Proteolytic Enzymes

The Practical Approach Series

Related **Practical Approach** Series Titles

Protein-Ligand Interactions: structure and spectroscopy
Protein-Ligand Interactions: hydrodynamic and calorimetry
Fmoc Solid Phase Peptide Synthesis
Post-Translational Modification
Gel Electrophoresis of Proteins 3/e
Protein Function 2/e
DNA and Protein Sequence Analysis
Protein Structure Prediction
Protein Blotting
Enzyme Assays
Protein Purification Methods

Please see the **Practical Approach** series website at
http://www.oup.co.uk/pas
for full contents lists of all Practical Approach titles.

Proteolytic Enzymes

Second Edition

A Practical Approach

Edited by

Robert Beynon

Department of Veterinary Preclinical Sciences,
University of Liverpool, U.K.

and

Judith S. Bond

Department of Biochemistry and
Molecular Biology,
The Pennsylvania State University, U.S.A.

OXFORD
UNIVERSITY PRESS

Great Clarendon Street, Oxford OX2 6DP

Oxford University Press is a department of the University of Oxford. It furthers the University's objective of excellence in research, scholarship, and education by publishing worldwide in

Oxford New York

Athens Auckland Bangkok Bogotá Buenos Aires Calcutta Cape Town Chennai Dar es Salaam Delhi Florence Hong Kong Istanbul Karachi Kuala Lumpur Madrid Melbourne Mexico City Mumbai Nairobi Paris São Paulo Singapore Taipei Tokyo Toronto Warsaw

with associated companies in Berlin Ibadan

Oxford is a registered trade mark of Oxford University Press in the UK and in certain other countries

Published in the United States by Oxford University Press Inc., New York

First Edition First Published 1989
Reprinted 1990 (with corrections)
Second Edition First Published 2001

A catalogue record for this title is available from the British Library

Library of Congress Cataloging in Publication Data

Proteolytic enzymes : a practical approach / edited by Rob Beynon and Judith S. Bond.–3rd ed.
(The practical approach series ; 247)
Includes bibliographical references and index.
1. Proteolytic enzymes. I. Beynon, R. J. (Robert J.) II. Bond, Judith S. III. Series.

QP609.P78 P77 2001 572′.76–dc21 00-063678

1 3 5 7 9 10 8 6 4 2

ISBN 0 19 963663 X (Hbk.)
ISBN 0 19 963662 1 (Pbk.)

Typeset in Swift by Footnote Graphics, Warminster, Wilts
Printed in Great Britain on acid-free paper
by The Bath Press, Bath, Avon

Preface

In the decade since the first edition of this book there has been a dramatic expansion in the identification and structural characterization of proteolytic enzymes complemented by knowledge of the functional and regulatory aspects. We have an increasing understanding of mechanisms and the critical roles that proteases play in developmental, physiological, and pathological processes. The information explosion concurrent with the sequencing of multiple genomes, including the human genome, will reveal even more proteases and lead to the discovery of additional functions, expression patterns, and regulatory roles of this group of enzymes. The ability of proteases to bring about an irreversible change in a target protein leads to regulatory roles for these hydrolases, as well as a role in manipulation of the dynamic state of cellular and extracellular protein content.

United only in their ability to hydrolyze the amide linkage between two amino acids, proteinases elicit an astonishing range of physiological and pathophysiological effects. Indeed, in the postgenome era it is evident that we will need new experimental strategies to discern the subtle roles of the new enzymes that will be identified through their sequence. Notwithstanding this diversity of function, there are a surprising number of common themes in the characterization and manipulation of this class of enzymes. This book aims to reinforce those common themes. In addition to be being studied in their own right, proteases are also essential tools in the armamentarium of the protein chemist. The ability to fragment a protein has innumerable applications including the generation of functionally active derivatives, or the creation of a reproducible set of peptides which when coupled with knowledge of their accurate mass turns out to be a powerful identification tool.

The wealth of information that already exists and the rapid expansion of that knowledge can be overwhelming for a scientist entering the realm of 'proteolysis'. This book attempts to present some fundamental information about how enzymes are classified, purified, assayed, and inhibited (Chapters 1, 2, 3, 5). Most endoproteases fit into one of five Classes of proteases (Serine, Threonine, Cysteine, Aspartic, Metallo), and Chapter 4 discusses the basic mechanistic characteristics of these Classes. Proteinaceous inhibitors are clearly important for the regulation of many inhibitors *in vivo*, and Chapter 6 discusses examples and strategies for identifying such inhibitors. For those who wish to use proteases as tools to study proteins, chapters on mass spectrometry, Edman Sequence analysis, probing native structures and peptide synthesis (Chapters 7–11) should be useful. The authors have tried to provide practical methods and strategies in addition to discussing the principles underlying the procedures. For many, the main interest in proteolysis will be to prevent unwanted hydrolysis by contaminating proteases. The chapters on protease inhibition (Chapter 5 and 9) and prevention of proteolysis are essential reading for those with this particular problem.

This book provides a base of fundamental information for a wide variety of scientists who are entering investigations of proteases and their interactions. It is intended to be used with other sources such as Protein Protocols, MEROPs (www.Merops.co.uk/Merops/merops.htm) and other databases Online. Because of the complexity of proteases (such as the proteasome, caspases, membrane-bound proteases) and proteolytic systems (e.g., those of blood coagulation, antigen processing, tissue remodeling and angiogenesis), no one source can provide sufficient information for the scientist at the lab bench. We refer readers to http//: www.protease.org the web page of the International Proteolysis Society where current and relevant links and resources will be maintained.

October 2000

Contents

Protocol list

Abbreviations

3,4-DCI	3,4-dichloroisocoumarin
A	absorbance
α_2M	α_2-macroglobulin
α_2M	α_2-macrogobulin
ACE	angiotensin-converting enzyme
ADAM	*a d*isintegrin *a*nd *m*etalloprotease
AEBSF	4-(2-aminoethyl)-benzenesulphonyl fluoride
AOMK	(acyloxy)methyl ketone
APMSF	4-amidinophenylmethanesulphonyl fluoride
ATAA	*N*-acetyl-L-Tyr-L-Arg-NH_2
ATEE	*N*-acetyl-L-Tyr-OEt
BODIPY	4,4-difluoro-4-bora-3a,4a-diazo-s-indacine
BSA	bovine serum albumin
CA074	*N*-(L-3-*trans*-propylcarbamoyloxirane-2-carbonyl)-L-Ile-L-Pro
CANP	calcium-activated neutral protease
Cbz-Phe	(*N*-carbobenzoxy-L-phenylalanine)
CID	collision-induced dissociation
CT	α-chymotrypsin
DFP	diisopropylphosphofluoridate
DHB	2,5-dihydroxybencoic acid
Dip-F	diisopropylphosphofluoridate
DMSO	dimethylsulphoxide
dpi	dots per inch
DTT	dithiothreitol
E-64	*N*-(L-3-*trans*-carboxyoxiran-2-carbonyl)-L-leucyl]-amido(4-guanidino)butane
EDC	1-ethyl-3-(3-dimethylaminopropyl) carbodiimide
EDTA	ethylenediaminetetraacetic acid
ELISA	enzyme-linked immunosorbent assay
ESI	electrospray ionization
ESI-MS	electrospray ionization mass spectrometry
EST	expressed sequence tag
Fa	*N*-furylacryloyl

FMK	fluoromethyl ketones (CH_2F)
FP	fluorescence polarization
FPLC	fast protein liquid chromatography
FRET	fluorescence resonance energy transfer
GFP	green fluorescent protein
HCCA	4-hydroxy-α-cyanocinnamic acid
HEPES	*N*-2-hydroxyethylpiperazine-*N*′-2-ethanesulfonic acid
HPLC	high-pressure liquid chromatography
IC_{50}	the molar concentration of the inhibitor that gives 50% inhibition of the target enzymatic activity
ID	internal diameter
IEF	isoelectric focusing
IUBMB	International Union of Biochemistry and Molecular Biology
LC	liquid chromatography
MALDI	matrix-assisted laser desorption/ionization
MALDI-TOF	matrix-assisted laser desorption ionization time of flight
MMP	matrix metalloproteases
MPC	multicatalytic endopeptidase complex
MS	mass spectrometry
MS-MS	tandem mass spectrometry
m/z	mass to charge ratio
NC-IUBMB	Nomenclature Committee of the International Union of Biochemistry and Molecular Biology
NRDB	non-redundant protein sequence database
ORF	open reading frames
PAGE	polyacrylamide gel electrophoresis
PAU-PAGE	phenol–acetic acid–urea polyacrylamide gel lectrophoresis
PBS	phosphate-buffered saline
PEEK	poly ether ether ketone
PHMB	*p*-hydroxy-mercuribenzoate
PMSF	phenylmethylsulphonyl fluoride
PVDF	polyvinylidene fluoride
SBTI	soybean trypsin inhibitor
SDS	sodium dodecyl sulphate
SDS-PAGE	sodium dodecyl sulphate polyacrylamide gel electrophoresis
TAME	tosyl arginine methyl ester
TCA	trichloroacetic acid
TIMP	tissue inhibitor of metalloprotease
TLC	thin-layer chromatography
TLCK	tosyl-lysine chloromethyl ketone
TOF	time-of-flight
TPCK	tosylamido-2-phenylethylchloromethyl ketone
UV	ultraviolet
WWW	World Wide Web.
Z	benzyloxycarbonyl, carbobenzoxy (= CBZ)

Chapter 1
Proteolytic enzymes: nomenclature and classification

Alan J. Barrett
MRC Peptidase Laboratory, Department of Immunology, The Babraham Institute, Babraham, Cambridgeshire CB2 4AT, UK

1 Introduction

A landmark in the development of any field of study is the appearance of a sound system of nomenclature and classification for the objects with which it deals. The introduction of the Linnaean system for naming and classifying organisms in the eighteenth century and the invention of a system of nomenclature for enzymes in the 1950s were such key events, and their value has been obvious. Both nomenclature and classification are vitally important for information-handling, allowing people to communicate efficiently, knowing that they are talking about the same thing, and to store and retrieve information unambiguously. A good system will also serve to highlight important questions, and thus prompt new discoveries.

The nomenclature and classification become increasingly necessary as a field of study grows, and the field of proteolytic enzymes is clearly in a stage of rapid expansion. This is seen as the accelerating rate of discovery of new proteolytic enzymes and the increase in what is known about them. These factors undoubtedly increase the need for sound nomenclature and classification, but happily they can also help to provide a basis for it.

The purpose of this chapter is first to review the terminology that is used in relation to proteolytic enzymes, and then to look at two systems that exist for their classification, the EC system of the IUBMB and the *MEROPS* system of peptidase clans and families. We shall consider what these systems do and how they can help the student or research worker.

2 Terminology and nomenclature

2.1 Peptidase and related terms

The term protease appeared in the German literature of physiological chemistry in the latter part of the nineteenth century in reference to proteolytic enzymes, and was used as a general term embracing all the hydrolases that act on proteins, or further degrade the fragments of them. Around 1930, the need to distinguish separate kinds of protease activity began to be recognized and two sets of terms were independently proposed, one in Germany and one in the United States. In Germany, Grassmann and Dyckerhoff (1) started to call the enzymes that act on proteins 'proteinases', and those that act preferentially on oligopeptides 'peptidases'. The reason for the ability of some proteolytic enzymes to act preferentially on small peptides became much clearer when peptide substrates with and without terminal blocking groups could be synthesized. It was found that the free end-groups were not required, and frequently were not tolerated, in the specificity sites of the proteinases, so that the enzymes acted well on long chains, away from the ends, but acted much less well on the oligopeptide products. In contrast, the enzymes that acted well on oligopeptides generally required at least one free terminus close to the scissile bond, so that they had little action on intact proteins, in which exceedingly few of the peptide bonds are close to a terminus. Bergmann and Ross (2) then used the term *peptidase* as a general term for peptide bond hydrolase, not in the restricted sense that had been proposed by Grassmann, and extended it to form the terms 'endopeptidase' and 'exopeptidase'. The use of the term peptidase in both the narrow and the broad senses has caused confusion, and I favour restricting it to the broad sense, as is recommended by IUBMB. For the narrower meaning of a peptide bond-hydrolysing enzyme that acts only or preferentially on 'peptide' substrates (i.e. oligopeptides, or small polypeptides), the best terms are exopeptidase if the specificity is a requirement for free terminal groups (3) (*Figure 1*), or oligopeptidase if the specificity is substrate size-dependent. We return to these and related terms, including aminopeptidase and carboxypeptidase which are defined in the EC system, in Section 3.1.

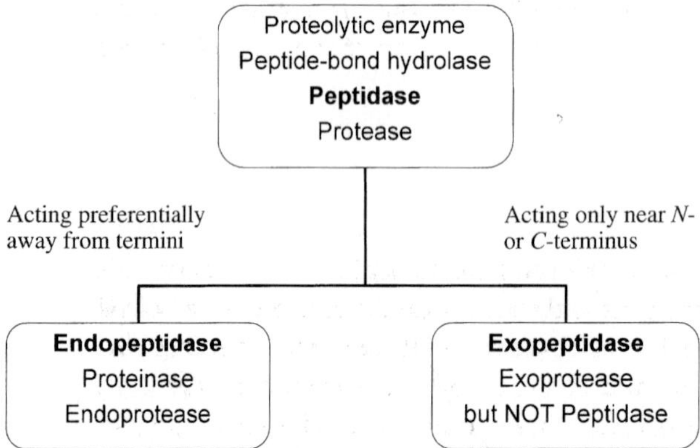

Figure 1 The synonymous terms for some kinds of proteolytic enzymes. The terms within each box presently have identical meanings. The terms peptidase, endopeptidase and exopeptidase (printed in bold type) are those that are recommended by IUBMB and they have a clear logic. The other synonyms are not particularly troublesome except for peptidase when used as a synonym for exopeptidase, since this is the general term for proteolytic enzymes.

2.2 Specificity-subsite terminology

Crystallographic structures show that the active site of a peptidase is commonly located in a groove on the surface of the molecule between adjacent structural domains, and the substrate specificity is dictated by the properties of binding sites arranged along the groove on one or both sides of the catalytic site that is responsible for hydrolysis of the peptide bond. Accordingly, the specificity of a peptidase is described by use of a conceptual model in which each specificity subsite is able to accommodate the side-chain of a single amino acid residue. The sites are numbered from the catalytic site, S_1, S_2...Sn towards the N-terminus of the substrate, and S'_1, S'_2 ...Sn' towards the C-terminus. The amino acids they accommodate are numbered P_1, P_2...P_n, and P'_1, P'_2...P'_n, respectively, as follows:

$$\text{Substrate:} - P_3 - P_2 - P_1 + P'_1 - P'_2 - P'_3 -$$

$$\text{Enzyme:} - S_3 - S_2 - S_1 \;{}^{*}\; S'_1 - S'_2 - S'_3 -$$

Here, the catalytic site of the enzyme is marked *, and the peptide bond that is cleaved (the scissile bond) is indicated by the symbol '+'. This system (4) is based on that of Berger and Schechter (5), but for typographical convenience, the original system is modified in that the subsite numbers are not subscripted, and are followed rather than preceded by the prime characters, when needed.

2.3 Catalytic type

It was first pointed out by Hartley (6) that proteolytic enzymes could be grouped according to the chemical nature of the catalytic site. Hartley recognized four distinct types of catalytic site, and minor refinements of the names for these in the light of subsequent discoveries have led to the present recognition of serine, cysteine, aspartic and metallo- types of peptidases (7,8). Most recently, the group of threonine peptidases has been discovered. These groupings contribute to both current systems of peptidase nomenclature and classification, the EC system and the *MEROPS* system.

2.4 Homology

In discussing the evolutionary and structural relationships of peptidases, the concept of homology is important. As is the case with other proteins, a peptidase is correctly described as being homologous with another when the two amino acid sequences are so similar that the resemblance cannot be attributed to chance; this is most readily explained by a common evolutionary origin. The concept of homology is thus absolute, not relative (9). The effect of this is that one may correctly say that two proteins are homologous, or that they are not homologous, but it is not correct to say that two proteins are more homologous than two others.

3 The EC classification of peptidases

3.1 What is the EC system?

The internationally recognized scheme for the classification and nomenclature of all enzymes, including peptidases, is that of the IUBMB. The Nomenclature Committee of the IUBMB is the successor to the Enzyme Commission, and the index numbers that it assigns to enzymes are still described as 'EC' numbers (4,10). The latest full publication of the EC list is in *Enzyme Nomenclature 1992* (4), but this has been updated by several supplements relating specifically to peptidases (11-14). The up to date text for the part of the EC list dealing with peptidases can be found on the Internet at http://www.chem.qmw.ac.uk/iubmb/enzyme/EC34/. This should be distinguished from abbreviated versions of the full EC list that are located at other sites on the Internet.

In the EC system all enzymes are divided between six classes: (1) Oxidoreductases, (2) Transferases, (3) Hydrolases, (4) Lyases, (5) Isomerases, (6) Ligases.

The hydrolases of peptide bonds, the peptidases, form subclass 3.4. There is an important difference between the way in which the great majority of enzymes are treated in the EC list and the way in which peptidases are dealt with, however. Throughout most of the EC list, an 'enzyme' is defined by the fact that it catalyses a single reaction. This has the important implications that (i) several quite different proteins are all described as the same enzyme, under a single EC number, if they catalyse the same reaction; (ii) conversely, a single protein that catalyses more than one reaction is treated as more than one enzyme, with more than one EC number; and (iii) a satisfactory name for any enzyme (catalysing a single reaction) can be formed directly from the description of the reaction it catalyses. However, a decision has been taken to treat peptidases differently from the other enzymes. The are several reasons for this. (i) The reaction catalysed is not a satisfactory basis for the classification of most peptidases. For all of the peptidases, the basic reaction is the same, hydrolysis of a peptide bond. The specificities differ, but for the endopeptidases, and for many of the exopeptidases, the specificity is extremely difficult to determine in any rigorous way and, if determined, is almost impossible to describe simply, or to use as the basis of a name. (ii) The value of the classification of peptidases by catalytic type was quickly recognized, and accordingly, this system was introduced into the EC list in 1972 (15). (iii) It was felt that it would be useful to scientists working in the field of proteolytic enzymes to be free to regard different proteins as different peptidases, even when they express similar or identical peptidase activities. An example would be the distinction of pancreatic elastase (EC 3.4.21.36) and leukocyte elastase (EC 3.4.21.37). It happens that these enzymes have slightly different specificities, but even if they did not, the genetic and biological differences between them would mean that they needed to be distinguished in biomedical science. The decision not to classify peptidases purely on the basis of the reactions they catalyse created the need for special systems for both classifying and naming them.

The peptidases in the EC list are divided between 13 sub-subclasses (*Table 1*). Two major sets of these sub-subclasses are recognized, comprising the exopeptidases (3.4.11–19) and the endopeptidases (3.4.21–24 together with 3.4.99). The exopeptidases act only near the ends of polypeptide chains. Those acting at a free N-terminus liberate a single amino acid residue (aminopeptidases, 3.4.11), or a dipeptide or a tripeptide (dipeptidyl-peptidases and tripeptidyl-peptidases, 3.4.14). Those acting at a free C-terminus liberate a single residue (carboxypeptidases, 3.4.16–18) or a dipeptide (peptidyl-dipeptidases, 3.4.15). The carboxypeptidases are allocated to three groups on the basis of catalytic groups: serine-type carboxypeptidases (3.4.16), metallocarboxypeptidases (3.4.17) and cysteine-type carboxypeptidases (3.4.18). Other exopeptidases are specific for dipeptides (dipeptidases, 3.4.13), or remove terminal residues that are substituted, cyclized or linked by isopeptide bonds (peptide linkages other than those of α-carboxyl to α-amino groups) (omega peptidases, 3.4.19).

The terms 'dipeptidyl-peptidase' and 'peptidyl-dipeptidase', for exopeptidases liberating dipeptides from N- or C-terminus, respectively, deserve a word of explanation. Historically, names involving aminopeptidase and carboxypeptidase have sometimes been used for these peptidases, but it is important to maintain the distinction between the enzymes that release single amino acids, and those that release dipeptides. The systematic basis for the term dipeptidyl-peptidase is that the substrate is thought of as a dipeptide linked to the N-terminus of another peptide (i.e. a dipeptidyl-peptide), and it is the linking peptide bond that is hydrolysed, so the enzyme is a dipeptidyl-peptide hydrolase or dipeptidylpeptidase (the hyphen indicating the scissile bond). Similar reasoning gives rise to the term peptidyldipeptidase.

Table 1 The EC system of classification of peptidases. The data apply to the latest full publication of the list, *Enzyme Nomenclature 1992* (4) as amended by four supplements (11–14)

Sub-subclass	Type of peptidase	Number of entries
3.4.11	Aminopeptidases	19
3.4.13	Dipeptidases	12
3.4.14	Dipeptidyl-peptidases	8
3.4.15	Peptidyl-dipeptidases	3
3.4.16	Serine-type carboxypeptidases	4
3.4.17	Metallocarboxypeptidases	19
3.4.18	Cysteine-type carboxypeptidases	1
3.4.19	Omega peptidases	11
3.4.21	Serine endopeptidases	75
3.4.22	Cysteine endopeptidases	26
3.4.23	Aspartic endopeptidases	32
3.4.24	Metalloendopeptidases	69
3.4.99	Endopeptidases of unknown type	2
	Total	281

Endopeptidases cleave internal bonds in polypeptide chains. In the EC list, the endopeptidases are divided into sub-subclasses on the basis of catalytic mechanism, and the specificity is used only to identify individual enzymes within the groups. The enzymes of sub-subclass 3.4.21 (serine endopeptidases) have an active centre serine involved in the catalytic process, those of 3.4.22 (cysteine endopeptidases) have a cysteine in the active centre, those of 3.4.23 depend on an aspartic acid residue (or commonly two) for their catalytic activity (aspartic endopeptidases), and those of 3.4.24 use a metal ion (often, but not always, Zn^{2+}) in the catalytic mechanism (metalloendopeptidases). Recently, sub-subclass 3.4.25 has been added for threonine peptidases. The initial assignment of an endopeptidase to a catalytic type commonly depends upon the results of tests with inhibitors, as is described elsewhere in the present volume (see Chapter 4).

Generally, the polypeptide chain hydrolysed by an endopeptidase may be of any length, so that proteins are among the substrates, but a subset of endopeptidases is recognized that are restricted to acting on substrates smaller than proteins; these are termed 'oligopeptidases'. Among the first oligopeptidases to be clearly recognized as such was thimet oligopeptidase (EC 3.4.24.15). This endopeptidase hydrolyses oligopeptides and polypeptides of up to 17 amino acids, but has no action on larger polypeptides or on proteins (16). Prior to the general awareness of the existence of the oligopeptidase subset of endopeptidases, a pitfall was repeatedly encountered in studies designed to identify endopeptidases specific for hydrolysis of particular peptide bonds in protein substrates, such as the β-secretase cleaving the Alzheimer's amyloid precursor protein (17). For experimental convenience, oligopeptide substrates were synthesized with sequences spanning the scissile bond of interest. These model substrates were then used in the search for the enzyme that cleaves the protein, but on several occasions it was actually an irrelevant oligopeptidase that was discovered (18). Oligopeptidases include prolyl oligopeptidase, oligopeptidase A, oligopeptidase B, pitrilysin, insulysin, neprilysin and others (19).

3.2 What information does the EC list contain?

The value of the EC list for peptidases lies in providing standard terminology for the various types of peptidase activity, and especially in the assignment of a unique identification number and a recommended name to each peptidase. The special strengths of the EC system are thus in the area of nomenclature rather than classification.

The EC entry for a peptidase contains the 'EC number', the 'Recommended name', a description of the 'Reaction', and a list of 'Other names' that have been used. There are also 'Comments', providing a brief summary of what is known about the enzyme (including the peptidase family number in the *MEROPS* system, for all new entries), links to other databases, and up to five 'References', complete with titles and inclusive page numbers (*Figure 2*).

3.4.22.36

Recommended name: Caspase-1

Reaction: Release of interleukin 1 by specific cleavage at -Asp116+Ala- and -Asp27+Gly- bonds in precursor. Also hydrolyses the small-molecule substrate, Ac-Tyr-Val-Ala-AspNHMec

Other names: Interleukin 1-converting enzyme

Comments: From mammalian monocytes. Inhibited by Ac-Tyr-Val-Ala-Asp-CHN_2. Type example of peptidase family C14

Links to other databases: BRENDA, EXPASY, MEROPS

References

1. Howard, A., Kostura, M.J., Thornberry, M., Ding, G.J.F., Limjuco, G., Weidner, J., Salley, J.P., Hogquist, K.A., Chaplin, D.D., Mumford, R.A., Schmidt, J.A. & Tocci, M.J. (1991) IL-1 converting enzyme requires aspartic acid residues for processing of the IL-1 precursor at two distinct sites and does not cleave 31-kDa IL-1. J. Immunol. 147, 2964-2969

2. Thornberry, N.A., Bull, H.G., Calaycay, J.R., Chapman, K.T., Howard, A.D., Kostura, M.J., Miller, D.K., Molineaux, S.M., Weidner, J.R., Aunins, J., Elliston, K.O., Ayala, J.M., Casano, F J., Chin, J., Ding, G.J.-F., Egger, L.A., Gaffney, E.P., Limjuco, G., Palyha, O.C., Raju, S.M., Rolando, A.M., Salley, J.P., Yamin, T.-T. & Tocci, M.J. (1992) A novel heterodimeric cysteine protease is required for interleukin-1 processing in monocytes. Nature 356, 768-774

3. Thornberry, N.A. (1994) Interleukin-1 converting enzyme. Methods Enzymol. 244, 615-631

4. Alnemri, E.S., Livingston, D.J., Nicholson, D.W., Salvesen, G., Thornberry, N.A., Wong, W.W. & Yuan, J.Y. (1996) Human ICE/CED-3 protease nomenclature. Cell 87, 171

5. Margolin, N., Raybuck, S.A., Wilson, K.P., Chen, W.Y., Fox, T., Gu, Y. & Livingston, D.J. (1997) Substrate and inhibitor specificity of interleukin-1-converting enzyme and related caspases. J. Biol. Chem. 272, 7223-7228

Figure 2 An example of an entry in sub-subclass 3.4 of the EC list as it appears on the World Wide Web. Five references is the maximum number normally allowed; titles and full page numbers are included.

3.3 When and how can a newly-discovered peptidase be added to the EC list?

In view of the valuable role that the EC list has to play in the terminology of peptidases, it is important that newly discovered peptidases are added to the list as soon as they have been characterized in sufficient detail. The present section summarizes the requirements that normally apply, and describes the procedure by which a new entry in the EC list may be proposed.

3.3.1 A distinctive peptidase

Clearly, a peptidase must be shown to be distinct from any already in the EC list before a new entry can be created for it. In the EC list, as a whole, the requirement would be simply that the type of reaction or its specificity are unlike those of any enzyme already included. These criteria are also sufficient for a peptidase, but it follows from what has been said above about the special handling of peptidases in the EC list that other factors also can justify the new entry. One such factor would be a difference in catalytic type, so that two peptidases that might have indistinguishable specificity would merit separate entries if they were of different catalytic types. Biological differences, such as encoding by separate genes in the same organism, or expression in different cell types, have also justified new entries. The over-riding consideration is that it is of practical usefulness to a significant number of scientists to be able to make unambiguous reference to the peptidase as a distinct entity.

3.3.2 Purification and characterization

The properties of a peptidase cannot be determined unambiguously without the use of a homogeneous preparation of the enzyme. Methods for the purification of naturally occurring forms of the enzymes are constantly improving (see Chapter 2), as are the methods for cloning and expression, which may lead to simpler purification. A proposal for addition of a peptidase to the EC list should normally include a reference to a refereed publication describing the isolation to homogeneity, commonly shown as a single zone of protein in gel electrophoresis or column chromatography which is associated with the enzymatic activity.

3.3.3 Evidence of catalytic type

In view of the importance that is placed on catalytic type in the EC classification of peptidases, it is expected, with rare exceptions, that the catalytic type will have been determined for any peptidase proposed for addition to the list. As was explained above, the catalytic types are the serine, cysteine, threonine, aspartic and metallo-peptidases, and the sub-subclasses of carboxypeptidases and endopeptidases depend directly upon catalytic type (*Table 1*). This gives rise to just seven sub-subclasses of these enzymes, since there are no known aspartic-type carboxypeptidases (or indeed exopeptidases of any kind).

The type of catalytic site is usually revealed most clearly by the effects of class-specific inhibitors. The use of these is described elsewhere (see Chapter 4 and ref. 20), but I should like to emphasize the usefulness of analysing any data showing partial inhibition by taking account of the effect of time on the degree of inhibition, in conjunction with the type of inhibition (irreversible or reversible) that is expected. For example, if using phenylmethane sulfonyl fluoride as a diagnostic inhibitor of many serine peptidases, one might note that since this is a slow-reacting, irreversible inhibitor, the expectation is that there will be time-dependent inhibition progression towards completion, with a rate that is first-order in inhibitor concentration. Partial inhibition that cannot be shown to have the expected time-dependence should be regarded with suspicion. Conversely, a reversible inhibitor such as soybean trypsin inhibitor will generally give apparently instantaneous inhibition under practical conditions, and when inhibition is partial, it will most probably depend not on time, but on inhibitor concentration that is in accord with a K_i value. The point is simply that it is known for all the catalytic-type-specific inhibitors that are commonly used what kind of inhibition is to be expected, and any partial inhibition obtained should be examined for dependence upon time of incubation and concentration of inhibitor that is consistent with this. Deviations from expectation may indicate that the enzyme preparation is inhomogeneous. Many published sets of data clearly have not been examined in this way and may be suspect.

If the amino acid sequence of a peptidase is known and allows the peptidase to be placed in one of the structure-based families (see Section 4.1) then it can safely be assumed that the catalytic type will be that already known for the family, and biochemical tests may be unnecessary.

3.3.4 Amino acid sequence data

It is now uncommon for a peptidase to be added to the EC list without its amino acid sequence being known. Although knowledge of the amino acid sequence is not a formal requirement it is clearly of great value, not least in allowing the compilers of the list to know that the enzyme being put forward really is distinct from all those already in the list. It usually allows the peptidase to be assigned to one of the known families (see Section 4.1), and confirms its catalytic type. The number of the family to which a peptidase belongs is included in the Comments section of new entries to the EC list. If the protein has not been cloned and sequenced, but has been purified to homogeneity, then limited chemical microsequencing of *N*-terminus and internal fragments may well yield sufficient data to allow at least tentative assignment to a family.

3.3.5 A satisfactory name

One of the most important parts of the entry for a peptidase in the EC list is the Recommended name. Those compiling the list see it as part of their responsibility to help ensure that satisfactory names are available for peptidases, and it is not uncommon for the inclusion of a peptidase to be delayed because none of the names in use can be recommended. Even if a peptidase were not to be included in the EC list, it would be important that it had a satisfactory name. The literature contains many sound reports of peptidases described only by a term such as 'beef brain acid proteinase'; such a term is not easily remembered or stored in indexes, so the report tends to sink into obscurity, to the disadvantage of science and the scientist concerned. It is therefore wise to put forward a distinctive name for any newly discovered peptidase at an early stage. If it should happen that the enzyme proves to be one for which a name already exists, that is not difficult to deal with later. Although a new name will initially sound strange, it is surprising how soon it becomes a natural and useful part of one's thinking and communication.

Creating a name for a newly-discovered peptidase is seldom easy. For most enzymes a name can be constructed from the reaction catalysed, but that is valid only because those enzymes are defined purely by the reaction catalysed so that the enzyme name does not have to distinguish a unique protein. This is not the case for the names of peptidases as was explained in Section 3.1.

I should like to offer some suggestions about what to do, and what not to do, in naming a peptidase.

3.3.5.1 Do not mistake a descriptive term for a name

The difficulty of inventing a good name has often been avoided by using a descriptive term in its place. The difference between a descriptive term and a true name is very important, but is not widely recognized. To take an example from the existing nomenclature of peptidases, the serine endopeptidase found in the acrosome of the sperm, for which the recommended name is acrosin, could equally well have been termed acrosomal neutral proteinase. I should describe acrosin as a true name, whereas acrosomal neutral proteinase is a descriptive phrase. The name, while often implying some information about the enzyme, does not explicitly state it, and is therefore more robust in surviving further discoveries about the enzyme. If acrosin is found to occur elsewhere than in the acrosome, the name will still be valid, at least reflecting the history of its discovery, but the descriptive designation acrosomal neutral proteinase would become untenable: clearly one could not say that one was working on acrosomal neutral proteinase from the ovary! Similarly, if another neutral proteinase were discovered in the acrosome, it could create a problem for the descriptive term, but not for the name, since it would not be acrosin.

Much the same considerations apply to naming on the basis of a source organism. One should assume that any peptidase will be found to occur in more than one organism. Imagine that a cysteine endopeptidase has been isolated from a fruit fly. The discoverer might consider calling the enzyme either drosophilain or *Drosophila melanogaster* cysteine proteinase. If species variants of the same enzyme are found to exist in other flies, the name drosophilain would still be satisfactory (reflecting the history of discovery of the enzyme), but the descriptive phrase would not. True names are also more convenient than descriptive phrases, because they are shorter and can contain unique strings of characters that will be valuable in computer searches and indexing.

3.3.5.2 Do not try to use the name of its gene as the name for a peptidase

Gene names are species-specific and, since we assume that any peptidase is likely to occur in more than one species, it will be encoded by genes of more than one name. Thus, *kex2* protease would not be a good designation for the processing enzyme from *Saccharomyces cerevisiae*, because genes with other names encode the same enzyme, even in other yeasts. However, the name kexin works well.

3.3.5.3 Base the name of an exopeptidase on the type of reaction

A look at the existing EC list will show that the names of exopeptidases almost invariably have been based on the type of activity—aminopeptidase, dipeptidase, etc.—qualified by a specificity descriptor (e.g. leucyl aminopeptidase) or a serial identifier (e.g. carboxypeptidase A). This is generally proving satisfactory.

3.3.5.4 For an endopeptidase, consider inventing a novel, one-word name

The majority of endopeptidases have one-word names, good examples of which would include trypsin, pepsin, acrosin and stromelysin. Biological similarities have sometimes led to the establishing of a series of names with a single root, such as the cathepsins and the granzymes. It is uncommon for the specificity of an endopeptidase to be so distinctive and so easily described as to form a satisfactory basis for a name. However, a term such 'glutamyl endopeptidase' has been found to need the addition of further qualifiers (e.g. glutamyl endopeptidase II) as other enzymes with similar specificity are discovered.

3.3.5.5 For the name of an endopeptidase, try to use a meaningful ending

The existing names of many cysteine peptidases end in '-ain', and it makes sense to try to follow this convention. Similarly, the ending '-epsin' is good for aspartic endopeptidases and the ending '-lysin' works well for metallo-endopeptidases. Take care with the ending '-ase', though: this is recommended to be reserved for the construction of names from those of substrates (e.g. collagenase), but even then only with caution, since it may tend to suggest a greater degree of specificity, or a primary biological function, that the enzyme does not prove to show.

3.3.5.6 Avoid using a set of capital letters as a name

Descriptive phrases are much easier to coin than real names, but apart from their other disadvantages described above, they are usually too long for regular use. As an attempt to get around this problem, they have sometimes been condensed to sets of initial capital letters. A string of capital letters does not make a good name, however, because it looks like an abbreviation and seems to need to be defined whenever it is used. Try changing it to lower case, and adding a few letters, as necessary, to make it into a novel and pronounceable name.

3.3.5.7 Do not make the name for your new peptidase into an advertisement

Occasionally, scientists have chosen to name a peptidase after their institute, for example. This may have pleased the chairman but generally it is a bad idea since it tends to delay acceptance by other scientists.

3.3.5.8 Some other points

It is important to invent a name that is easy to use, in speech, in writing and at a computer keyboard. A one-word name is usually best and, if the sequence of letters is unique, it will work well in a computer search. Avoid Greek letters and typographical complications such as the bar characters that are used in the notation of the complement system.

It is important to be absolutely sure that the new name really is novel and not already in use, so an essential step is to search the literature databases for it. This will also show how easy or difficult such a search will be for others later.

Even if the planned new name is not yet in the literature, it may be on its way. Serial names (a root name with a letter or number, such as cathepsin K) are particularly likely to be duplicated. At any given time, many scientists feel that they know what the 'next' letter or number should be, but they have no way of knowing whether someone else is also adopting that same designation for their own new peptidase. When the publications come out it will be too late to avoid a muddle, so it is worth making enquiries and perhaps using a special interest news group on the Internet to announce the intention of using the name at an early stage.

3.3.6 Procedure for proposing a new or amended EC entry

When the description of the isolation and specificity of the new peptidase has been accepted for publication in a refereed journal, and the sequence has been deposited, it is time to propose the EC entry. Proposals for inclusion of new entries in *Enzyme Nomenclature*, as well as for amendment of existing entries, should be sent to Prof. K. F. Tipton, Trinity College, Dublin. A simple form for this purpose can be found on the WWW in the Enzyme folder of the ExPASy site (http://www.expasy.ch/).

4 The *MEROPS* system for the classification of peptidases

As we have seen, the EC system places each of the peptidases in one of just 13 sub-subclasses, and only five of the sub-subclasses are available for the very numerous endopeptidases (comprising 204 of the total of 281 entries). This inevitably lumps together peptidases that are very different. Most importantly, it takes no account of structural groupings of peptidases that would reflect evolutionary relationships. About 1992, N. D. Rawlings and the present author started to develop a new form of classification of peptidases that was designed to group them in a way that would reflect essential structural features and evolutionary relationships, taking advantage of the wealth of amino acid sequence data that was becoming available. The beginnings of the system were described in 1993 (21); it was then further developed as the organizational basis for collections of articles in *Methods in Enzymology* (22, 23), and in 1996 the system was published on the WWW as the *MEROPS* database (URL: http://www.merops.co.uk). The name *MEROPS* has no special significance, being simply the name of a bird whose social behaviour prompted our adoption of the term 'clan', but it now forms a convenient label for reference to the system of peptidase families

Table 2 The three levels of the *MEROPS* classification. The identifiers used for clans, families and peptidases in the *MEROPS* classification each start with a letter indicating the catalytic type of peptidases in the group. These letters are: A, aspartic; C, cysteine; M, metallo; S, serine; T, threonine, or U, unclassified; the letter P is also used, to identify two clans that contain peptidases of more than one of the types C, S and T. At the time of writing, no clan is yet recognised that contains peptidases only of type T or of type U. The identifier of a clan is completed with a second capital letter, assigned sequentially (*e.g.* CD), and the identifier for a family is completed with a number (*e.g.* C14). The *MEROPS* identifier for each individual peptidase starts with the family identifier and is completed with a decimal number (*e.g.* C14.001). All of the identifiers are stable, being changed only under very special circumstances, and never re-used

Level	Description	Catalytic types recognized in forming identifiers
Clan	A clan is a set of families in which all the peptidases have evolved from a single ancestor. Families in the same clan are best recognised by the similar protein folds of their peptidases.	A, C, M, P, S
Family	A family contains peptidases that can be proved to be homologous by comparison of amino acid sequences.	A, C, M, S, T, U
Peptidase	An individual peptidase is distinguishable from all others by differences in activity, structure or genetics.	A, C, M, S, T, U

and clans whether it be in the form of the World Wide Web database or printed versions. Cross-references to the *MEROPS* system will be found in the SWISS-PROT database (at the ExPASy site, http://www.expasy.ch/) and in the EC list (URL: http://www.chem.qmw.ac.uk/iubmb/enzyme/index.html).

MEROPS is a three-layered system in which we allocate the peptidases to families and group these as far as possible into clans (*Table 2*). The summary of the resulting classification is presented in *Table 3*. We shall now look at each of the three levels of *MEROPS* in a little more detail.

4.1 Families

The families of peptidases are constructed by comparisons of amino acid sequences, with the use of strict statistical criteria such that there is no realistic possibility of two peptidases being placed in the same family when they are not truly homologous. Each family is built around a 'type example' peptidase, and every member of the family shows a statistically significant relationship in amino acid sequence to at least one other member of the family in the part of the molecule that is responsible for peptidase activity. The point that we require that the relationship be demonstrable in the catalytically active part of the molecule is an important one, necessitated by the existence of peptidases that are chimeric proteins. For example, procollagen I N-endopeptidase is a chimeric protein that contains a catalytic domain related to that of astacin, but also contains segments that are clearly homologous with non-catalytic parts of the complement components C1r and C1s, which are in the chymotrypsin family

Table 3 Clans and families of peptidases. The clans and families of peptidases, together with the type example one example from each family. The clan PA contains protein nucleophile peptidases of more than one catalytic type, previously assigned to clans SA and CB, as described in the text. The miscellaneous groups of families that cannot yet be assigned to clans are listed as if in clans SX, CX, AX, MX and UX for convenience. The *MEROPS* classification is under constant review, so the reader is advised to consult the Web version at www.merops.co.uk for the latest information

Clan	Family	Example
(a) Serine, threonine and cysteine peptidases		
PA	S1	Trypsin
	S2	Streptogrisin A
	S3	Togavirin
	S6	IgA1-Specific serine endopeptidase
	S7	Flavivirin
	S29	Hepatitis C virus NS3 polyprotein peptidase
	S30	Potyvirus P1 proteinase
	S31	Pestivirus polyprotein peptidase p80
	S32	Equine arteritis virus serine endopeptidase
	S35	Apple stem grooving virus protease
	C3	Poliovirus picornain 3C
	C4	Tobacco etch virus NIa endopeptidase
	C24	Feline calicivirus endopeptidase
	C30	Mouse hepatitis coronavirus picornain 3C-like endopeptidase
	C37	Southampton virus processing peptidase
SB	S8	Subtilisin
SC	S9	Prolyl oligopeptidase
	S10	Carboxypeptidase C
	S15	X-Pro dipeptidyl-peptidase
	S28	Pro-X carboxypeptidase
	S33	Prolyl aminopeptidase
	S37	PS-10 peptidase (*Streptomyces lividans*)
SE	S11	D-Ala-D-Ala carboxypeptidase A
	S12	D-Ala-D-Ala carboxypeptidase B
	S13	D-Ala-D-Ala peptidase C
SF	S24	Repressor LexA
	S26	Signal peptidase I
	S41	Tail-specific protease
SH	S21	Assemblin
TA	T1	Proteasome
CA	C1	Papain
	C2	Calpain
	C6	Tobacco etch virus HC-proteinase
	C7	Chestnut blight virus p29 endopeptidase
	C8	Chestnut blight virus p48 endopeptidase
	C9	Sindbis virus nsP2 endopeptidase
	C10	Streptopain
	C12	Deubiquitinating peptidase Yuh1
	C16	Mouse hepatitis virus endopeptidase
	C19	Isopeptidase T
	C21	Turnip yellow mosaic virus endopeptidase
	C23	Blueberry scorch carlavirus endopeptidase
	C27	Rubella rubivirus endopeptidase

Table 3 Continued

Clan	Family	Example
	C28	Foot-and-mouth disease virus L proteinase
	C29	Mouse hepatitis coronavirus papain-like endopeptidase 2
	C31	Porcine respiratory and reproductive syndrome arterivirus α-endopeptidase
	C32	Equine arteritis virus PCP α-endopeptidase
	C36	Beet necrotic yellow vein furovirus papain-like endopeptidase
CD	C11	Clostripain
	C13	Legumain
	C14	Caspase-1
	C25	Gingipain R
CE	C5	Adenovirus endopeptidase
SX	S14	Endopeptidase Clp
	S16	Endopeptidase La
	S18	Omptin
	S19	Chymotrypsin-like protease (*Coccidioides*)
	S34	HflA protease
	S38	Chymotrypsin-like protease (*Treponema denticola*)
CX	C15	Pyroglutamyl peptidase I
	C26	γ-Glutamyl hydrolase
	C33	Equine arterivirus Nsp2 endopeptidase
	C40	Dipeptidyl-peptidase VI
	C41	Hepatitis E cysteine proteinase
Aspartic peptidases		
AA	A1	Pepsin
	A2	HIV 1 retropepsin
	A3	Cauliflower mosaic virus endopeptidase
	A9	Simian foamy virus polyprotein peptidase
	A10	*Schizosaccharomyces* retropepsin-like transposon
	A15	Rice tungro bacilliform virus protease
AB	A6	Nodavirus endopeptidase
AX	A4	Scytalidopepsin B
	A5	Thermopsin
	A7	Pseudomonapepsin
	A8	Signal peptidase II
	A11	*Drosophila* transposon copia peptidase
	A12	Maize transposon bs1 peptidase
Metallopeptidases		
MA	M1	Membrane alanyl aminopeptidase
	M2	Peptidyl-dipeptidase A
	M4	Thermolysin
	M5	Mycolysin
	M6	Immune inhibitor A
	M7	*Streptomyces* small neutral endopeptidase
	M8	Leishmanolysin
	M9	*Vibrio* collagenase
	M10	Interstitial collagenase
	M11	Gametolysin
	M12	Astacin
	M13	Neprilysin
	M30	Hyicolysin
	M36	Fungalysin

Table 3 Continued

Clan	Family	Example
MC	M14	Carboxypeptidase A
MD	M15	Zinc D-Ala-D-Ala carboxypeptidase
ME	M16	Pitrilysin
MF	M17	Leucyl aminopeptidase
MG	M24	Methionyl aminopeptidase
MH	M18	Aminopeptidase I
	M20	Glutamate carboxypeptidase
	M25	X-His dipeptidase
	M28	Aminopeptidase Y
	M40	Carboxypeptidase (*Sulfolobus sulfataricus*)
	M42	Glutamyl aminopeptidase (*Lactococcus*)
MX	M3	Thimet oligopeptidase
	M19	Membrane dipeptidase
	M22	*O*-Sialoglycoprotein endopeptidase
	M23	β-Lytic endopeptidase
	M26	IgA-specific metalloendopeptidase
	M27	Tentoxilysin
	M29	Aminopeptidase T
	M32	Carboxypeptidase Taq
	M34	Anthrax lethal factor
	M35	Penicillolysin
	M37	Lysostaphin
	M38	β-Aspartyl dipeptidase
	M41	FtsH endopeptidase
	M45	D-Ala-D-Ala dipeptidase (*Enterococcus*)
Unclassified peptidases		
UX	U3	Endopeptidase gpr
	U4	Sporulation sigma E factor processing peptidase
	U6	Murein endopeptidase MepA
	U7	Protease IV
	U9	Prohead proteinase (bacteriophage T4)
	U12	Prepilin type IV signal peptidase
	U26	vanY D-Ala-D-Ala carboxypeptidase
	U27	ATP-dependent protease (*Lactococcus*)
	U28	Aspartyl dipeptidase
	U29	Cardiovirus endopeptidase 2A
	U32	Microbial collagenase (*Porphyromonas)*
	U39	Hepatitis C virus NS2-3 protease
	U40	Protein P5 murein endopeptidase (bacteriophage phi6)
	U43	Infectious pancreatic necrosis birnavirus endopeptidase

(24). The procollagen endopeptidase is placed in the family of astacin (M12), and not that of chymotrypsin (S1).

A further aspect of the way families are assembled in *MEROPS* that has sometimes given rise to confusion is the use of 'linking sequences'. The essential characteristic of peptidases within a single family is that they are evolutionarily related. If peptidase A is related to peptidase B, and peptidase B is related to peptidase C, then peptidase A must obviously be related to peptidase C, even if they have rather different structures. In these circumstances it is the existence of peptidase B that proves the relationship of peptidases A and C, and it can therefore be termed a 'linking sequence'. As more sequence information is obtained, we are discovering new linking sequences and from time to time these allow families to be merged. The new, enlarged families may, however, contain individual peptidases that differ greatly from each other.

At the time of writing, there are about 140 families in the *MEROPS* database. A few of the families contain two or more rather distinct groups of peptidases, as shown by a deep divergence in the dendrogram. For these, subfamilies are recognized.

The naming of the families follows the system introduced by Rawlings and Barrett (21) in which each family is named with a letter denoting the catalytic type (S, C, T, A, M or U, for serine, cysteine, threonine, aspartic, metallo- or unknown, respectively), followed by a sequentially assigned number. The threonine-type peptidases are the most recently discovered catalytic type (25). When a family disappears, usually because it is merged with another, the family name is not reused. For this reason, there are interruptions in the numerical sequences of families that are of no current significance.

4.2 Clans

Although the families are the largest groupings of peptidases that can be proven rigorously to be homologous, there are persuasive lines of evidence that many of the families do share common ancestry with others. That is to say, there are sets of families in which all of the proteins have diverged from a single ancestral protein, but they have diverged so far that we can no longer prove their relationship by comparison of the primary structures. The term clan is used to describe such a group of families (21, 26). The best kind of evidence to support the formation of a clan is similarity in three-dimensional structures when the data are available, but the arrangement of catalytic residues in the polypeptide chains and limited similarities in amino acid sequence around the catalytic amino acids are also taken into account, so that the assignment of a family to a clan normally depends upon several lines of evidence. Sometimes, the assignment can be made with a high level of confidence but on other occasions it has to be considered provisional.

The name of each clan is formed from the letter for the catalytic type of the peptidases it contains (as for families) followed by an arbitrary second capital letter (*Table 3*). As with families, if a clan disappears, the name is not reused. A clan that contains families of more than one of the C, S and T types is treated as being of 'P' type. An example is clan PA that contains trypsin and many other serine peptidases but also contains cysteine peptidases from RNA viruses (27) (see also *Table 3*). About 32 clans are recognized at the time of writing.

4.3 Individual peptidases

The *MEROPS* system—in the form of version 2 on the WWW, and the databanks tables in the *Handbook of Proteolytic Enzymes* (28)—includes data for all the peptidases that are recognized in the EC system and for which sequence data are available. A good many other peptidases are included that are not yet in the EC system. All of the available amino acid sequences of peptidases are included, after removal of 'redundant' items that are essentially identical. Each individual peptidase is given an identifier that acts much like an accession number. This is formed by concatenation of the three-character identifier of the family to which the peptidase belongs, a point, and a three-figure number.

4.4 Uses of the *MEROPS* system

The literature on peptidases contains many sweeping statements about the properties of peptidases of serine, cysteine or metallo- type that should perhaps be restricted to the enzymes of particular families. The 30 and more clans that can already be recognized indicate that there are at least this number of quite distinct lines of peptidase that have had different evolutionary origins and have very different structures and chemistry. These merit separate consideration, and the classification in *MEROPS* provides for this.

The existence of distinct 'families' of proteins is well recognized, but the exact meaning of a family has had no single, clear definition, so the concept has been a vague one. Within the *MEROPS* system there is a rigorous definition of family that is applied across the hundreds of known peptidases and, most importantly, it provides a way of dealing with the problem of mosaic proteins by restricting the comparison of amino acid sequences to the part of each enzyme that is responsible for its catalytic activity.

Until now, there has been no unambiguous way to name families of peptidases, and a family has generally been referred to by the name of one of its members. For example, one might speak of the 'prolyl oligopeptidase family' or the 'dipeptidyl-peptidase IV family'. It happens that both of these would be references to the same family, S9 in the *MEROPS* system, since both peptidases named are in this family. One of the useful features of *MEROPS* is therefore that it provides an identifier that can be used as a concise and unambiguous way to refer to any family or clan of peptidases. As the *MEROPS* family names have been increasingly introduced into the records of the sequence databases and the EC list, they have become effective ways of searching for information about the family as a whole.

The WWW version of the *MEROPS* database forms a convenient mode of access to a great deal of the information about peptidases that is available on the WWW (*Figure 3*). At the family level, there is access to the SCOP, MIPS, HSSP and PROSITE databases. These provide sequence alignments, secondary structure predictions, dendrograms, and conserved sequence motifs. *MEROPS* itself contains family alignments, family trees and search facilities. At the level of individual peptidases there is access to the EC list. At the level of sequences there is access to the annotated amino acid and nucleic acid sequence data of SWISS-PROT, PIR and EMBL, as well as the data on human genetics in OMIM. Data for three-dimensional structures are available from the Brookhaven Protein Databank

Caspase-1

MEROPS classification			
Clan: CD	Family: C14	Subfamily: -	MEROPS ID: C14.001
IUBMB classification			
3.4.22 Cysteine endopeptidases		EC Number: EC 3.4.22.36	
Nomenclature			
Recommended Name	Caspase-1		
Other names	Interleukin 1-beta-converting enzyme		
Other information			
For a review, see *Handbook of Proteolytic Enzymes* chapter 248			
Catalytic type	Cysteine		
Links:	Functional relevance		3D Image

Human Genetics			
Gene names	***Chromosome location***	***GDB***	***OMIM***
IL1BC; IL1BCE; CASP1	11q22.2-q22.3		147678

Protein Sequence Data		
Species (and isoform)	***SwissProt***	***PIR***
Homo sapiens	P29466	A42677
Mus musculus	P29452	A46495
Rattus norvegicus	P43527	I53300

Nucleic Acid Sequence Data				
	Comment	***EMBL***	***GenBank***	***CDS***
Gallus gallus				
	cDNA	AF031351	AF031351	G2642241
Homo sapiens				
	cDNA	M87507	M87507	

Tertiary Structure Data					
	Comment	***Resolution***	***PDB***	***DSSP***	***RASMOL***
Homo sapiens					
	mature peptidase	2.6 A	1ICE	1ICE	1ICE

Figure 3 One view of the *MEROPS* database as it appears on the World Wide Web. This view of the output for an individual peptidase is just a sample of what is contained in the database. The figure shows only the format of the data for caspase-1; many sequence identifiers have been omitted to save space. The clickable links to other parts of *MEROPS* and to other databases can be seen.

(PDB), and *MEROPS* provides for viewing and manipulation of the three-dimensional images by use of RASMOL (29). Bibliographic data are accessible through SWISS-PROT and EMBL.

5 Steps one might take on discovering a new peptidase

Clearly, the systems of nomenclature (EC list) and classification (*MEROPS*) for peptidases can work well only if they are kept up to date. It is therefore important that newly discovered peptidases are included when the time is right. I would therefore urge anyone who feels that they have discovered a new peptidase to consider putting it forward for an EC entry, as described above. Once the sequence has been deposited in the databanks, it should find its way into the *MEROPS* system automatically. However, the mass of data is such that this can take longer than it should, and the curator of the WWW version of *MEROPS* (see below) would be happy to be informed of new peptidases that should be included.

Acknowledgement

My colleague Neil D. Rawlings (neil.rawlings@bbsrc.ac.uk) has made essential contributions at all stages of the development of the *MEROPS* system, and is curator of the World Wide Web database. I thank him for his advice during the writing of the present chapter.

References

1. Grassmann, W. and Dyckerhoff, H. (1928). *Hoppe-Seyler's Z. Physiol. Chem.* **179,** 41.
2. Bergmann, M. and Ross, W. F. (1936). *J. Biol. Chem.* **114,** 717.
3. Barrett, A. J. and McDonald, J. K. (1985). *Biochem. J.* **231,** 807.
4. NC-IUBMB (Nomenclature Committee of the International Union of Biochemistry and Molecular Biology) (1992). *Enzyme Nomenclature 1992. Recommendations of the Nomenclature Committee of the International Union of Biochemistry and Molecular Biology on the nomenclature and classification of enzymes.* Academic Press, Orlando.
5. Berger, A. and Schechter, I. (1970). *Philos. Trans. R. Soc. Lond. Ser. B Biol. Sci.* **257,** 249.
6. Hartley, B. S. (1960). *Annu. Rev. Biochem.* **29,** 45.
7. Barrett, A. J. (1980). In *Protein degradation in health and disease. Ciba Foundation Symposium 75* (ed. Evered, D. and Whelan, J.), p. 1. Excerpta Medica, Amsterdam.
8. Barrett, A. J. (1986). In *Proteinase inhibitors* (ed. Barrett, A. J. and Salvesen, G.), p. 3. Elsevier Science Publ., Amsterdam.
9. Reeck, G. R., de Haën, C., Teller, D. C., Doolittle, R. F., Fitch, W. M., Dickerson, R. E. *et al.* (1987) *Cell* **50,** 667.
10. Webb, E. C. (1993). *FASEB J.* **7,** 1192.
11. NC-IUBMB (Nomenclature Committee of the International Union of Biochemistry and Molecular Biology) (1994). *Eur. J. Biochem.* **223,** 1.
12. NC-IUBMB (Nomenclature Committee of the International Union of Biochemistry and Molecular Biology) (1995). *Eur. J. Biochem.* **232,** 1.
13. NC-IUBMB (Nomenclature Committee of the International Union of Biochemistry and Molecular Biology) (1996). *Eur. J. Biochem.* **237,** 1.
14. NC-IUBMB (Nomenclature Committee of the International Union of Biochemistry and Molecular Biology) (1997). *Eur. J. Biochem.* (in press).
15. CBN (Commission on Biochemical Nomenclature) (1973). *Enzyme Nomenclature. Recommendations (1972) of the Commission on Biochemical Nomenclature on the nomenclature and classification of enzymes together with their units and the symbols of enzyme kinetics.* Elsevier Scientific Publ., Amsterdam.
16. Knight, C. G., Dando, P. M., and Barrett, A. J. (1995). *Biochem. J.* **308,** 145.
17. Selkoe, D. J., Yamazaki, T., Citron, M., Podlisny, M. B., Koo, E. H., Teplow, D. B., and Haass, C. (1996). *Ann. N.Y. Acad. Sci.* **777,** 57.

18. Brown, A. M., Tummolo, D. M., Spruyt, M. A., Jacobsen, J. S., and Sonnenberg-Reines, J. (1996). *J. Neurochem.* **66,** 2436.
19. Monnet, V. (1995). In *Methods in enzymology* (ed. Barrett, A. J.), Vol. 248, p. 579. Academic Press, San Diego.
20. Bieth, J. G. (1995). In *Methods in enzymology* (ed. Barrett, A. J.), Vol. 248, p. 59. Academic Press, San Diego.
21. Rawlings, N. D. and Barrett, A. J. (1993). *Biochem. J.* **290,** 205.
22. Barrett, A. J. (ed.) (1994). *Methods in enzymology*, Vol. 244. Academic Press, San Diego.
23. Barrett, A. J. (ed.) (1995). *Methods in enzymology*, Vol. 248. Academic Press, San Diego.
24. Rawlings, N. D. and Barrett, A. J. (1990). *Biochem. J.* **266,** 622.
25. Seemüller, E., Lupas, A., Stock, D., Löwe, J., Huber, R., and Baumeister, W. (1995). *Science* **268,** 579.
26. Barrett, A. J. and Rawlings, N. D. (1995). *Arch. Biochem. Biophys.* **318,** 247.
27. Allaire, M., Chernaia, M.M., Malcolm, B.A., and James, M.N.G. (1994). *Nature* **369,** 72.
28. Barrett, A. J., Rawlings, N. D, and Woessner Jr, J. F. (eds) (1998) *Handbook of proteolytic enzymes*. Academic Press, London.
29. Sayle, R. A. and Milner-White, E. J. (1995) *Trends Biochem. Sci.* **20**, 374.

Chapter 2
Purification of proteolytic enzymes

Sherwin Wilk
Department of Pharmacology, Mount Sinai School of Medicine, New York, NY 10029-5205, USA

1 Introduction

Proteolytic enzymes are purified according to well established and extensively reviewed procedures for purification of proteins in general. However, there are aspects of the purification of some proteolytic enzymes which may require special attention. Many intracellular proteinases are under tight regulation to prevent uncontrolled cellular proteolysis. Tissue disruption may release potent inhibitors, which, if tightly bound, can pose serious problems in purification. In addition, unlike other proteins, proteinases are susceptible to autolysis. The first section of this chapter discusses issues that must be dealt with prior to initiating purification, and also discusses potential problems and pitfalls that are associated with the purification of this class of enzymes in particular. The second section of this chapter deals with the purification procedure in general. Particular attention is paid to affinity chromatography as it relates to proteolytic enzymes. The third section deals with optimizing the purification protocol and judging homogeneity.

In this chapter, it will be assumed either that the enzyme of interest has not previously been purified or that if it has, the existing purification procedure has serious deficiencies and therefore requires modification. Clearly, once an enzyme has been cloned and large amounts are required for further characterization or crystallization, expression systems, especially for low abundance enzymes, may be methods of choice. For example, secreted peptidyl-dipeptidase A (angiotensin-I converting enzyme; EC 3.4.15.1) from *Drosophila melanogaster* was expressed in the yeast *Pichia pastoris* at a level of 160 mg/l culture medium supernatant (1). The secreted recombinant protein was apparently homogeneous on SDS-PAGE without purification. Traditionally, a protein is first purified, sequence information is obtained and cloning follows. With the rapid expansion of the protein and expressed sequence tag databases, and the refinement of the techniques of molecular biology, cloning often precedes purification. Examples of the latter approach can be found in the expanding list of interleukin-1-β-converting enzyme-like cysteine proteinases (caspases) (2). Specialized methods for the expression and purification of recombinant proteins will not be considered in this chapter.

2 Prelude to purification

2.1 Assay

2.1.1 The problem of identity

The starting point of any purification procedure is the establishment of a convenient assay. Ideally, the assay used should be specific for the enzyme under investigation, but this is rarely achieved (see below). Since proteolytic enzymes can cleave peptide bonds in peptides, proteins, and synthetic peptide substrates, apparently 'novel' proteolytic enzymes frequently prove to be identical to enzymes that had been previously characterized with other substrates. For example, the enzyme now known as thimet oligopeptidase (EC 3.4.24.15) has appeared over the years under several different guises. Early studies on collagen metabolism led to the description of an enzyme called Pz-peptidase (3). This name derives from the ability of the enzyme to cleave Pz-peptide (phenylazobenzyloxycarbonyl-Pro-Leu-Gly-Pro-D-Arg), a substrate designed to detect enzymes with collagenase-like activity. The use of bradykinin as substrate (Arg-Pro-Pro-Gly-Phe-Ser-Pro-Phe-Arg) led to the purification of an enzyme termed 'endo-oligopeptidase A' which cleaved this biologically active peptide at the Phe-Ser bond (4). Use of the synthetic substrate α-*N*-benzoyl-Gly-Ala-Ala-Phe-*p*-aminobenzoate, led to the purification of the neuropeptide-degrading enzyme endopeptidase 24.15 (5). All three enzymes are now recognized as identical (6,7). It therefore seems reasonable to propose that the substrate specificity of the proteinase of interest once purified should be characterized using a variety of natural and synthetic substrates before concluding that it is a new enzyme. Molecular cloning has led to an accelerated deposition of the primary structures of proteinases in the protein data bank. Partial sequence analysis may provide an unequivocal answer to most questions of molecular identity.

2.1.2 The problem of overlapping specificities

Many proteolytic enzymes may be capable of cleaving a common substrate. For example, many proteinases have a 'trypsin-like' specificity, i.e. these enzymes can cleave substrates containing arginine or lysine in the P_1 position—the amino acid residue in the *N*-terminal direction of the scissile peptide bond is designated P1 and the residue in the *C*-terminal direction is designated P_1' according to the nomenclature of Schechter and Berger (8). These include enzymes of the clotting cascade, pro-hormone convertases, kallikreins, tryptase, acrosin and some cathepsins. Therefore, the use of a relatively non-selective substrate can complicate the early stages of purification and facilitate the possibility of purifying the wrong enzyme. This is particularly true when one seeks to purify a novel proteolytic enzyme armed only with the knowledge of the peptide bond cleaved in a particular endogenous substrate. For example, the early work of Steiner and colleagues established that hormones are generated from their precursors (pro-hormones) by a trypsin-like cleavage which occurs generally after a pair of basic amino acids (9). A variety of synthetic substrates containing Arg in the P_1 position have been used in studies extending over 20 years in an attempt to purify this elusive activity. It is now known that pro-hormone convertases belong to the family of subtilisin-like enzymes, i.e. Ca^{2+}-activated serine proteinases (10). The earlier literature however is strewn with many false leads.

2.1.3 Autolysis

Autolysis of a homogeneous enzyme preparation is a problem unique to proteinases. Autolysis can be established by gel electrophoretic analysis of the protein as a function of the time of its storage. If autolysis is suspected, enzyme preparations should be periodically examined by gel electrophoresis. The problem can be minimized by storage of the enzyme at −80 °C or storage in the presence of an inhibitor. A specialized solution to this problem is exemplified by trypsin where autolysis is prevented by Ca^{2+}. Storage of the enzyme in the presence of an inhibitor can be inconvenient as the inhibitor must be removed before working with the enzyme. Moreover the use of dialysis to effect inhibitor removal may in some cases facilitate autolysis (11). It should be noted that autolysis does not always lead to inactivation. For example, autolytic changes in the multicatalytic proteinase complex (proteasome, EC 3.4. 99.46) will activate the hydrolysis of β-casein (11). The autolysis of calpain *in vivo* has been proposed as a mechanism involved in its activation (12).

2.1.4 Proteinase inhibitors

Many endogenous inhibitors of proteolytic enzymes have been described and characterized. One of the problems occasionally encountered in the purification of these enzymes is loss of activity due to a proteinase–proteinase inhibitor interaction. Since many proteinase inhibitors are present in blood, a good precaution is to wash the tissue free of blood before homogenization. Tissue perfusion to remove all blood is even more effective. If inhibitor and proteinase occur in different cellular compartments, the initial homogenate can be made in a 0.25 M sucrose solution. Differential centrifugation may provide a subcellular fraction enriched in the proteinase and lacking the proteinase inhibitor. There are cases in which the presence of a proteinase is masked by the inhibitor and is revealed only in later steps of purification. For example, a calcium-activated proteinase can be detected in unfractionated extracts of skeletal muscle but not in unfractionated extracts of cardiac muscle. After DEAE-cellulose chromatography of a cardiac muscle extract, an endogenous inhibitor is separated and a peak of proteinase activity is detected (13).

3 Initial considerations

3.1 Source

Purification will be facilitated by using a source rich in the enzyme of interest. In many cases, however, the experimental goal is to purify the enzyme from a specific tissue and/or species. If the enzyme is present in only very low concentrations in the tissue of interest but is abundant in another tissue, purification from the abundant source is often a good way to initiate studies. The chromatographic behaviour of the enzyme from the abundant source can be characterized and the results applied to the tissue of interest. Moreover, the enzyme purified from the abundant source could be used for antibody production. An immunoaffinity column could then be prepared and this column employed in a subsequent purification from the tissue of interest.

3.2 Buffer composition

Buffer composition is frequently critical for successful enzyme purification. Factors such as pH, ionic strength and nature of the buffer can all contribute to protein instability. Although selection of the optimal buffer composition is often an empirical process, some rational judgements can be made. For example, cysteine proteinases generally require the inclusion of a sulphydryl reducing agent such as DTT or 2-mercaptoethanol in the buffers. Because heavy metals will readily inactivate cysteine proteinases, it is essential that the highest quality water and reagents be used for buffer preparation. As an added precaution, it is a good idea to also include EDTA. It should be noted that some serine proteinases and even metalloproteinases contain a sulphydryl group whose integrity is necessary for expression of enzyme activity. Therefore, inclusion of thiol-reducing agents in buffers is not always limited to cysteine proteinases. For example, prolyl oligopeptidase (EC 3.4.21.26), a serine proteinase and thimet oligopeptidase (EC 3. 4.24.15) a metalloproteinase, were previously considered by some investigators to be sulphydryl proteinases on the basis of their sensitivity to thiol-blocking reagents. Since metal ion chelators will inhibit metalloproteinases, buffers used for the isolation of cysteine proteinases are generally incompatible with buffers used for isolation of metalloproteinases. Finally, some proteolytic enzymes are particularly labile and special procedures such as inclusion of glycerol in buffer solutions must be used to stabilize them.

3.3 Membrane-bound or soluble?

The next question to be addressed is whether the proteinase of interest is membrane-bound or soluble. A clear answer to this question is frequently difficult since a highly specific assay is required for an unambiguous interpretation of results obtained from crude tissue fractions. Moreover, some proteinases may be found in both fractions. It will be assumed in the following discussion that the proteinase to be purified is either predominantly present in one of these fractions or that the investigator has set out initially to purify either a membrane-bound or soluble enzyme.

3.4 Membrane-bound enzymes

A true membrane-bound enzyme will resist extensive washings of membrane preparations with high ionic strength buffers. This is, however, not an absolute criterion. Occasionally, a fraction of a proteinase will remain firmly associated to membranes and the membrane-associated activity will resist repeated washings, despite the absence of a membrane-spanning domain in the primary sequence of the proteinase. For example, ≈ 20% of the total activity of thimet oligopeptidase in rat brain homogenates remains tightly associated with membrane fractions (14). The enzyme cannot be removed by repeated washings with 0.32 M sucrose, hypotonic buffers or buffers containing 0.5 M NaCl. However, the enzyme can be completely solubilized by treatment with Triton X-100. A similar situation is seen with the puromycin-sensitive cytosol alanyl aminopeptidase (EC 3.4.11.6). It has been proposed (15) that this enzyme may belong to the class of amphitropic proteins as described by Burn (16). Such proteins interact with lipids, thus facilitating their association with membranes.

3.4.1 Solubilization of membrane-bound proteins with detergents

Detergent treatment is the most commonly used approach for solubilizing membrane proteins. The major concern with this method is a detergent-induced loss of enzymatic activity. Although Triton X-100 remains a popular first choice because of cost, many other ionic and non-ionic detergents are available if Triton X-100 treatment inactivates the enzyme. Another and frequently more troublesome problem is being able to remove the detergent from the solubilized enzyme without causing enzyme precipitation. In many cases, it is possible to lower the detergent concentration as the purification proceeds, eventually entirely omitting the detergent in the final steps.

3.4.2 Proteolytic solubilization of membrane-bound enzymes

Proteolytic cleavage of the enzyme at its membrane anchoring site is frequently employed to solubilize membrane-bound enzymes. In this method, a membrane suspension is incubated with a proteolytic enzyme such as trypsin or papain. In some cases, autolysis by cathepsins is used for solubilization. For autolytic solubilization, the pH of the membrane suspension is lowered to the acidic range and the suspension incubated at 37 °C.

During the course of proteolytic solubilization, aliquots of the suspension are removed, centrifuged and the supernatant tested for enzymatic activity. The goal is to minimize both the concentration of proteolytic enzyme used for solubilization and the incubation time (or temperature) while maximizing the yield of soluble enzyme. The major drawback of this approach is proteolysis of the target enzyme. Limited proteolysis of the target enzyme is not always easy to detect since it may not necessarily lead to loss of enzymatic activity if non-covalent forces maintain the tertiary structure of the protein. Retention of enzymatic activity is therefore no guarantee that 'over-proteolysis' has not occurred. Under conditions of over-proteolysis, SDS-PAGE of a homogeneous preparation can yield several bands (17). If the target enzyme has never been characterized, the gel electrophoretic pattern may be incorrectly interpreted either in terms of protein subunit structure or purity of the preparation.

The amount of proteolytic enzyme necessary for solubilization can vary widely. For example, solubilization of pig kidney membrane alanyl aminopeptidase (EC 3.4.11.2) requires treatment for 1 h at 37 °C with 200 μg trypsin/g kidney (18), whereas solubilization of pyroglutamyl peptidase II from rat brain synaptosomal membranes requires only a 15-min incubation at 4 °C with 50 μg trypsin/g membrane protein (17). More vigorous trypsinization of pyroglutamyl peptidase II cleaves the protein chain into several fragments (17, and own unpublished observations). After proteolysis has been completed, it is necessary to remove or inactivate the proteolytic enzyme employed for solubilization. When trypsin is used, soybean trypsin inhibitor can be added to terminate the reaction. Papain can be inactivated by a thiol-blocking reagent provided that the target enzyme is not sensitive to thiol blockers.

Various methods of solubilization may work for a given enzyme. *Table 1* lists solubilization procedures that have worked for the membrane- bound ecto-enzyme dipeptidyl peptidase IV (EC 3.4.14.5). Hopsu-Havu and Sarimo (19) compared various methods for the solubilization of dipeptidyl peptidase IV from rat liver membranes. The procedures compared were Triton X-100, deoxycholate, trypsin, autolysis, ultrasound, freezing and thawing (10×), and treatment with snake venom. The most effective method found for this enzyme was autolysis.

Table 1 Methods for the solubilization of dipeptidyl peptidase IV

Enzyme source	Solubilization method	Reference
Human kidney	Triton X-100	22
Human placenta	0.1% Nonidet P40	21
Human submaxillary gland	Autolysis	26
Porcine kidney	Autolysis	23,24
Rat intestine	Triton X-100	20
Rat kidney	Autolysis	25
Rat liver	Autolysis	19
	Sulphobetaine	28
	Triton X-100	27

Protocol 1

Autolytic solubilization of dipeptidyl peptidase IV from rat kidney membranes[a]

Equipment and reagents

- Rat kidneys, fresh or frozen
- 1 M HCl
- 1 M Trizma base
- Waring blender

Protocol 1 continued

Method

1 If frozen kidneys are used, allow them to partly thaw on ice. Slice each kidney in half.

2 Add 200 ml H_2O to 50 g kidneys and homogenize at 4 °C in a blender for 2 min at high speed.

3 Wait 5 min for the homogenate to cool to 4 °C and homogenize once more for 2 min at high speed.

4 Centrifuge at 1100 **g** for 15 min.

5 Collect the supernatant and acidify to pH 3.9 with 1 M HCl. Incubate overnight (18 h) at 37 °C.

6 Centrifuge at 18 000 **g** for 30 min. Collect the supernatant and immediately adjust to pH 7.2 with 1 M Trizma base.

[a] This protocol can also be used to solubilize the kidney brush border membrane proteinases glutamyl aminopeptidase (EC 3.4.11.7), membrane alanyl aminopeptidase (EC 3.4.11.2) and neprilysin (EC 3.4.24.11).

3.4.3 Solubilization of glycosyl phosphatidylinositol anchored proteinases

Some proteolytic enzymes are attached to the membrane by a glycosyl phosphatidyl inositol anchor (29). These proteins can be conveniently solubilized by treatment of membranes with phospatidylinositol-specific phospholipase C. For X-Pro aminopeptidase (aminopeptidase P, EC 3.4.11.9) solubilization was achieved by treatment with phospholipase from *Bacillus thurigensis* (30). Incubation of crude rat lung microsomes for 8–11 h at 39 °C with phospholipase C was sufficient to solubilize more than 90% of the enzyme.

3.4.4 Other solubilization techniques

n-Butanol has been used to solubilize X-Pro aminopeptidase (31). The authors speculate that this glycosyl phosphatidyl inositol-anchored protein is actually released by cellular phospholipase D during the process of butanol treatment.

3.5 Endogenous inhibitors and activators

Many proteolytic enzymes are under tight cellular regulation because uncontrolled proteolysis could be catastrophic to a cell (for a review see ref. 32). Endogenous inhibitors and activators are commonly revealed in the early stages of purification. Their existence is suspected if, during purification, one encounters either a sudden increase or decrease in enzymatic activity. However, although an increase in recovery greater than 100% in any step may suggest removal of an inhibitor, it is also possible that activity has increased merely as a result of the removal of proteins that compete with the substrate for binding to the enzyme. Similarly, although a precipitous loss of activity may suggest the removal of an endogenous activator, there are other possibilities. These include destabilization of the enzyme in more dilute protein solutions, instability in a new buffer solution or irreversible binding to the chromatographic support. Once the enzyme is obtained in a highly purified state, earlier fractions can be added back to it to test for the presence of endogenous modulators. This method has been applied for the detection and purification of the proteasome activator PA28 (33). Dialysis or ultrafiltration will readily reveal if the effector is a macromolecule. The stability of the effector to heating can provide information on its nature.

4 General scheme for purification of proteolytic enzymes

4.1 Initial steps

The initial stages of proteinase purification generally contain one or more crude fractionation steps to remove bulk protein and prepare the material for subsequent chromatography. Pilot studies must be performed to determine which procedure is most suitable and the experimental conditions to be used. The most commonly employed initial step is ammonium sulfate fractionation (34). Ammonium sulfate of high purity designed for protein purification should be used, since trace contaminants of heavy metals can inactivate many proteinases. Batch ion-exchange chromatography is a reasonable alternative provided that the protein of interest binds to the exchanger. In this method, the ion-exchanger is equilibrated with the same buffer used to prepare the tissue extract, mixed with the extract and the suspension poured into a Büchner funnel. Unbound protein is removed by washing with the equilibrating buffer until the absorbance at 280 nm of the effluent reaches baseline. The protein of interest is eluted by the equilibrating buffer containing NaCl. A limitation of this method is its cost. Another approach is the use of polyethylene glycol for fractional precipitation. This technique has the advantage of being a fairly mild treatment (35). Other procedures that have been used include acid precipitation, heat inactivation and fractionation with organic solvents (34); these are harsher treatments and some enzymes will not readily withstand them. Cathepsins which are optimally active at an acidic pH may more readily tolerate a low-pH fractionation step. For example, an initial step in the purification of porcine spleen cathepsin D is an acid precipitation at pH 3.7 (36). Heat inactivation may facilitate autolysis or proteolysis by other proteinases present in the crude extract. Following initial fractionation, the protein of interest is either resuspended in or dialysed against the buffer to be used in the next step.

Protocol 2

Partial purification of prolyl oligopeptidase (EC 3.4.21.26) by acid precipitation and ammonium sulfate fractionation

Equipment and reagents

- Rat brains, fresh or frozen (Pel Freez)
- 25 mM acetic acid
- Tris-EDTA buffers, 10 mM, pH 7.5 and pH 8.3
- 1 M Trizma base
- Ammonium sulfate (SigmaUltra)
- Magnetic stirring plate

Method

A. Acid precipitation

1 Add 250 ml Tris-EDTA buffer, pH 7.5–50 g of rat brains and homogenize at 4 °C for 1 min in a blender at high speed. Wait 5 min and repeat.[a]

2 Centrifuge for 30 min at 20 000 g at 4 °C and collect the supernatant.

3 Add the supernatant to a beaker containing a magnetic stir bar. Place the beaker in an ice-filled container and stir at medium speed on a magnetic stirring plate.

4 Add 25 mM acetic acid slowly and dropwise until the pH reaches 5.0. Try to avoid allowing the pH to fall below 5.0. Continue stirring for 5 min and determine if the pH has stabilized. If necessary, add more acetic acid.

5 Centrifuge at 20 000 **g** for 30 min. Transfer the supernatant an ice-cooled beaker and immediately adjust to pH 7.2 with 1 M Trizma base.

B. Ammonium sulfate fractionation

1 Measure the volume of the pH 7.2 supernatant. Add solid ammonium sulfate to 50% saturation (313 g/l) over 30 min to the cooled and efficiently stirred solution. Break up any lumps of ammonium sulfate before addition.[b] Continue stirring for a further 30 min.

2 Centrifuge for 20 min at 20 000 **g**. Remove the supernatant and measure its volume.

3 Add ammonium sulfate to 80% saturation (214 g/l) over a 30-min period. Stir for a further 30 min and centrifuge for 20 min at 20 000 **g**. Discard the supernatant and dissolve the precipitate in a minimal amount of 10 mM Tris-EDTA buffer, pH 8.3 with the aid of a glass rod.

[a] It is important to avoid foaming at this step.

[b] It is necessary to avoid a local high concentration of ammonium sulfate which could result in precipitation of unwanted proteins.

4.2 Intermediate and final steps

4.2.1 Column chromatography

Protein purification relies on exploiting as many of the physical properties of the protein as are necessary for separation from all other proteins. Most commonly, these steps will include methods that separate by molecular weight (size exclusion chromatography), charge (ion exchange chromatography), hydrophobicity (hydrophobic interaction chromatography), adsorptive properties (hydroxylapatite chromatography) and group-specific interactions (affinity chromatography). In general, ion exchange chromatography precedes size exclusion chromatography in sequence because of the limitation of sample size in the latter procedure. Ion exchange chromatography will generally remove enough protein to allow the sample to be concentrated to a small volume sufficient to apply to a gel-filtration column. In most cases, a series of standard column chromatographic steps will be sufficient to yield a homogeneous enzyme

4.2.2 FPLC

FPLC provides excellent speed and resolution, but is limiting with respect to sample size. Large-scale preparations generally employ standard column chromatographic techniques as intermediate steps and may be followed by FPLC separations as final steps. In cases where only moderate amounts of material are

Table 2 Some affinity supports used for proteinase purification

Support	Proteinase	Source or reference[a]
Affi-Gel 501 (organomercurial)	Cysteine	Bio-Rad
Amastatin-AH Sepharose 4B	Glutamyl aminopeptidase	39
Arginine Sepharose 4B	Trypsin-like	Amersham Pharmacia Biotech
Bacitracin-Sepharose 4B	Glutamyl endopeptidase	40
Benzamidine Sepharose 4B	Trypsin-like	Amersham Pharmacia Biotech
Bestatin-Sepharose 4B	Microsomal aminopeptidase	41
Chelating Sepharose	Metallo-	Amersham Pharmacia Biotech
D-Trytophanmethyl ester-CH Sepharose	Chymase	47
Gelatin-Sepharose 4B	Gelatinases	Amersham Pharmacia Biotech
Glutathione Sepharose 4B	Cysteine	Amersham Pharmacia Biotech
Gly-Phe-glycinal-semicarbazone-Sepharose 4B	Cathepsin B	42
Gly-Pro-aminoacetonitrile-CH-Sepharose 4B	Papain, papaya proteinase	43
Gly-Pro-Sepharose 4B	Dipeptidyl peptidase IV	44
Hydroxamate-Sepharose 4B	Collagenase	45
Lysinopril-Sepharose CL 4B	Peptidyl-dipeptidase A	37
Pro-Leu-Gly-AH Sepharose 4B	Prolyl aminopeptidase	46
Thiopropyl Sepharose 6B	Cysteine	Amersham Pharmacia Biotech

[a] Commercial source is listed if available. In other cases, the original reference should be consulted for preparation.

used, e.g. purification of a proteinase from a cell culture, a series of FPLC steps can rapidly and efficiently provide homogeneous enzyme. The speed of FPLC is of particular advantage when working with a labile enzyme.

4.2.3 Affinity chromatography

The most efficient methods employ affinity chromatography. For example, the potent angiotensin converting enzyme inhibitor lisinopril when coupled to epoxy-activated Sepharose CL-4B provides a means of obtaining electrophoretically homogeneous angiotensin-converting enzyme (EC 3.4.15.1) from a crude tissue homogenate in a single step (37). In this case, active enzyme is eluted from the column by an equilibrating solution containing free lisinopril. Affinity supports for proteinase purification have employed both substrates and inhibitors as ligands. Some of the affinity supports used in proteinase purification are listed in *Table 2*. Note that affinity chromatography can also be used to remove unwanted proteolytic enzymes. For example, it was found to be necessary to remove trypsin and chymotrypsin during the preparation of carboxypeptidases A and B from pancreatic juice. This was achieved by passage through an immobilized soybean trypsin inhibitor column (38).

5 Specialized techniques for proteolytic enzymes

5.1 Peptidyl aldehyde affinity chromatography

Peptidyl aldehydes inhibit serine and cysteine proteinases by forming transition-state analogue complexes with the active site residue (48,49). The interaction of peptidyl aldehydes with some proteinases has formed the basis of inhibitor affinity chromatography. The synthesis of these columns proceeds by the coupling of a protected aldehyde (e.g. semicarbazone or acetal) to the matrix, and removal of the protecting group prior to use. Since aldehydes tend to bind tightly to proteinases, elution can be problematic. Trypsin bound to a Gly-Gly-Argininal column is eluted either with a 50 mM HCl/1 M NaCl solution or with the argininal inhibitor leupeptin (50). Semicarbazide-containing buffers have been used to elute some trypsin-like proteinases (51). Differences in subsite specificities among the trypsin-like enzymes can be exploited for more specific purification by judicious choice of the ligand. For example, pyroglutamyl-Lys-Leu-argininal was coupled to agarose by the ε-amino group of lysine and the resulting affinity column used for the purification of urokinase (52). Elution was achieved with a 0.2 M solution of citric acid, resulting in a recovery of 90%. Aldehyde affinity chromatography has also been applied more recently for the purification of interleukin-1-β-converting enzyme (53). The inhibitor Ac-Tyr-Val-Ala-Asp-CHO, protected as its dimethyl acetal, was coupled to Sepharose 4B. Enzyme was eluted by free inhibitor and essentially homogeneous enzyme was obtained following chromatography of a crude cellular lysate. Interleukin-1-β-converting enzyme is a cysteine proteinase. Active enzyme was recovered from the eluate by treatment with hydroxylamine and glutathione disulfide, followed by reduction with DTT.

The weaker inhibition of some serine and cysteine proteinases by semicarbazones has also been applied for affinity purification. In this case, elution is much simpler (54).

5.2 Affinity columns for cysteine proteinases

The properties of the active site cysteine residue have been exploited for the affinity purification of this class of enzymes (see ref. 55 for a more complete discussion). Two classes of affinity columns are available for cysteine proteinase purification. Organomercurial agarose will form a covalent mercaptide bond with the active site cysteine residue. Advantage is taken of the low pK_a of the active site cysteine ($\approx$ 4) (56). Application of the sample at low pH will favour binding of cysteine proteinases over non-selective binding of protein thiol groups. Elution is achieved with a 10 mM solution of DTT or 2-mercaptoethanol. Since cysteine proteinases are usually stored in a buffer containing a thiol-reducing agent, the reducing agent must be removed prior to chromatography. Rapid gel filtration chromatography over columns designed for desalting or ultrafiltration can be used for buffer exchange. Note that particularly labile cysteine proteinases may not withstand prolonged storage in the absence of a reducing agent. An example of the use of organomercurial-Sepharose is found in a procedure for the purification of cysteine-type carboxypeptidase (cathepsin B2; EC 3.4.18.1) (57). After Sephadex G-100 chromatography, the partially purified enzyme was incubated with 5 mM DTE to maximize binding, and then dialysed against a buffer composed of 0.05 M sodium acetate, pH 4.6, 0.2 M NaCl, and 1 mM EDTA to remove the DTE. After application onto the organomercurial column, unbound protein was removed by washing with the same buffer, and elution was achieved with 10 mM cysteine. It was noted that better results could be obtained by stopping the column for 15–20 min after one column bed of cysteine-containing buffer had entered the column.

The second type of affinity column is represented by supports which contain a covalently bound thiol group. Cysteine proteinases will form a mixed disulfide with the thiolated matrix. These commercially available supports are supplied in a stable disulfide form. The protecting group is removed by a reducing agent prior to use. After binding of cysteine proteinases to the affinity column, unbound proteins are removed by washing. Cysteine proteinases are then eluted by reduction of the disulfide bond by solutions of DTT, 2-mercaptoethanol or cysteine. An example of the use of this technique is the purification of cathepsins B and H from bovine spleen on a thiopropyl-Sepharose 6B column (58). The enzymes were co-eluted by 20 mM cysteine and resolved on a carboxymethyl cellulose column.

5.3 Affinity columns for trypsin-like enzymes

Many serine proteinases have a trypsin-like specificity, i.e. they cleave peptide bonds on the carboxyl side of arginine residues. A commercially available arginine-Sepharose column can be used for affinity chromatography of this class of enzymes. For example, arginine-Sepharose (*Table 2*) was used to purify a trypsin-like serine proteinase 91-fold from a cytolytic T cell line (59). The enzyme was eluted by a Tris buffer containing 0.5 M arginine hydrochloride. It should be noted that some cysteine proteinases, such as cathepsin B, also have a trypsin-like specificity.

Benzamidine is a reversible inhibitor of trypsin-like enzymes and a benzamidine-Sepharose support is commercially available (*Table 2*). The use of *p*-aminobenzamidine as an affinity ligand was first described in 1973 and used in the purification of thrombin and trypsin (60). Benzamidine-Sepharose has been quite effective as the major step in the purification of the trypsin-like enzyme acrosin from boar semen (61). This step is conducted very rapidly since acrosin undergoes autolysis at the pH of adsorption to the affinity column.

5.4 Affinity columns for metalloproteinases

Commercially available affinity supports have been used in the purification of some metalloproteinases (*Table 2*). Examples are gelatin-Sepharose 4B for the purification of gelatinases (62) and chelating-Sepharose for the purification of matrilysin (63). Inhibition of metalloenzymes by hydroxamates has served as the basis of affinity purification of collagenase (45).

5.5 Other affinity columns

The use of cyclic peptides such as bacitraicin and gramicidin and of immobilized dyes as affinity ligands for proteinase purification has been extensively reviewed (64). The immobilized antibiotics appear to be particularly useful for the purification of proteolytic enzymes from microbial sources. Some excellent results have been obtained with chromatography of proteinases on immobilized dye-containing supports. This technique is not specific for proteolytic enzymes and is used for protein purification in general. The authors of the review article (64) recommend screening a panel of immobilized dyes to find one which selectively interacts with the target proteinase or with a difficult to remove contaminant. Protein binding to immobilized dyes is believed to be due to both hydrophobic and ionic interactions between the dye and the protein. The predominant interaction of the protein of interest will determine the nature of the mobile eluent. An example of this procedure is the use of Orange Sepharose in the purification of human PMN leukocyte collagenase (65); the Orange Sepharose column was essential for obtaining a homogeneous preparation.

Protocol 3

Coupling a peptide to an affinity support via an amino group[a]

Equipment and reagents

- Sintered glass funnel
- End-over-end mixer

Protocol 3 continued

- CH Sepharose 4B (Amersham Pharmacia Biotech)[b]
- Peptide for coupling
- Coupling buffer: 0.1 M $NaHCO_3$, 0.5 M NaCl, pH 8.0
- 1 mM HCl
- Low pH wash buffer: 0.1 M Na-acetate, 0.5 M NaCl, pH 4.0
- High pH wash buffer: 0.1 M Tris-HCl, 0.5 M NaCl, pH 8.0

Method

1 Weigh out the required amount of activated CH Sepharose 4B (1 g lyophilized material will give 3 ml of gel, and 1 ml of gel will couple 6–16 μmol peptide).[c]

2 Suspend lyophilized powder in ice-cold 1 mM HCl (200 ml 1 mM HCl/g lyophilized powder) and wash on a sintered glass filter for 15 min. Begin coupling within 30 min of swelling.

3 Dissolve peptide ligand in coupling buffer at a concentration several-fold in excess of active ester groups.

4 Mix ligand solution with the gel in an end-over-end mixer for 1 h at room temperature.

5 Wash gel on a sintered glass funnel sequentially with the high and low pH wash buffers to remove excess ligand and deactivate active ester groups.

6 Store at 4 °C in the presence of a bacteriostatic agent such as sodium azide.

[a] Adapted from the manufacturer's instructions (Amersham Pharmacia Biotech).

[b] Activated CH Sepharose 4B is formed by esterification of a carboxyl group of the support with *N*-hydroxysuccinimide (active ester).

[c] Since coupling is to a free primary amino group, both the α-amino group of the peptide and the ε-amino group of a lysine side-chain can react. Reaction of α-amino groups is favoured by coupling at a lower pH. Ligands can be coupled in the pH range of 5–10, with optimal reaction at pH 8.0. Alternatively, a peptide can be designed in which either of these groups is blocked.

Protocol 4

Coupling a peptide to an affinity support via a carboxyl group[a]

Equipment and reagents

- Sintered glass funnel
- end-over-end stirrer
- Affi Gel 102 (Bio Rad).
- EDC (Sigma)
- 1 N HCl
- Low pH wash buffer: 0.1 M Na-acetate, 0.5 M NaCl, pH 4.0
- High pH wash buffer: 0.1 M Tris-HCl, 0.5 M NaCl, pH 8.0

Protocol 1 continued

Method

1 Remove the appropriate amount of pre-swollen gel (each ml of drained gel should couple 15 μmol ligand)[b] and dilute with an equal volume of distilled water.

2 Add 10–50 μmol ligand in appropriate solvent[c]/ml gel and mix in an end-over-end shaker. Adjust the pH to 4.7–5.0 with 1 N HCl.

3 Gradually add 2–10 mg EDC with continued shaking.

4 Immediately readjust the pH to 4.7–5.0 with 1 N HCl and continue shaking overnight at room temperature.

5 Wash and store gel as described in Protocol 3.

[a] Adapted from the manufacturer's instructions (BioRad).

[b] Since coupling is to a free carboxyl group, the α-carboxyl group of the peptide as well as carboxyl groups of aspartyl and glutamyl residues can react.

[c] If the ligand is insoluble in water, coupling can be achieved under anhydrous conditions in 100% DMSO, 2-propanol or ethanol.

6 Optimization of the purification protocol

The initial purification scheme can be optimized for efficiency. This is most dramatically illustrated by the purification of pyroglutamyl peptidase II (EC 3.4.19.6), as described by Bauer (17). This membrane-bound ectoenzyme cleaves the pyroglutamyl-histidyl bond of thyrotropin-releasing hormone (*p*-Glu-His-Pro-NH_2). Pyroglutamyl peptidase II is present in very low concentrations in a brain synaptosomal membrane preparation. A protocol consisting of eight column chromatographic steps was carefully optimized to arrive at an overall 200 000-fold purification at a 20% yield.

It is essential that all steps of the purification protocol be carefully monitored for yield and purity. A table of purification should be constructed with a format as in *Table 3*. This will make it possible to readily identify efficient and inefficient steps. Final evaluation will be based on electrophoretic analysis of the preparation.

Once a highly purified proteinase preparation is obtained, it can be unambiguously characterized with respect to proteinase class, substrate specificity and sensitivity to inhibitors. This will allow for a rational choice of affinity column for subsequent purification.

A tedious purification procedure may be followed by a much simpler protocol.

Table 3 Table of purification

Step	Volume	Protein	Activity	Specific activity	Fold purification	Total activity	Yield
n	ml	mg/ml	units/ml	units/mg	*n*	units	%

For example, the separation of rat kidney membrane alanyl aminopeptidase (EC 3.4.11.2) and glutamyl aminopeptidase (EC 3.4.11.7) is very difficult (66): the two enzymes share significant sequence homology and cannot totally be resolved by conventional gel filtration or ion-exchange chromatography. The enzymes can be partially separated on a membrane alanyl aminopeptidase substrate affinity column (Leu-Gly-Sepharose) and repeated chromatography of the glutamyl aminopeptidase-enriched fractions eventually yields essentially homogeneous glutamyl aminopeptidase. We used the purified glutamyl aminopeptidase for antibody production and the antibodies in turn were used to prepare a glutamyl aminopeptidase immunoaffinity column. Subsequent purifications were efficiently achieved by immunoaffinity chromatography, with elution effected by a 2 M solution of $MgCl_2$.

7 Determination of homogeneity

The first judgement made about homogeneity should derive from the elution pattern of the enzyme after the last chromatographic step. The peak for enzymatic activity should be totally coincident with the protein peak. Any skewing can be taken as evidence that the preparation is not homogeneous. Traditionally, gel electrophoresis is used to confirm the homogeneity of enzyme preparations. A single band on a non-dissociating gel is, however, not unequivocal evidence of homogeneity: two proteins may migrate at the same position on the gel. Further confirmation of homogeneity is obtained by electrophoresis at several different pH values. It is also possible that the enzyme could be only a trace protein in the preparation but be mistaken for the major protein band. Enzymatic activity should be determined directly on the gel, if possible, or the gel should be sliced and activity determined in the gel slices to demonstrate that the activity is coincident with the protein band. The enzyme should then be analysed by SDS-PAGE. A single band obtained following both non-dissociating gel electrophoresis and SDS-PAGE is good evidence of apparent homogeneity. A tryptic digest of the protein can also be subjected to analysis by MALDI mass spectrometry. The masses obtained can be used for data base searching for protein identification. An enzyme composed of non-identical subunits frequently presents a difficult problem of interpretation. This problem is considered in the following section on the purification of the multi-catalytic proteinase complex.

8 Purification of the MPC (proteasome; EC 3.4.99.46)

The MPC or 20S proteasome, because of its highly unique properties, presents an interesting and somewhat extreme problem in proteinase purification and evaluation of homogeneity. The properties of the constitutive eukaryotic enzyme briefly summarized below have been reviewed in a compendium of publications (67). MPC is a 28 subunit-containing protein composed of four rings of seven subunits each. The two outer rings are each made up of subunits structurally related to the α-subunit of the enzyme from *Thermoplasma acidophilum*

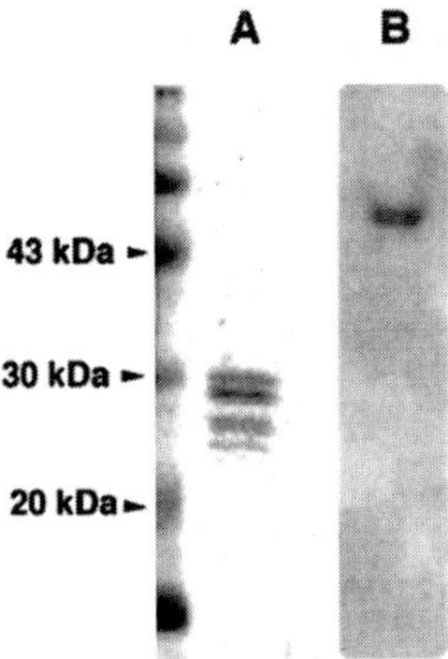

Figure 1 PAGE of purified bovine pituitary multi-catalytic proteinase complex. (A) SDS-PAGE run on a 12.5% gel in a 0.25 M Tris, 1.92 M glycine, 1% SDS buffer, pH 8.3. Molecular weight markers are ovalbumin (43 kDa), carbonic anhydrase (30 kDa), and soybean trypsin inhibitor (20.1 kDa). Proteins stained with Coomassie Brilliant Blue. (B) Non-dissociating gel electrophoresis run on a 5% gel in a 0.25 M Tris, 1.92 M glycine buffer, pH 8.3. Proteins stained with Coomassie Brilliant Blue.

and are referred to as α-type subunits. The inner rings are composed of subunits structurally related to the β-subunit of *T. acidophilum* and are referred to as β-type subunits (68). The enzyme is arranged as a symmetrical dimer of composition $\alpha_7\beta_7\beta_7\alpha_7$. The seven α-type subunits of the eukaryotic enzyme, although structurally related to each other, are distinct, as are the seven β-type subunits. Therefore, the eukaryotic enzyme contains 14 distinct subunits which can be resolved from each other by two-dimensional gel electrophoresis. One-dimensional SDS-PAGE of the enzyme reveals a series of closely spaced bands with M_r values of about 22 000–32 000 (*Figure 1A*). A unique feature of this enzyme is that it contains multiple distinct active sites. Electron microscopy has revealed that the active sites are present in the central channel of this barrel shaped structure (69). Site-directed mutagenesis identified the active site residue as an *N*-terminal threonine present on some β-type subunits (70).

Some fundamental problems of proteinase purification and characterization are illustrated when one reviews the evidence supporting the multi-catalytic nature of this macromolecule (71). The use of synthetic chromogenic substrates demonstrated that the preparation could cleave peptide bonds after basic amino acids, hydrophobic amino acids and acidic amino acids. The purified enzyme migrates as a single band following one dimensional non-dissociating gel electrophoresis (*Figure 1B*) but when examined by SDS-PAGE reveals multiple bands (*Figure 1A*). Since it is possible for unrelated proteins to migrate identically on non-dissociating gels yet be resolved by SDS-PAGE, and since enzymatic assays uncovered more than one type of proteolytic cleavage (there was no precedent for a multi-catalytic proteinase at that time), it was necessary to rule out the possibility that the preparation was an aggregate of unrelated proteinases perhaps in complex with a proteinase inhibitor such as α_2-macroglobulin.

It was first noted that the characteristics of repeated preparations were very

Table 4 Separation techniques used in the purification of the multicatalytic endopeptidase complex (20S proteasome)

Procedure	Reference
Agarose-hexylamine chromatography	79
Anion exchange chromatography	75
Arginine-Sepharose chromatography	80
DEAE-Affi Gel Blue chromatography	78
Gel filtration chromatography	75
Glycerol gradient centrifugation	76
Heparin-Sepharose chromatography	78
Hydroxylapatite chromatography	77
Immuno-affinity chromatography	82
Phenyl Sepharose chromatography	81

reproducible. Moreover, the three catalytic activities were monitored at each step of the purification procedure and found to co-purify and reside in a single high molecular weight molecule. By slicing the gel and measuring enzymatic activities in each gel slice it was found that the three catalytic activities were confined to the observed single protein band. After ammonium sulfate fractionation, the ratios of the catalytic activities remained constant throughout the remainder of the purification procedure. When the preparation was incubated in buffers of high ionic strength, all three activities declined in parallel, suggesting that they were in some way related. Dissociation by SDS led to loss of all three activities. Thus integrity of the macromolecule appeared to be necessary for expression of all activities.

Further confirmation of the homogeneity followed later when virtually identical preparations were isolated from other sources (see *Table 4*). Electron microscopy confirmed the homogeneity of the preparation and established its quaternary structure (72) and molecular cloning was used to establish the primary structures of each of the subunits (73). The enzyme can be purified to virtual homogeneity by alternating ion-exchange chromatography and gel filtration steps (74).

Protocol 5

Purification of the multi-catalytic proteinase complex from bovine pituitaries

Equipment and reagents

- Peristaltic pump
- Fraction collector
- Ultrafiltration cell
- Spectrophotometer
- Frozen bovine pituitaries (Pel Freez)
- Tris-EDTA, 10 mM, pH 7.5
- Tris-EDTA, 300 mM, pH 7.5
- Tris-EDTA 10 mM, pH 8.3
- Tris-EDTA, 400 mM, pH 8.3
- DEAE-Sephacel (Pharmacia)
- Ultrogel AcA 22 (Biosepra Inc., Marlborough, MA)

Protocol 5 continued

Method

A. Preparation of homogenate and ammonium sulfate fractionation

1. Place 200 g of frozen bovine pituitaries in a beaker of water and allow to partly thaw on ice. Wash the pituitaries free of blood and remove connective tissue with a sharp scissors.
2. Homogenize for 1 min in 200 ml 10 mM Tris-EDTA, pH 7.5 at 4 °C in a blender set at high speed. Repeat two more times, waiting 10 min before each homogenization.
3. Centrifuge at 9000 **g** for 30 min. Retain the supernatant and re-homogenize the pellet for 1 min in 400 ml of the same buffer. Centrifuge at 9000 **g** for 30 min and combine the supernatant fractions.
4. Fractionate with ammonium sulfate (38–60% saturation) following protocol 2 in part B. Dissolve the precipitate in the Tris-EDTA buffer, 10 mM, pH 7.5 and dialyse for 18 h against 7 l of the same buffer.

B. Column chromatography

1. Equilibrate DEAE-Sephacel with the same buffer. Pack a 20 × 2.5 cm column with the anion exchanger. Pass the dialysed enzyme over the column and wash with the same buffer until the absorbance at 280 nm falls below 0.1. With the aid of a peristaltic pump set at 60 ml/h elute the enzyme with a linear gradient of 250 ml 10 mM Tris-EDTA buffer, pH 7.5 vs 250 ml 300 mM Tris-EDTA buffer, pH 7.5.[a] Collect 6-ml fractions and pool all tubes containing enzymatic activity. Concentrate by ultrafiltration through a PM-10 membrane to 10–20 ml.
2. Apply the concentrated enzyme to a 5 × 50 cm AcA 22 gel filtration column equilibrated with 10 mM Tris-EDTA, pH 8.3. Elute the enzyme with the same buffer at 40 ml/h. Collect 12-ml samples and combine active fractions.
3. Apply the enzyme to a 2.5 × 10 cm DEAE-Sephacel column equilibrated with 10 mM Tris-EDTA, pH 8.3. Wash with 100 ml of the same buffer. Elute with a linear gradient established between 150 ml of the same buffer and 150 ml of 400 mM Tris-EDTA, pH 8.3. Elute the enzyme at a speed of 90 ml/h and collect 6-ml fractions. The enzyme elutes near the end of the gradient. Combine the active fractions and concentrate to 10–20 ml.
4. Apply the concentrated enzyme to a 5 × 50 cm AcA 22 gel filtration column equilibrated with 10 mM Tris-EDTA, pH 7.5. Elute the enzyme at a speed of 40 ml/h and collect 12-ml fractions. Pool the active fractions, concentrate to 0.2–0.3 mg protein/ml and store 1-ml aliquots at −70 °C. Generally, 30–40 ml of the concentrated enzyme is obtained.[b]

[a] The bovine pituitary enzyme is inhibited by NaCl. Elution is therefore achieved by increasing the buffer concentration rather than by using a commonly employed NaCl gradient.

[b] This procedure used with freshly prepared DEAE-Sephacel and AcA 22 will give an apparently homogenous preparation as shown in Figure 2.1. Repeated use of these columns lowers their efficiency, resulting in preparations that may contain higher molecular weight contaminants visible after SDS-PAGE. In this case, a phenyl Sepharose chromatographic step will remove these contaminants (81).

9 Conclusions: many roads lead to Rome

There is no fixed sequence or nature of chromatographic steps that must be followed to yield a homogeneous preparation of a proteolytic enzyme. Although the properties of the molecule isolated may suggest likely choices, protein purification is in most cases an empirical process. The high molecular weight of the multicatalytic proteinase complex pointed to filtration on a gel with a high molecular weight exclusion volume as a fundamental step. This enzyme has now been purified by many investigators from varied sources. Virtually all procedures for the purification of this enzyme also use anion exchange chromatography. The selection of other chromatographic steps varies. *Table 4* lists some of the separation techniques that have been used to purify this enzyme.

References

1. Williams, T. A., Michaud, A., Houard, X., Chauvet, M-T., Soubrier, F., and Corvol, P. (1996). *Biochem. J.* **318**, 125.
2. Fernandes-Alnemri, T., Armstrong, R. C., Krebs, J., Srinivasula, S. M., Wang, L., Bullrich, F. *et al.* (1996). *Proc. Natl. Acad. Sci.* USA **93**, 7464.
3. Morales, T. I. and Woessner Jr, J. F. (1977). *J. Biol. Chem.* **252**, 4855.
4. Camargo, A. C.M., Shapanka, R., and Greene, L. J. *Biochemistry* **12**, 1838.
5. Orlowski, M., Michaud, C., and Chu, T. G. (1988). *Eur. J. Biochem.* **135**, 81.
6. Tisljar, U., Camargo, A. C. M., da Costa, C. A., and Barrett, A. J. (1989). *Biochem. Biophys. Res. Comm.* **162**, 1460.
7. Barrett, A. J. and Tisljar, U. (1989). *Biochem. J.* **261**, 1047.
8. Schechter, I. and Berger, A. (1967). *Biochem. Biophys. Res. Comm.* **27**, 157.
9. Nolan, C., Margoliash, E., Peterson, J. D., and Steiner, D. F. (1971). *J. Biol. Chem.* **246**, 2780.
10. Smeekens, S. P. (1993). *Bio/Technol.* **11**, 182.
11. Yu, B., Pereira, M. E., and Wilk, S. (1993). *J. Biol. Chem.* **268**, 2029.
12. Suzuki, K., Sorimachi, H., Yoshizawa, T., Kinbara, K., and Ishiura, S. (1995). *Biol. Chem. Hoppe-Seyler* **376**, 523.
13. Waxman, L. and Krebs, E. G. (1978). *J. Biol. Chem.* **253**, 5888.
14. Acker, G. R., Molineaux, C., and Orlowski, M. (1987). *J. Neurochem.* **48**, 284.
15. Dyer, S. H., Slaughter, C. A., Orth, K., Moomaw, C. R., and Hersh, L. B. (1990). *J. Neurochem.* **54**, 547.
16. Burn, P. (1988). *Trends Biochem. Sci.* **13**, 79.
17. Bauer, K. (1994). *Eur. J. Biochem.* **224**, 387.
18. Pfleiderer, G. (1970). *Methods Enzymol.* **19**, 514.
19. Hopsu-Havu, V. K. and Sarimo, S. R. (1967). *Hoppe-Seyler's Z. Physiol. Chem.* **348**, 1540.
20. Bella, A. M., Erickson, R. H., and Kim, Y. S. (1982). *Arch. Biochem. Biophys.* **218,** 156.
21. Puschel, G., Mentlein, R., and Heymann, E. (1982). *Eur. J. Biochem.* **126,** 359.
22. Brownlees, J., Williams, C. H., Brennan, G. P., and Halton, D. W. (1992). *Biol. Chem. Hoppe-Seyler* **373,** 911.
23. Kenny, A. J. (1976). *Biochem. J.* **155,** 169.
24. Seidl, R. and Schaefer, W. (1991). *Prep. Biochem.* **21**, 141.
25. Li, J., Wilk, E., and Wilk, S. (1995). *Arch. Biochem. Biophys.* **323,** 148.
26. Kojima, K., Hama, T., Kato, T., and Nagatsu, T. (1980). *J. Chromatog.* **189,** 233.
27. Ogata, S., Misumi, Y., Tsuji, E., Takami, N., Oda, K., and Ikehara, Y. (1992). *Biochem.* **31,** 2582.

28. Hartel, S., Hanski, C., Kreisel, W., Hoffmann, C., Mauck, J., and Reutter, W. (1987). *Biochim. Biophys. Acta* **924,** 543.
29. Hooper, N. M. and Turner, A. J. (1988). *FEBS Lett.* **229**, 340.
30. Simmons, W. H. and Orawski, A. T. (1992). *J. Biol. Chem.* **267**, 4897.
31. Vergas Romero, C., Neudorfer, I., Mann, K. and Schafer, W. (1995). *Eur. J. Biochem.* **229:** 262.
32. Twining, S. S. (1994). *Crit. Rev. Biochem. Mol. Biol.* **29**, 315.
33. Chu-Ping, M., Slaughter, C. A. and DeMartino, G. N. (1992). *J. Biol. Chem.* **267**, 10515.
34. Englard, S. and Seifter, S. (1990). In *Methods in Enzymology* (ed. Deutscher, M. P.), Vol. 182, p. 285. Academic Press, San Diego.
35. Ingham, K. C. (1990). In *Methods in Enzymology* (ed. Deutscher, M. P.), Vol. 182, p. 301. Academic Press, San Diego.
36. Takahashi, T. and Tang, J. (1981). In *Methods in Enzymology* (ed. Lorand, L.), Vol. **80**, p. 565. Academic Press, San Diego.
37. Bull, H. G., Thornberry, N., and Cordes, E. H. (1985). *J. Biol. Chem.* **260**, 2963.
38. Reeck, G. R., Walsh, K. A., and Neurath, H. (1971). *Biochemistry* **10**, 4690.
39. Tobe, H., Kojima, F., Aoyagi, T., and Umezawa, H. (1980). *Biochim. Biophys. Acta* **613,** 459.
40. Svendsen, I. and Breddam, K. (1992). *Eur. J. Biochem.* **204,** 165.
41. Kurauchi, O., Mizutani, S., Okano, K., Narita, O., and Tomoda, Y. (1986). *Enzyme* **35,** 197.
42. Rich, D. H., Brown, M. A., and Barrett, A. J. (1986). *Biochem. J.* **235,** 731.
43. Buttle, D. J., Kembhavi, A. A., Sharp, S. L., Shute, R. E., Rich, D. H., and Barrett, A. J. (1989). *Biochem. J.* **261:** 469.
44. Fukasawa, K. M., Fukasawa, K., and Harada, M. (1978). *Biochim. Biophys Acta* **535:** 161.
45. Tschesche, H. (1995). In: *Methods in enzymology* (ed. Barrett, A. J.), Vol. 248**,** p.431. Academic Press, San Diego.
46. Turzynski, A. and Mentlein, R. (1990). *Eur. J. Biochem.* **190,** 509.
47. Yurt, R. and Austen, K. F. (1977). *J. Exp. Med.* **146,** 1977.
48. Westerik, J. C. and Wolfenden, R. (1972). *J. Biol. Chem.* **247**, 8195.
49. Thompson, R. C. (1973). *Biochemistry* **12**, 47.
50. Nishikata, M., Kasai, K-I., and Ishii, S-I. (1981). *Biochim. Biophys. Acta* **660**, 256.
51. Patel, A. H., Ahsan, A., Suthar, B. P., and Schultz, R. M. (1983). *Biochim. Biophys. Acta* **748**, 321.
52. Someno, T., Saino, T., Katoh, K., Miyazaki, H., and Ishii, S-I. (1985). *J. Biochem.* **97**, 1493.
53. Thornberry, N. A., Bull, H. G., Calycay, J. R., Chapman, K. T., Howard, A. D., Kostura, M. J., Miller, D. K. *et al.* (1992). *Nature* **356**, 768.
54. Basak, A., Yuan, X-W., Seidah, N. G. Chretien, M., and Lazure, C. J. (1992). *J. Chromatog.* **581**, 17.
55. Buttle, D. J. (1994). In *Methods in Enzymology* (ed. Barrett, A. J.), Vol. 244, p. 639. Academic Press, San Diego.
56. Storer, A. C. and Menard, R. (1994). In *Methods in enzymology* (ed. Barrett, A. J.), Vol. 244, p. 486. Academic Press, San Diego.
57. Singh, H. and Kalnitsky, G. (1983). *J. Biol. Chem* . **253**, 4319.
58. Willenbrock, F. and Brockelehurst, K. (1985). *Biochem. J.* **227**, 511.
59. Simon, M. M., Simon, H. G., Fruth, U., Epplen, J., Muller-Hermelink, H. K., and Kramer, M. D. (1987). *Immunology* **60**, 219–230.
60. Hixson, H. F. and Nishikawa, A. H. (1973). *Arch. Biochem. Biophys.* **154**, 501.
61. Muller-Esterl, W. and Fritz, H. (1981). In *Methods in enzymology* (ed. Lorand, L.), Vol. 80, p. 621. Academic Press, San Diego.

62. Murphy, G. and Crabbe, T. (1995). In *Methods in enzymology* (ed. Barrett, A. J.), Vol. 248, p. 470. Academic Press, San Diego.
63. M. Woessner Jr, F. J., (1995). In *Methods in enzymology* (ed. Barrett, A. J.), Vol. 248, p. 485. Academic Press, San Diego.
64. Ibrahim-Granet, O. and Bertrand, O. (1996). *J. Chromatog. B* **684**, 239.
65. Knauper, V., Kramer, S., Reinke, H., and Tschesche, H. (1990). *Biol. Chem. Hoppe-Seyler* **371**, 295.
66. Wilk, S. and Healy, D. P. (1993). *Adv. Neuroimmunol.* **3**, 195.
67. *Enzyme and Protein* (1993). **47**, 189.
68. Zwickl, P., Grziwa, A., Puhler, G., Dahlmann, B., Lottspeich, F., and Baumeister, W. (1992). *Biochemistry* **31**, 964.
69. Lowe, J., Stock, D., Jap, B., Zwickl, P., Baumeister, W., and Huber, R. (1995). *Science* **268**, 533.
70. Seemuller, E., Lupas, A., Stock, D., Lowe, J., Huber, R. and Baumeister, W. (1995). *Science* **268**, 579.
71. Wilk, S. and Orlowski, M. (1983). *J. Neurochem.* **40**, 842.
72. Kopp, F., Steiner, R., Dahlmann, B., Kuehn, L., and Reinauer, H. (1990). *Biochim. Biophys. Acta* **872**, 253.
73. Tanaka, K., Tamura, T., Yoshimura, T., and Ichihara, A. (1992). *New Biologist* **4**, 173.
74. Orlowski, M. and Michaud, C. (1989). *Biochemistry* **28**, 9270.
75. Wilk, S. and Orlowski, M. (1980). *J. Neurochem.* **35,** 1172.
76. Hough, R., Pratt, G., and Rechsteiner, M. (1987). *J. Biol. Chem.* **262,** 8303.
77. Ray, K. and Harris, H. (1985). *Proc. Natl. Acad. Sci USA* **82,** 7545.
78. Tanaka, K., Ii, K., Ichihara, A., Waxmann, L., and Goldberg, A. L. (1986). *J. Biol. Chem.* **261,** 15197.
79. Rivett, A. J. (1989). *J. Biol. Chem.* **264,** 12215.
80. Mykles, D. (1992). *Arch. Biochem. Biophys.* **274,** 216.
81. Pereira, M. E., Yu, B., and Wilk, S. (1992). *Arch. Biochem. Biophys.* **294**, 1.
82. Hendil, K. and Uerkvitz, W. (1991) *J. Biochem. Biophys. Methods* **22,** 159.

Chapter 3
Protease assay methods

Gautam Sarath
Department of Biochemistry and Center for Biotechnology, University of Nebraska-Lincoln, Lincoln, NE 68588-0664, USA

Michael G. Zeece
Department of Food Science, University of Nebraska-Lincoln, Lincoln, NE 68583-0919, USA

Alan R. Penheiter
Department of Biochemistry, Mayo Clinic, Rochester, MN 55905, USA

1 Introduction

Peptidases have continued to receive significant attention in the last decade. Although some novel means to assay these enzymes have been published, many of the older protocols for the assay of proteases are still in use. One obvious change has been brought about by the revolution in recombinant protein expression technology. This ability to engineer proteins has added a new dimension to the study of proteases. Other changes have arisen from an understanding of the complexity of the proteolytic machinery that govern many critical cellular processes (e.g. caspases), and the development of instrumental means to automate these analyses. Thus, in addition to several standard protocols, this chapter describes some of these newer assays that can be readily imported into most laboratories. Other, more complex methods are described briefly. It is expected that the reader will adapt the protocols presented to meet their specific needs. The substrate nomenclature used in this chapter is shown in *Table 1*.

2 Assays with natural substrates

2.1 Endopeptidase assays

Endopeptidase assays can be used to detect a specific enzyme or evaluate a number of enzymes, such as those present in a crude extract. In general, these assays involve the incubation of a protein substrate containing several or many scissile sites of undefined kinetic properties with a protease or a mixture of proteases. Substrates such as gelatin, casein, haemoglobin, etc., are commonly used.

Table 1 Common peptidase substrate[a] nomenclature (substituent-NH-PI-P I'-COO-substituent)

NH_2 substituents		COOH substituents	
Abbreviation	**Group**	**Abbreviation**	**Group**
Bz	Benzoyl	NHNap	2-Naphthylamide
Z	Benzoyloxycarbonyl	NHNan	4-Nitroanilide
Ac	Acetyl	NHMec	7-Amido-4-methylcoumarin
Suc	Succinyl	OMe	Methylester
Abz	*O*-Aminobenzoyl		
Fa	Furylacryloyl		

[a] There are many commercial sources of substrates (See Core List of Suppliers).

With the advent of recombinant methods to produce proteins, very low abundance proteins can be readily generated for such studies, provided that their cDNAs are available. Quantification of proteolytic activity depends upon the nature of the substrate, the type and amount of activity being detected and the sensitivity and precision needed.

Protocols 1–5 provide procedural details for some general endopeptidase assays in which protein substrates are used. In the following subsections, experimental and theoretical considerations pertaining to these and related assay techniques are described.

2.1.1 Choice of the protein substrate

The greatest problem in assaying endopeptidases is the choice of substrate. If the enzyme of interest is a well-characterized protease, a number of synthetic substrates are usually known, and information concerning the kinetics of hydrolysis of these materials can be obtained. However, the endopeptidase to be assayed is often either not well characterized or, in many instances, is only inferred to exist from general experimental results. Under these conditions, the development of an assay is more involved.

In a few instances, the protein substrate best suited for the assay is determined by the enzyme to be assayed. For example, collagen is the protein of choice for the assay of collagenase. In other instances, however, the detection of the protease precedes the identification of its natural substrate(s). Under these conditions, a protein substrate must be selected so that good results are obtained with the detection methodology being used. From a historical perspective, haemoglobin and casein have received wide use in assay systems because they are inexpensive and easily obtained in highly purified form. Procedures using these substrates are still commonly employed. The method given in *Protocol 1* provides a simple and inexpensive assay suitable for experiments in which crude materials are screened for general proteolytic activity. In a related method, the traditional Coomassie dye-binding assay for quantifying proteins (1) has been employed to assay proteases (2,3). Coomassie Brilliant Blue tends to bind to larger polypeptides and thus can be used to detect the amount of substrate protein in the presence of smaller polypeptides released by proteolytic activity (*Protocol* 4).

Chromogenic derivatives of proteins, notably azocasein and azoalbumin, are commercially available. Azogelatin has also recently been synthesized (4; *Protocol* 2) and used in assays to detect proteases in germinated barley. Azogelatin is soluble in buffers from pH 3.0 to 9.0, a range suitable for the analyses of most types of proteases (4). Detection of protease activity using chromogenic protein substrates is convenient because it depends upon the release of soluble, low molecular weight coloured peptides in a supernatant fluid after precipitation of large fragments of the azoprotein. These azopeptides have an absorbance maximum in the visible region, which permits measuring enzymatic activity by even a low-cost colorimeter (*Protocol* 3).

Protocol 1

Endopeptidase assays with protein substrates.

Equipment and reagents

- Constant-temperature water bath
- UV spectrophotometer
- Matched quartz cuvettes
- Protein substrate: generally haemoglobin or casein is used. These are prepared as follows:
- Prepare 2% (w/v) haemoglobin substrate by dissolving, with gentle stirring, 5 g of haemoglobin substrate powder (Worthington Biochemical Corp.) in a solution of 80 g of urea in 80 ml of water. After incubation at 37 °C for 60 min, add 50 ml of 0.25 M buffer, adjust the pH to the desired value (7.0–7.5)[a] then dilute to 250 ml with water. When casein or another protein is used, the substrate can be prepared without urea.
- With casein, the protein is dissolved in water with the addition of concentrated NaOH. The solution is diluted to 125 ml with water, then diluted two-fold with buffer. The pH is adjusted to the desired value. With casein substrate, pH values must be kept above 6.0 to prevent precipitation of the protein.

Method

1. Prior to the assay, equilibrate all solutions at 25 °C[b].
2. To start the assay, add 5 ml of temperature-equilibrated substrate solution to 1 ml of enzyme solution, and mix thoroughly, but gently to prevent frothing. Incubate the mixture at the desired temperature for a fixed period of time not exceeding 10 min[c]. Multiple assays can be performed by adding substrate to successive tubes in 30-s intervals. The reactions will be terminated in the same sequence.
3. Terminate the reaction at the end of the incubation period by adding 10 ml of 5%

Protocol 1 continued

trichloroacetic acid to the assay mixture[c]. Prepare a blank by combining the TCA with the enzyme, then adding the substrate. Allow the assay mixture and blank to stand for 30 min.

4. Remove precipitated material from each mixture by filtering it through Whatman 3 filter paper. Centrifugation in a bench centrifuge can also be used to separate the TCA-soluble products, but this is not part of the 'standard' procedure.
5. Measure the absorbance of the filtered solution at 280 nm, with reference to the blank. One unit of enzyme activity is defined as the amount of enzyme required to cause a unit increase in absorbance at 280 nm across a 1 cm path length, under the conditions of the assay.

[a] Phosphate buffer is commonly used, although the choice of buffer will be dictated by stability and activity requirements of the protease being assayed. Other recommended buffer salts include Tris, Tricine, and HEPES.

[b] Temperatures at which protease assays are performed typically lie in the range of 25–37 °C; 25 °C is preferable in order to provide consistency between research groups.

[c] Times exceeding 10 min may cause extensive denaturation of the enzyme, owing to the high urea concentration. If denaturation is very severe, urea can be removed from the substrate prior to the assay by dialysis; however, this generally entails some precipitation of the substrate. If urea is not present in the substrate, prolonged incubation times are feasible.

More accurate absorbance measurements are made using plate readers or recording spectrophotometers. High sensitivity is also afforded by preparing fluorescent conjugates of polypeptides and using these conjugates as substrates for proteases (5 but see also 45) (*Protocol 5*). The present need for high-throughput analysis of biological samples, has resulted in robust methods to assay proteases in microwell plates. Detection in such systems are based on small chromogenic substrates or immobilizing biotinylated-proteins in the wells (discussed later; see *Protocol 14*)

Occasionally, the native substrate for the endopeptidase is easily obtained and can be used for assays. As an example, in many germinating seeds, specific proteases have been identified for specific seed proteins (6). Other examples of endopeptidase assays based on the use of native proteins as substrates include those of enteropeptidase (7) and signal (processing) peptidases (8–10). Signal peptidases, have been cumbersome to assay and, consequently, to purify. The low abundance of these enzymes and the lack of small synthetic substrates have been hurdles to their purification. To assay such enzymes, the mRNA for the preproteins is isolated or generated *in vitro*, translated *in vitro* using radioactive amino acids to produce preproteins with high specific radioactivity. Incubation of the preproteins (employed in vanishingly small molar concentrations) with enzyme leads to proteolytic processing. Proteolytically processed polypeptides are separated by SDS-PAGE and subsequently detected by imaging the labelled products (8–10).

New assays have recently been developed to assay signal peptidases. These protocols use recombinant methods to generate fluorescent derivatives of the GFP, originally isolated from a jellyfish (11). In a very elegant assay, two GFP derivatives with different fluorescent characteristics are produced as fusion products separated by linker that contains a scissile site. Cleavage of the site by a signal peptidase releases the two GFP molecules with concomitant increase in fluorescence (12). This class of assay is also known as FRET assay. Peptide substrates capable of FRET have been designed and utilized for the assay of proteases (13,14).

The other assay for signal-/sequence-specific proteases also involves recombinant means to generate polypeptides fused to GFP (10). In this assay, the polypeptide domain containing the scissile site is *N*-terminal to the GFP and expressed using a vector that contains an affinity handle of six histidines. Recombinant expression of these constructs yields fluorescent products that can be purified by metal-chelate affinity chromatography. Incubation of the protein substrates with a signal peptidase, results in the cleavage of the substrate domain and the release of free GFP. Unreacted substrate and non-GFP polypeptides containing the polyhistidine linker are removed by incubation with metal-chelate resin, followed by centrifugation. The fluorescence of the supernatant is then measured and quantified (*Figure 1*). This method is robust and will be useful for the assay of multiple samples, such as those generated during enzyme purification.

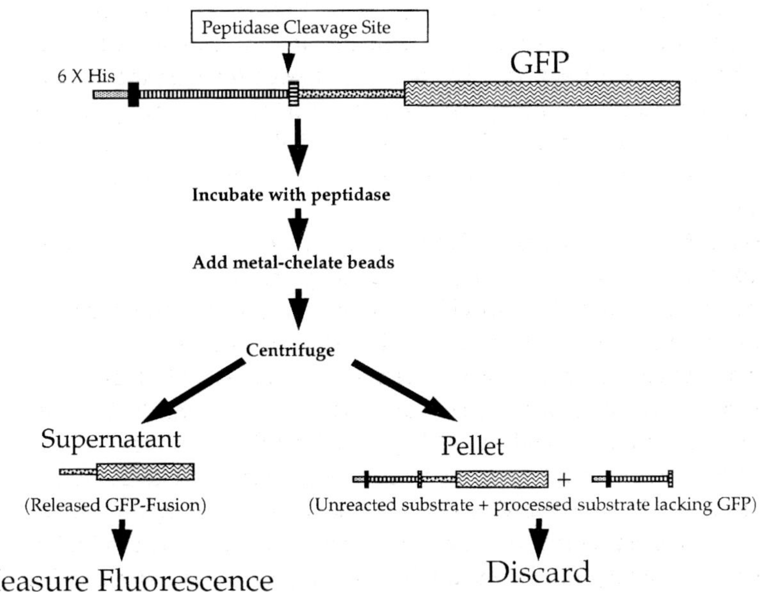

Figure 1 Method for expression of GFP-fusion protein. Printed with permission of Pehrson, J. C., Weatherman, A., Markwell, J., Sarath, G., and Schwartzbach, S. D. (1999), *Biotechniques* **27**: 28–32.

Protocol 2

Synthesis of azogelatin[a]

Equipment and reagents

- Water bath
- Lyophilizer
- Waring Blender
- Azogelatin is prepared as follows:
  - Solution A is prepared by suspending 20 g of porcine skin gelatin in 275 ml of water containing 4 g $NaHCO_3$ and raising the temperature to 54 °C
  - Solution B is prepared by dissolving 0.01 mol of sulphanilic acid in 30 ml of water containing 0.01 mol of NaOH; next, add 0.01 mol of $NaNO_2$, followed by 0.02 mol of HCl with stirring for 2 min or until the solution turns cloudy

Method

1. Add 0.02 mol of NaOH to solution B, stir for 5 s and add with brisk stirring to Solution A.
2. The reddish orange solution is stirred for 5 min, dialysed against 3 × 4 l of 0.01% sodium azide and freeze dried.
3. The freeze-dried material is reduced to a fine powder using a blender and stored at 4 °C. Azogelatin is soluble in buffers in a pH range from 3 to 9. This azoprotein substrate can be used as indicated in *Protocol 3*.

[a] Method adapted from Jones *et al.* (4)

Protocol 3

Endopeptidase assays with azoprotein substrates.

Equipment and reagents

- Spectrophotometer or colorimeter; a simple colorimeter capable of absorbance measurements between 400 and 700 nm is adequate
- Constant-temperature water bath
- Microfuge or centrifuge capable of processing many small tubes simultaneously
- Azoprotein substrate (2% w/v azocasein, 2% w/v azoalbumin or 1% w/v azogelatin) is prepared by dissolving azoprotein in 50–100 mM buffer of the desired pH. The solubility of azocasein becomes limiting at pH values below 6.0. After the substrate[a] has dissolved, clarify it by centrifugation at 12 000 **g** for 10 min

Methods

1. Equilibrate substrate and enzyme solutions at 25 °C (see *Table 2*, note b).
2. Pipette 250 μl of substrate into a 1.5 ml microcentrifuge tube. Initiate the assay by pipetting 150 μl enzyme into the substrate. Mix the reaction gently, but thoroughly. Incubate the reaction mixture at 25 °C for a set time period (15–30 min)[b]. Prepare a

Protocol 3 continued

reagent blank by replacing the enzyme solution with buffer in the above procedure. It is advisable to run each sample in duplicate or triplicate.

3 Terminate the assay by adding 1.2 ml of 10% TCA to each microfuge tube. Prepare an enzyme blank by mixing buffer, enzyme, trichloroacetic acid and substrate in that order. Mix the contents of the microcentrifuge tubes thoroughly and allow them to stand for 15 min to ensure complete precipitation of the remaining azoprotein and azoprotein fragments.

4 Centrifuge the samples at about 8000 **g** for 2–5 min and transfer 1.2 ml of the supernatant fluid to a test tube containing 1.4 ml of 1.0 M NaOH[c].

5 Determine the absorbance of this solution at 440 nm. One unit of protease activity is defined as the amount of enzyme required to produce an absorbance change of 1.0 in a 1 cm cuvette, under the conditions of the assay.

[a] Azocasein yields lower colour in reagent blanks after precipitation than does azoalbumin; however, azocasein is insoluble in buffers with pH values below 6.0. The choice of buffer depends on the enzyme to be assayed. Buffer salts such as Tris, HEPES, phosphate and Tricine are all satisfactory.

[b] Because the assay conditions are non-denaturing, incubations can be for prolonged periods. In this laboratory, we have incubated assay mixtures for up to 18 h to test for trace levels of endopeptidases. Under these conditions, the enzyme–substrate mixture and blank should be filter-sterilized to prevent hydrolysis resulting from microbial growth.

[c] The rate of centrifugation and the volume of NaOH solution given here were chosen for convenience and may be altered to satisfy limitations imposed by available equipment.

Table 2 Solubility characteristics of selected peptidase substrates

Substrate[a]	Solvents[b]	Solubility (mM)	Concentration[c] (mM)
N-Bz-AA-X	Me_2SO, EtOH, MeOH, HOAc	0.2–3.0	0.05–0.5
AA-X	Me_2SO, MeOH, H_2O, HOAc	1.0–5.0	0.1–1.0
N-t-Boc-AA-X	Me_2SO, MeOH	0. 1–2.0	0.05–0.5
N-Suc-AA-X	MeOH, MeCN	0.05–2.0	0.02–1.0
N-Ac-AA-X	MeOH, Me_2SO	0.05–5.0	0.03–3.0
AA-AA, etc.	H_2O, Me_2SO	At most < 5	0.05–2.0
Fa-AA	Me_2SO, H_2O	At most < 5	1.0
Abz-AA	MeCN, Me_2SO	0.02–2.0	0.01–0. 1

[a] For substrates, X represents 2-naphthylamide, 4-nitroanilide, or 7-amido-4-methylcoumarin (Nmec); AA denotes the aminoacyl moiety.

[b] Solvents are MeZSO, dimethyl sulphoxide; MeOH, methanol; EtOH, ethanol; HOAc, acetic acid; MeCN, acetonitrile. Dissolving the substrate in one of these solvents, then diluting with buffer or water to the specified final concentration often speeds the preparation of the substrate stock. Generally, an organic solvent concentration of 5–10% in the assay does not inhibit enzyme activity; however, it is always advisable to check enzyme activity in assays with a buffer-soluble substrate and various concentrations of organic solvents.

[c] Concentration in aqueous solution for routine assays. These values are approximate solubilities for different substrates; actual solubilities will vary with the identity of the amino acid residue(s) in the substrate.

2.1.2 Separation of products from substrates

Product peptides will possess the same general chemical properties as the substrate from which they were derived. Thus, peptide products must usually be separated from the protein substrate prior to quantification. For large polypeptide substrates, the most commonly used technique is to precipitate the remaining intact substrate and large fragments with trichloroacetic acid and remove them by filtration or centrifugation. Perchloric acid can be substituted for TCA. Changes in the levels of peptides liberated from test samples and from controls (prepared by denaturing either substrate or enzyme with prior addition of TCA) can be assessed by direct measurement at 210 nm or by colorimetric or fluorimetric procedures (*Protocols 1–5*). In many instances, product peptides and intact substrate can be separated and analysed by electrophoretic methods. Microassays have been developed in which protein degradation can be monitored in microtitre plates or by instrumental means. These, and other related methods are discussed in more detail in the later sections.

Protocol 4

Coomassie blue dye binding micro-assays using casein[a]

Equipment and reagents

- Substrate. This is prepared by dissolving casein at 4 mg/ml in a suitable buffer by gentle boiling, filtering through muslin cloth and storing at 4 °C for up to 2 weeks. Most pure proteins that are readily soluble in buffers should also work
- Spectrophotometer
- Constant temperature water bath
- Vortex mixer

Method

1. Prepare a series of tubes containing 250 μl of 4 mg/ml casein in the appropriate buffer. Add the enzyme to be assayed in 250 μl so as to obtain a final volume of 500 μl and a final substrate concentration of 2 mg/ml.
2. Incubate at the desired temperature for the desired length of time (frequently under 30 min for pure enzymes, but this can be extended for extracts containing low proteolytic activities).
3. Remove two 25-μl aliquots from each reaction to test tubes containing 775 μl of water. Add 200 μl of Coomassie dye reagent (Bio-Rad Inc.), vortex and wait 10 min.
4. Read absorbance at 595 nm. Plot change in $A_{595\ nm}$ against appropriate controls.

[a] Method from Buroker-Kilgore and Wang (2) and Bickerstaff and Zhou (3). This method can be adapted for larger volumes (2), more dilute substrate (2,3) and for microtitre plate assays (2). It can be adapted for most protein substrates.

Protocol 5

Fluorescent haemoglobin derivatives as substrates for proteases[a,b]

Equipment and reagents

- Spectrofluorimeter interfaced to a microcomputer (or to a chart recorder)
- Vortex mixer
- Microfuge
- Controlled temperature water bath/incubator
- BODIPY-haemoglobin: Prepared in a manner similar to Oregon Green-haemoglobin (*Protocol 11*). FITC-casein (see refs 5 or 37) or other fluorescently labelled proteins can be synthesized and used in such assays.

Method

1 Place 12.5 μl BODIPY-haemoglobin and 12.5 μl of 0.2 M sodium formate buffer, pH 3.5, into a 500 μl microfuge tube.

2 Preincubate the mixture for 5 min at 37 °C.

3 Include a control by adding 10 μl of 10 μM pepstatin to the solution.

4 Start the reaction in all tubes by adding 50 μl of supernatant and continue incubating at 37 °C.

5 Stop the reaction after 30 min by the addition of 50 μl 10% TCA. Place the tube on ice for 5 min and then centrifuge at 11 000 **g** for 5 min.

5 Take a 50-μl aliquot of the supernatant and add to 4.0 ml of 25 mM sodium borate buffer, pH 9.2.

6 Read the fluorescence intensity of this solution using 488 nm excitation and 520 nm emission.

7 Express the activity due to cathepsin D[c] in the sample as fluorescence intensity of the sample, minus the fluorescence intensity of the pepstatin control.

8 Cathepsin D activity determined above can also be expressed per μg of protein in the sample (specific activity).

9 The concentration of cathepsin D in samples can be calculated from a standard curve of fluorescence intensity versus enzyme concentration (generated with purified cathepsin D).

[a] Adapted from M. G. Zeece, unpublished data. This method can detect nanomolar levels of Cathepsin D.

[b] Twinning (5) first demonstrated the utility of FITC conjugates for protease assays (see ref. 37).

[c] This protocol was designed to detect Cathepsin D activity in extracts. It can be adapted for other proteolytic activities.

2.1.3 Advantages and disadvantages of the use of proteins as substrates

Of the many substrates for the assay of an endopeptidase, each is limited in its utility, and if used inappropriately will generate misleading kinetic data. This is especially true for whole proteins or mixed peptides as substrates. The problem lies in the fact that proteins and peptides are collections of substrates (unless only a single bond is cleaved), not just a single substrate. As an example, suppose that lysozyme, with 129 amino acid residues, is chosen as the substrate to measure an endopeptidase in a tissue extract. Further suppose that the lysozyme has been completely denatured so that all 128 peptide bonds are available for cleavage. Under these initial conditions, there is not a single substrate available for hydrolysis; rather, there are 128 of them, each with different kinetics of hydrolysis. Moreover, as hydrolysis proceeds and the lysozyme is fragmented, the 'substrate' becomes poorly defined. In the instant that a hydrolytic event takes place in a natural substrate assay, the substrate is no longer chemically the same as it was initially; instead, two new substrate molecules are ordinarily generated from the original intact protein. Because of this, initial velocity, as we usually construe it, has no comparative or kinetic meaning and those values that we derive from measurements of initial velocity, such as K_m and V_{max}, are usually not interpretable. Even if K_m is most carefully determined from kinetic data, what does 'the concentration of substrate when the velocity is half maximal' mean when the number of susceptible bonds in the substrate (and therefore the actual substrate concentration) is unknown?

A similar difficulty arises when comparing the results of assays with proteases from different sources, even when no attempt is made to calculate kinetic parameters rigorously. A protease from one source may cleave an entirely different set of bonds in the substrate than a protease from another source, so that for all practical purposes, the enzymes are acting on two different substrates. While velocity data from such comparative experiments may have physiological meaning if the two proteases are suspected of acting on the chosen substrate molecule *in vivo*, there is no information of enzymological interest in the comparison when this physiological relationship is lacking.

It is also important to understand that when intact proteins are used as endopeptidase substrates in pH optimum determinations, changes in velocity with pH may provide more information about the susceptibility of the substrate to proteolysis than about the ability of the enzyme to act on the substrate (see Chapter 10). This is because protein substrates may contain tens or hundreds of ionizable groups, each of which is potentially capable of influencing the enzyme's ability to act on the substrate.

The use of a single protein as a substrate for a proteolytic enzyme has great analytical utility. Under a defined set of conditions (pH, buffer salt and temperature) an increase in substrate hydrolysis rate is directly correlated with enzyme activity, and this information is extremely useful while a protease is being purified and characterized. However, for the reasons described above, this sort of information is not, in most instances, comparable to that obtained under other conditions, with other protein substrates, or from experiments with other proteases.

The notable exception to these rules is the rare instance in which a protease hydrolyses a single bond in a protein substrate. In this instance, a further caution is in order: care should be taken when measuring velocity with a substrate, [S], in concentrations so dilute that they approach the concentration of the enzyme, [E], a condition that can easily arise when very large protein substrates are used. Under these conditions, the assumptions to derive the velocity equation must include a substrate conservation term of the form $[S]_{total} = [S]_{free} + [ES]$, and the solution to the velocity equation becomes considerably more complicated.

2.2 Exopeptidase assays

Natural substrates are rarely employed in the routine assay of exopeptidases, because of the diversity of synthetic substrates and the ease with which an elegant assay which uses modified substrates can be developed. Exceptions are exopeptidases involved in the processing of prohormones and other metabolically active peptides.

3 Assays with synthetic substrates

Synthetic substrates are constructed around the chemistry of amino acids and peptides. Novel substituents are added to either the α-NH_2 or α-COOH groups of amino acids for several reasons. First, the group may block an α-NH_2 or α-COOH group on the peptide to produce a substrate susceptible to proteolytic cleavage. Second, the group may impart a chromogenic nature to the substrate, which allows for simple quantification. Finally, the group may actually provide the bond to be hydrolysed by a peptidase.

The simplest substrates are single amino acids coupled through their α-COOH to a chromogenic amine, such as 4-nitroaniline or 2-naphthylamine. Substrates with highly fluorescent, substituents such as 7-amino-4-methylcoumarin are also available commercially. Based on the blocking groups employed, substrates can be classified into three categories, namely endopeptidase, aminopeptidase and carboxypeptidase substrates. Endopeptidase substrates will have no free amino- or carboxy-termini, while aminopeptidase substrates will have a free amino-terminus, and carboxypeptidase substrates will have a free carboxy-terminus. Classification of peptidases as endopeptidases or exopeptidases is based on the results of studies that determine the class of substrate hydrolysed by a given enzyme. It must be emphasized that certain endopeptidases and exopeptidases can act on substrates belonging to different classes. Therefore, with newly isolated enzymes, a variety of substrates must be tested before the enzyme can be assigned to specificity categories.

3.1 Endopeptidase and aminopeptidase substrates

The simplest endopeptidase substrate consists of a doubly blocked aminoacyl residue in which the acyl group is amide-bonded to a group that yields a chromophore upon hydrolysis. A typical substrate is Bz-L-Arg-NHNan. A corresponding aminopeptidase substrate will be L-Arg-NHNan. The amino-terminus of an amino acid or peptide can be blocked with any of a number of reagents, of which those used in peptide synthesis predominate; some common examples of NH_2 blocking groups are Bz, Z, Ac, and Suc (see *Table 1*). A problem frequently encountered is the limited solubility of these substrates in aqueous solutions. *Table 2* lists the solubility characteristics of a few selected substrates, as well as suggested concentrations for assays. For both endopeptidase and aminopeptidase substrates, the COOH-terminal substituent generally determines the sensitivity and method of detection for the assay. Exceptions occur when an endopeptidase cleaves at the amino-side of a substrate, or when internally quenched peptide substrates are used. Fluorescent peptides composed entirely of α-amino acids have been synthesized and used to assay renin (15).

Esters prepared at the carboxyl end of the amino acid residue were among the first synthetic substrates. Assays with these substrates were normally potentiometric in nature, since they depended on the esterolytic properties of the peptidase (16). The introduction of 4-nitroaniline and 2-naphthylamine (17,18) improved the sensitivity of the assays, since both of these compounds, as well as diazotized derivatives of 2-naphthylamine, can be detected spectrophotometrically. Liberation of 2-naphthylamine from acyl-NHNap substrates can also be followed by fluorimetry. More recently, Whittaker *et al.* (19) have designed a microtitre plate assay, where the protease catalysed hydrolysis of an ester substrate is detected by including a pH indicator in the assay mix. In aqueous solutions, ester hydrolysis results in the production of an alcohol with subsequent release of the product containing a free carboxyl and a proton. Interaction of the proton with the dye causes a change in absorbance that is reflective of the extent of proteolytic activity. This assay is useful in cases where chromogenic substrates are not appropriate.

3.2 Spectrophotometric assays

Spectrophotometric assays rely on the difference in molar absorptivity between the substrates and products. Substrates and products with large differences in molar absorptivities in the visible range are more desirable than those absorbing in the UV range, where absorbance due to enzymes and components of crude extracts can cause interference. In an interesting paper, Cathers and Schloss (20) describe an enzyme-coupled assay for proteases. Peptide and amino acid products of protease hydrolysis are incubated with a coupling solution containing carboxypeptidase A, amino acid oxidase and horseradish peroxidase. Free amino acids released by proteases, or generated from proteolytically derived peptides by carboxypeptidase A, are substrates for amino acid oxidase, which generates peroxide during oxidation of the free amino acids. The amount of peroxide generated in the assay is detected colorimetrically using peroxidase and a suitable substrate. This assay was linear between 0 and 25 μM phenylalanine in 96-well microtitre plates.

3.2.1 Substrates conjugated to 4-nitroaniline

N-Acylated derivatives of 4-nitroaniline have low absorbance at wavelengths around 400 nm, whereas the free amine is highly absorbant at these wavelengths. The difference in molar absorptivity for many 4-nitroanilides and 4-nitroaniline at 400 nm is about 10 000/M.cm at pH 8.0 (17); however, minor variations have been reported. These may be due to assays at different wavelengths and pH values. It is advisable to calculate the molar absorptivity under the conditions of the specific assay for optimal precision. A representative protocol is given in *Protocol 6*. Many nitroanilide substrates will hydrolyse spontaneously in basic solutions (the rate of autolysis increases with pH); thus, preparation of a stock solution in water or in an organic solvent usually facilitates the long-term storage of these substrates. Alternatively, fresh substrate can be prepared daily.

Protocol 6

Protease assays based on continuous spectrophotometric recording

Equipment and reagents

- A recording spectrophotometer, equipped with thermoregulated cuvette holder or housing, and capable of monitoring absorbance (220–700 nm is preferable; in this example, 400 nm is used). The spectrophotometer output should be connected to variable-speed strip chart recorder capable of speeds of about 2.5 cm/min or to an on-line data capture system. Scale expansions of 0. 1–1.0 absorbance full scale are desirable
- Matched cuvettes, glass or quartz; preferably of 1.0 ml volume. This arrangement allows for assays which involve either an increase or decrease in absorbance.
- The substrate concentration for routine assay depends on the kinetic parameters of the protease being studied, but is usually in the millimolar range. This example represents a characteristic substrate for a peptidase capable of hydrolysing L-Leu-NHNan, or any other 4-nitroanilides. Dissolve 12.57 mg of L-Leu-NHNan in 25 ml of water (substrate concentration 2 mM)
- Buffer (100 mM), preferably Tris, Tricine or HEPES, of the desired pH

Method

1 Add from 0.05 to 0.48 ml substrate[a] to a 1 ml cuvette. Add 0.5 ml of buffer and enough water to give a final volume of 0.98 ml. When test reagents are to be incorporated in the substrate, they should be added in place of, or dissolved in, the water.

2 Place the cuvette in the spectrophotometer cuvette housing that is equilibrated at the temperature of the assay (see *Table 2*, note b). Check that the scale expansion is correct (0.1 *A* is preferred) and that the proper chart speed is selected (1–2 cm/min is recommended. Set the wavelength to 400 nm.

Protocol 6 continued

3 Add 10–20 μl of enzyme solution to the cuvette. This volume can be applied to the side of the cuvette above the solution in the form of a bead and thus be kept separate from the substrate until mixing. When the enzyme is very dilute its volume can be increased, and should be added partially in lieu of water so that the total volume is always 1 ml.

4 Place a small piece of Parafilm over the cuvette and invert it several times, then put the cuvette back into the spectrophotometer. Begin recording immediately. If the pen goes off scale due to the addition of enzyme solution, adjust the spectrophotometer to reposition the pen. The initial absorbance is not critical as long as the recorder scale expansion stays constant and there is adequate energy from the monochromator.

5 Record the reaction sufficiently long to ascertain a good linear trace. If the slope of the recorder trace exceeds about 60°, either dilute the enzyme or adjust the spectrophotometer to greater scale expansion or a faster chart speed.

6 Calculate velocity, *v*, from the rate of change of absorbance with respect to time, using the difference between molar absorptivity for L-Leu-NHNan and 4-nitroaniline (10 500/M.cm) using the following equation:

Note that velocity is not defined as a function of enzyme volume, weight or equivalents. The values *v*/(ml enzyme), *v*/(mg enzyme) or *v*/(mol enzyme) are expressions of specific activity. Other substrates that can be used with this general procedure and their spectrophotometric properties are given below:

Substrate	Wavelength to measure reaction (nm)	Direction of absorbance change	(/M.cm)
Fa-Gly-Leu-NH2	345	down	317
L-Leu-NHNap (and other-NHNaps)	340	up	1780
Bz-L-Arg OEt	253	up	1150
Leu-NH,	220	down	50

[a] The chosen substrate concentration should result in linear absorbance changes over the time of the assay. Usually, 0.25–1 mM final substrate concentration yields good results. Prepare only as much substrate solution as will be used during the day's experimentation, since nitroanilide substrates show varying degrees of autolysis during storage. This can be exaggerated in Tris and other amine buffers, which catalyse nucleophilic attack on the substrate.

3.2.2 Substrates conjugated to 2-naphthylamine

2-Naphthylamine possesses an absorption maximum at 340 nm, and the difference in molar absorptivity between this compound and *N*-acylated 2-naphthylamine is about 1700/M.cm. This difference is pH dependent and decreases to zero at pH values below 5.5 (21). Thus, rates of release of 2-naphthylamine during substrate hydrolysis can be monitored at 340 nm in alkaline or neutral buffers.

Protocols for spectrophotometric assays using 2-naphthylamide substrates are similar to those outlined for 4-nitroanilides.

Enzyme-catalysed release of 2-naphthylamine can also be determined by diazotization of free naphthylamine with *N*(1-naphthyl)-ethylenediamine dihydrochloride (22). Some investigators have used a stable diazonium salt, Fast Garnet GBC (23), to detect rates of release of 2-naphthylamine. The conjugated product is insoluble and requires the presence of detergents to maintain solubility. Note that 2-naphthylamides are carcinogenic and should be handled accordingly. Increasingly, naphthylamide substrates have been replaced with NHMec or Nan substrates. Although these substrates are convenient for solution assays, they are unsuited for histochemical localization. Only naphthylamide derivatives are well suited for this purpose.

3.2.3 Substrates containing amides and esters

Amino acid amides, esters of amino acids and *N*-blocked amino acids are aminopeptidase and endopeptidase substrates. The degree of sensitivity afforded by these substrates is often much lower than those containing 4-nitroaniline or 2-naphthylamine, unless the products of hydrolysis have a high molar extinction coefficient (24). Another potential problem with ester substrates is their susceptibility to hydrolysis by non-proteolytic enzymes, especially when crude extracts are being analysed. Amide substrate hydrolysis can be followed spectrophotometrically at 230 nm as a decrease in absorbance upon amide bond hydrolysis, or by measuring the amount of free amino acid released (see ninhydrin protocol for carboxypeptidases—*Protocol 8*). However, high concentrations of substrate and enzyme are needed, which may result in a high background absorbance. It is normally more convenient to measure enzyme activity with more sensitive chromogenic substrates.

3.2.4 Substrates *N*-acylated with furylacrylic acid

N-Furylacryloyl (Fa) substituted peptides were first introduced as substrates for neutral proteases, especially metallo-endopeptidases (25). Assays using Fa-peptides are followed by measuring the decrease in absorbance at 345 nm.

3.3 **Fluorimetric assays**

Fluorimetric assays employ either the inherent fluorescence of an amino acid or depend on the fluorescent properties of products. In such assays the spectrofluorimeter has to be standardized with known concentrations of the fluorescent substance, or standardized using secondary fluors. The choice of fluor is dependent on the type of substrate being utilized and the sensitivity necessary for the experiments.

For example, experiments designed to evaluate kinetic parameters of an enzyme will require standardization of the output of the instrument with known concentrations of the fluorophore being measured. When relative fluorescence, but not absolute activity, of samples is required, it is often easier to zero the

Table 3 Problems encountered in fluorimetric assays

Light scattering	Particulate matter in solution such as bacterial impurities in reagents; limited solubility of substrate	Centrifuge samples and/or filter through 0.45 mm filter. Lower the substrate concentration
Quenching (loss of light intensity generated by fluorescence)	Enzyme sample (especially if impure extract) contains constituents that absorb the emitted light. Substrate solution may contain components that absorb emitted light.	Dialyse enzyme to remove low molecular coloured materials. Partial purification of enzyme preparation? Check spectral properties of substrate solution
Rapid increase in fluorescence signal.	Fluorescence assays are often very sensitive, and a small degree of substrate hydrolysis can produce a large signal change.	If a large amount of the substrate is hydrolysed, dilute the enzyme preparation, or otherwise adjust the sensitivity of the fluorimeter
Very slow changes in fluorescence	Inadequate hydrolysis	Use more enzyme
No change in fluorescence signal, independent of concentration.		Inappropriate substrate? Stock enzyme solution of substrate already hydrolysed? Instrument malfunction ?

fluorimeter with a buffer sample or a substrate sample at the start of the assay, and then measure samples containing enzyme and the substrate. It should be borne in mind that not all fluorimetric assays work well with all enzymes, nor with all tissue types. Furthermore, assays involving crude homogenates might yield misleading results due to light scattering and quenching (*Table 3*).

3.3.1 Substrates containing 2-naphthylamine (NHNap)

2-Naphthylamine absorbs at 335 nm and emits light at 410 nm (26). The fluorescence of 2-naphthylamine is pH dependent and is linear in a concentration range from 0.25–100 ng/ml (27), which is equivalent to 1.75–698 nmol/ml. The sensitivity of this procedure is nearly three orders of magnitude greater than the corresponding assay relying on diazotization. Thus, much smaller quantities of enzyme can be readily detected.

3.3.2 Substrates containing 7 amino-4-methylcoumarin (NHMec)

Glutaryl-L-phenylalanyl-NHMec was first synthesized as a fluorogenic substrate for chymotrypsin by Zimmerman *et al.* (28). Since then, numerous peptide-NHMec substrates have been synthesized for specific enzymes. Rates of hydrolysis of NHMec substrates are measured at an excitation wavelength of 380 nm and an emission wavelength of 460 nm (*Protocol 7*). At these wavelengths, the free 7-amino-4-methylcoumarin group is approximately 500–700 times more fluorescent than the conjugated forms. For chymotrypsin the detection limit is 0.5 μg/ml at a substrate concentration of 0.2 mM, compared with 10 μg/ml enzyme for the corresponding 4-nitroanilide substrate, and the initial velocity of the reaction is proportional to enzyme concentration over a 100-fold concentration range of enzyme (28). A large number of NHMec substrates are available commercially.

Most NHMec substrates are water-soluble and show low rates of autolysis upon storage. Further, since they have high fluorescence yields, it is possible to assay dilute enzyme preparations at lower substrate concentrations.

Protocol 7

Fluorimetric assay for 7-amino-4-methylcoumarin substrates

Equipment and reagents

- Fluorimeter (variable wavelength) equipped with temperature-controlled cuvette housing
- Matched UV grade fluorescence cuvettes (examples here employ 1.0 ml cuvettes; however, 3 ml cuvettes can also be used by scaling up the procedure)
- Circulating water bath (connected to the cuvette housing).
- Computer or strip chart recorder for monitoring fluorescence output
- Substrate: aminoacyl-NHMec or N blocked aminoacyl-NHMec (0.1–0.5 mM) in buffer. Substrate can be diluted to the desired concentration
- Buffer: 50 mM solution of a suitable buffer of the desired pH

Method

1. Calibrate the fluorescence output with solutions of 7-amino-4-methylcoumarin of known concentrations. It is often convenient to use secondary standards, such as plastic blocks with inherent fluorescence, for daily calibration. This obviates the need to calibrate the instrument with 7-amino-4-methylcoumarin solutions every time the instrument is used. Measure the fluorescence of the secondary standards at the same time as the primary standard. Plot the fluorescent intensity of the primary standard against concentration. Using the standard curve, determine the 7-amino-4-methylcoumarin concentration corresponding to the fluorescence output of each block. Subsequently, the instrument can be standardized using only the secondary standards.
2. Place all solutions (substrate, buffer and enzyme) in a constant-temperature water bath set at 25 °C.
3. Pipette the substrate solution and buffer into a 1.0 ml cuvette to obtain the desired substrate concentration in 1.0 ml.
4. Place the cuvette in a thermostatted holder (25 °C) and allow it to equilibrate for about 5 min.
5. Set the excitation and emission wavelengths to 380 and 460 nm, respectively. The slit width should be kept to a minimum. If the slit setting is changed, the instrument should be recalibrated.

Protocol 7 continued

6 Initiate the reaction by adding 10 μl of enzyme. Place a piece of Parafilm over the cuvette and invert it three or four times. Place the cuvette in the fluorimeter and record the fluorescence emission for 2–5 min.

7 Calculate the protease activity from the slope of the recorder trace in terms of 7-amino-4-methylcoumarin produced from substrates using the standard curve relating fluorescent intensity to 7-amino-4-methylcoumarin concentration. One unit of protease activity is defined as the amount of enzyme required to release 1 nmol of 7-amino-4-methylcoumarin from the substrate per minute under the conditions of the assay.

3.4 Miscellaneous fluorimetric methods

Yaron *et al.* (29) has reviewed the design and synthesis of intramolecularly quenched fluorogenic substrates. These substrates contain a fluorophore quenched by another moiety present in the substrate molecule, but separated from the fluorophore by the scissile bond. The fluorophore is in a P position and the quenching group in a P′ position. The advantage of these substrates is that cleavage of a peptide bond between two amino acids, rather than a peptide analogue, occurs, which may be argued to represent a more 'natural' hydrolytic event. A typical substrate fulfilling these constraints is Abz-L-Gly-Phe(NO_2)-Pro, a substrate for angiotensin-1 converting enzyme. The Abz group is a fluorophore with high quantum yield, and is quenched in the substrate by the presence of the Phe(NO_2) group. Enzyme-catalysed hydrolysis of the Abz-Gly-Phe(NO_2) bond releases the fluorescent moiety, Abz-Gly. Fluorescence is monitored at an excitation wavelength of 310 nm and at an emission wavelength of 410 nm using a protocol similar to the one given in Protocol 7.

Bratovanova and Petkov (30) reported the synthesis of fluorogenic substrates from peptide 4-nitroanilides. *N*-Acylation of peptide-4-nitroanilides with Abz yielded *N*-Abz substrates that were internally quenched by the presence of the 4-nitroanilide moiety. On hydrolysis of the aminoacyl-4-nitroanilide bond, the highly fluorescent *N*-Abz peptide or amino acid was released.

Routine assays were performed at a substrate concentration of 0.05 mM in a total of 1 ml with an excitation wavelength of 340 nm and an emission wavelength of 425 nm. When Abz-L-Phe-NHNan was used as a substrate for chymotrypsin, linearity was observed over a concentration range of enzyme from 1 to 10 000 nM (30). Since then a whole range of internally quenched fluorescent substrates have been developed to assay many classes of proteases, including renin (15), viral proteases (31) and tail-specific protease from *Escherichia coli* (32).

In a novel development, Lawler and Snyder (33) described a viral protease assay based on the construction of plasmids containing the GAL4-DNA binding domain joined by a peptide linker to the GAL4 transactivating domain. The peptide linker contains a viral protease site. Plasmids containing this construct are cotransfected into mammalian cell lines along with a plasmid containing the GAL4 upstream activating sequences ligated to a luciferase gene. The protein product of the first plasmid binds to the GAL4 upstream activating sequence, resulting in the synthesis of luciferase. The amount of luciferase activity is subsequently measured by chemiluminescence. If the transfected cells were infected with a virus, the viral protease would cleave the linker peptide between the GAL4 DNA-binding domain and the GAL4 transactivating domain, leading ultimately to a loss in luciferase production, through a loss in the transcription of the luciferase gene. The amount of luciferase produced is expected to reflect the amount of viral protease produced *in situ*, and/or the extent of inhibition of viral multiplication. This unique assay has the capacity for high throughput analysis and can evaluate the efficacy of antiviral drugs under better physiological control. However, the amount of luciferase produced was found to be dependent on the cell line (33).

FP assays have been used to detect protease (34–36). These assays depend on a change in the FP of a given molecule in solution. The relative FP of a molecule increases with size, thus large polypeptides will have much greater FP than small peptides. FP measurements, however, require a special instrument designed for this purpose and a suitably labelled polypeptide or peptide substrate. Incubation of the substrate with a protease leads to a decrease in net FP and the change in FP is dependent on the amount of protease and the nature of the substrate. These measurements can be performed within minutes and can detect pure proteases in the 5–20 ng ml^{-1} range (35), or have been designed to detect a specific protease using peptide substrates (36).

Protocol 8

Ninhydrin method for carboxypeptidase assays.

Equipment and reagents

- Colorimeter or spectrophotometer (570 nm)
- Temperature-controlled water bath
- Buffer: a 50 mM solution of the buffer of choice adjusted to the desired pH. Substrate solution (2 mM) should be prepared in the buffer.
- Boiling-water bath
- Ninhydrin reagent: dissolve 200 mg of ninhydrin and 75 mg of hydrindantin in 7.5 ml of peroxide-free methyl cellusolve and dilute to 10 ml with 1.0 M acetate buffer at pH 5.5

Method

1. In a 13 × 100 mm test tube, combine 1.0 ml of substrate and 1.0 ml of the enzyme sample. Incubate at the desired temperature for 10–30 min. Terminate the reactions by adding 1.0 ml of the ninhydrin reagent.
2. Heat the test tubes in a boiling water bath for precisely 15 min, then cool them quickly in an ice-water bath. Add 5.0 ml of 50% ethanol to each tube and mix thoroughly.

Protocol 8 continued

3. Measure the absorbance of this solution at 570 nm.
4. Calculate the carboxypeptidase activity from the absorbance value using a standard curve prepared with known concentrations of leucine. One unit of peptidase activity is the amount of enzyme required to produce 1 μmol of L-Leu equivalent amino acid from the substrate per min under the conditions of the assay.

3.5 Carboxypeptidase substrates

By definition, carboxypeptidases require a free carboxyl group for activity, and hydrolyse peptide bonds from the *C*-terminal end of a peptide; thus, a carboxypeptidase substrate should possess an unblocked aminoacyl carboxylate.

Commonly used carboxypeptidases substrates require the quantification of released amino acids by secondary methods such as reaction with ninhydrin (*Protocol 8*), or qualitative measurements of released amino acids by TLC or HPLC (37).

Carboxypeptidase-catalysed hydrolysis can also be monitored potentiometrically or in a enzyme-coupled assay (20).

N-Dansylated peptides have been synthesized to study carboxypeptidase–substrate complexes with stopped-flow fluorimetry (38). While the method provides for a sensitive assay for all types of peptidases (39), highly concentrated enzyme preparations are needed and the instrument requirements are considerable and expensive.

Carboxypeptidases can also be assayed spectrophotometrically using *N*-acylated Fa-peptides and *N*-acylated Abz- or dansyl-peptides. Methods are similar to those outlined for endopeptidases or aminopeptidases. Many carboxypeptidases can catalyse the hydrolysis of ester substrates. These assays are useful with purified enzymes, but results obtained with crude homogenates are questionable because esterases and endopeptidases are usually present.

3.6 Radiometric assays

Radioactive peptide substrates containing a few amino acid residues are prepared by chemical coupling techniques, and are normally used when other non-radioactive substrates are either unavailable or involve cumbersome assay techniques (40). When assays are performed with small peptide substrates, it is necessary to separate the liberated amino acid from the residual intact substrate. This may involve solvent extraction and ion-exchange chromatography. In most instances, the sensitivity of these methods is high and a small amount of peptidase can be reliably assayed. However, as with most radioisotope studies, substrates are expensive, materials require special handling and proper disposal of radioactive wastes is necessary. However, many radioactive peptides can be conveniently synthesized by solid-phase methodologies. Large protein substrates can be prepared by iodination with ^{125}I (37) or reductively labelled with tritium

(41). In general, the current non-radioactive micro-methods provide suitable sensitivity for the assay of most peptidases and radiolabelled assays are only needed in special circumstances.

3.7 HPLC assays for peptidases

Several HPLC protocols are available to analyse the products of enzymic hydrolysis of peptidyl or protein substrates. An obvious advantage of HPLC protocols is that both chromogenic and proteinaceous substrates can be analysed at high sensitivities and, if proteins are used as substrates, it is possible to collect the digestion products for amino acid analyses. There are, however, several drawbacks to HPLC procedures, in addition to the cost of the equipment. These are: (i) only one sample can be analysed at a time; (ii) turn-around times for injections may be long and will vary with the type of column and solvents used; (iii) defining the working protocol may take a good deal of development time; (iv) contamination of the column and loss of resolution can occur if crude homogenates are used; and (v) analytical columns need to be replaced at regular intervals. Despite these drawbacks, HPLC methods are extremely useful for the assay of peptidases. A protocol for the analysis of NHNan substrates is provided in *Protocol 9*.

Protocol 9

HPLC assay using 4-nitroanilide substrates[a]

Equipment and reagents

- Isocratic HPLC (only one pump required) equipped with a flow-through spectrophotometric detector set at 314 nm.
- Column: Nucleosil 5 μm, C 18, 4.6 × 150 mm (Phenomenex)
- Peak integrator: preferably microcomputer-based (conventional manual procedures could be used)
- Amino acyl-NHNan (1 mM) of choice dissolved in a desirable buffer, e.g. 50 mM Tris-HCl, pH 7. 4
- 4-Nitroanilide standard (10 mM) in water or assay buffer
- HPLC solvent: sodium phosphate (10 mM, pH 3.0) and methanol in 1: 1 mixture
- HPLC operation: flow rate, 0.6 ml/min, 0.1 absorbance units full-scale expansion

Method

1. In a microfuge tube. combine 100 μl of substrate solution and 5–10 μl of enzyme solution: incubate at 37 °C for 30 min[b]. Stop the reaction by adding 10 μl of glacial acetic acid[c].
2. Inject 1–20 μl for HPLC analyses.
3. Calibrate peak areas by injecting known volumes of authentic standards of known concentrations.

Protocol 9 continued

[a] Method adapted from (42).

[b] Times of incubation have to be determined by experimentation for each substrate and enzyme combination.

[c] When crude extracts are assayed, 10 μl of 60% trichloroacetic acid can be used to precipitate protein. Precipitated proteins are sedimented by centrifugation at 10 000 **g** for 5 min, and 10 μl of the supernatant fluid is analysed by HPLC. If a tripeptide substrate such as Suc-Ala_3-NHNan is used, it is possible to follow both endopeptidase and aminopeptidase activity simultaneously. since the products of each enzyme, AlaNHNan and 4-nitroaniline are clearly resolved from each other (32). HPLC assays have been adapted for a number of protease substrates. The availability of photodiode array detectors can greatly facilitate the identification of compounds in the HPLC effluent. This is of importance in substrates containing chemically modified amino acids.

Numerous other protocols are available to detect peptides and amino acids by HPLC. Although quite specialized, many of these methods can be adapted for the study of proteases. Protein substrates for measuring femtomolar quantities of endopeptidases have been reported by Green (43). A protocol for this technique is given in *Protocol 10*. This procedure can be readily modified for other protease assays.

Protocol 10

HPLC assays using protein substrates[a]

Equipment and reagents

- HPLC: an HPLC (or LC) system capable of generating a gradient is required for this procedure. The instrument should be set to run linear gradients automatically. A flow-through photodiode array or UV/visible spectrophotometer are also required
- Column: C-18 or C-8 reversed-phased column
- Microfuge
- HPLC solvents: aqueous eluent, 0. 1% trifluoroacetic acid in water; organic eluent, 90% acetonitrile in water containing 0.08% trifluoroacetic acid. Use HPLC-grade solvents
- Protein substrate: e.g. a 10 μM solution of cytochrome *c* or *S*-carboxymethylated lysozyme, in an appropriate buffer of desired pH, is used as the substrate. Any protein will suffice; however, the proteins suggested here are of known sequence. Frequently, HPLC peaks containing peptides are analysed by mass spectrometry, automated protein sequencing or by amino acid analyses. All three methods will yield data on the sites of enzymatic cleavage. Thus, different protein substrates of known sequence can be used as an aid to determining enzyme specificity.

Method

1 Equilibrate the HPLC column with the aqueous eluent at a flow rate recommended

Protocol 10 continued

for the column. Monitor the absorbance of the effluent stream by recording absorbance at 206 nm, with a scale expansion of 0.05 or 0.1 absorbance units full scale.

2 Incubate 300–500 μ1 of substrate with the enzyme (1–100 fmol) at the desired temperature (see *Table 2*, note b). At specific time intervals, e.g. every 30 min, add 100 μ1 of the reaction mixture to 5 μl of 10% trifluoroacetic acid, and mix with a vortex mixer. Centrifuge the solution at 10 000 **g** for 5 min to remove any precipitate that may form.

3 Inject 50 μl of the sample for HPLC analysis. Elute the column with a linear gradient, from 100% aqueous eluent to 100% organic eluent, generated over a 60–120 min period. To regenerate the column before injection of the next sample, pump the organic eluent over the column for an additional 10 min, then re-equilibrate the column with the aqueous eluent.

[a] Adapted from Green (43).

Endoproteinases are routinely used in protein sequencing, and several HPLC procedures have been developed to isolate and quantify the released peptide fragments. Thus, it should be possible to adapt existing HPLC techniques to assay peptidases using unmodified or specifically modified protein substrates. One of the biggest change in the LC analysis of peptides has been the development of capillary LC and the ability to directly inject the LC effluent stream into a mass spectrometer. Data can be simultaneously obtained about peptide masses and peptide sequence. Many elegant studies published in the last decade using these techniques. Such instrumentation in combination with software to mine the sequence databases, has resulted in the new field of proteomics (44) (see Chapter 7).

3.8 Capillary electrophoretic analyses of proteases

CE is a very powerful analytical tool used to detect a broad range of analytes with high resolution and high sensitivity. One advantage is that the reactants can be present in high or low concentrations.

Several proteases have been analysed routinely by CE but for the purpose of this chapter, we present a protocol to determine the activity of cathepsin D by analysis of the products of its reaction with fluorescently-labelled haemoglobin (45) (*Protocol 11*; *Figure 2*). This method has advantages in the size of sample required for analysis and sensitivity of detection. Initial products of the reaction appeared very quickly ($<$ 5 min) and one of these peptides was identified as resulting from cleavage between Phe-32 andPhe-33 residues of haemoglobin α-chain by amino acid sequencing of the purified peptide. In addition, there was reasonable assurance that proteolytic activity observed in crude extracts, was due to an aspartyl proteinase because incubations were conducted at low pH (3.5) and the haemoglobin α 1–32 peptide product was not observed when control incubations were conducted with the inhibitor, pepstatin.

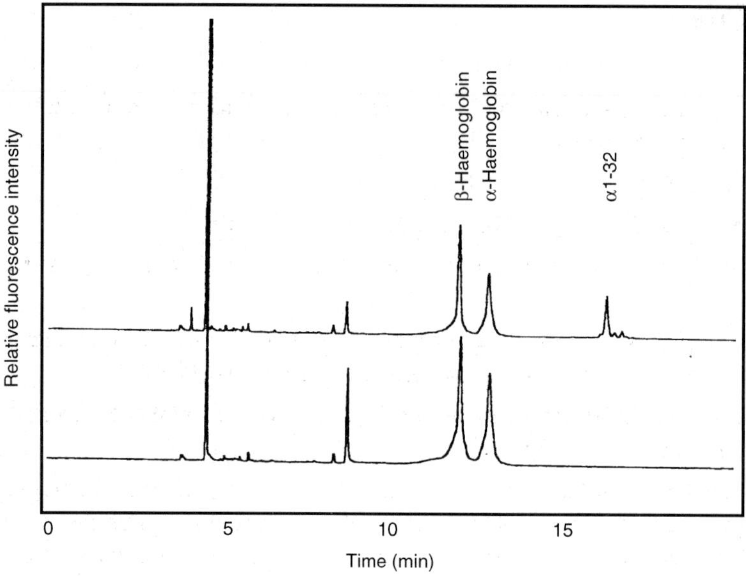

Figure 2 Electrophoretogram of cathepsin D activity with Oregon Green-haemoglobin. Cathepsin D (84 mg) was incubated with Oregon Green-haemoglobin, prepared as shown in *Table 14*. The lower trace is the control to which pepstatin was added before enzyme. The upper trace represents an incubation of 4 min of substrate + enzyme before the reaction was stopped with pepstatin. Peaks correspond to the α and β chains of haemoglobins and the α 1–32 fragment released as a consequence of cathepsin D action are labelled. Conditions for electrophoresis are as shown in *Protocol 11*.

Protocol 11

Capillary electrophoretic analysis of cathepsin D using Oregon Green-labelled haemoglobin[a]

Equipment and reagents

- Capillary electrophoresis, equipped with a laser-induced fluorescence detector and interfaced to a microcomputer
- Capillary columns
- Oregon Green-haemoglobin. The fluorescent substrate is prepared as follows: haemoglobin is dialysed for 2 days versus water with several changes and centrifuged to remove insoluble materials. The pH of the supernatant is adjusted to pH 8.5 with dilute NaOH, and the protein concentration adjusted to 25–35 mg/ml with water. To 3.5 ml of the haemoglobin solution containing 0.125 M sodium borate buffer, pH 8.5, 0.875 ml of a 5 mg/ml Oregon Green (Sigma Chemical Co.) solution in dry dimethylformamide is added. The reaction is incubated at room temperature for 5 min and then stopped by the addition of 0.5 ml of 0.5 M NH_2OH. This reaction mixture is passed over a Pharmacia G-25 Sephadex column and the labelled protein eluted in the void volume is collected, dialysed against water, lyophilized and stored at −70 °C. Stock solutions containing approximately 25–35 mg/ml of labelled haemoglobin are prepared fresh prior to use
- Capillary electrophoresis buffer: 75 mM SDS, 25 mM sodium borate, 50 mM sodium phosphate, pH 8.5

Protocol 11 continued

Method

1. Add 12.5 μl Oregon Green-haemoglobin and 12.5 μl 0.2 M sodium formate buffer, pH 3.5 in a 500 μl microfuge tube.
2. Preincubate the mixture for 5 min at 37 °C.
3. Include a control by adding 10 μl of 10 μM pepstatin to one tube.
4. Start the reaction in all tubes by adding of 50 μl of a Cathepsin D source, e.g. tissue extracts.
5. Stop the reaction after 30 min by the addition of 10 μl of 10 μM pepstatin. Place the tube on ice and remove any precipitate by centrifuging 5 min at 11 000 **g**.
6. Add 2 μl of 1.0 μM BODIPY-ethanol amine per 20 μl of sample as internal standard, just prior to electrophoretic analysis.
7. CE is performed with an appropriate instrument (e.g. Beckman P/ACE 5510) equipped with laser (Argon)-induced fluorescence detector. The excitation and emission wave lengths are 488 nm and 520 nm, respectively. The column material (Polymicro Technologies) used in these separations is untreated fused silica (50 μm ID, 27 cm in long). Separations are performed in a micellar mode containing 75 mM SDS, 25 mM Na borate, 50 mM Na phosphate, pH 8.5. Voltage for these separations is 18 kV, with the temperature maintained at 20 °C.
8. Pressure inject the sample for 3 s and separate as above.
9. Identify the peak corresponding to α-haemoglobin 1–32 in the separation. It occurs at about 15 min under these conditions (see *Figure 2*).
10. Determine the areas under the peaks corresponding to α-haemoglobin 1–32 and the internal standard. Express the area of the α-haemoglobin 1–32 peak as a ratio of the internal standard peak area.
11. The relative amount of cathepsin D activity in samples can be expressed as normalized α-haemoglobin 1–32 peak area/μg protein.
12. The concentration of cathepsin D in samples can be calculated from a standard curve generated with purified cathepsin D and normalization of the α-haemoglobin 1–32 peak area versus enzyme concentration.

[a] Adapted from (45) and M. G. Zeece, unpublished data). Oregon Green carries an extra carboxylate anion, the negative charge is of benefit in CE separations, since it diminishes interactions with the surface of the silica capillaries.

4 Solid-phase protease assays

Solid-phase assays, in which either enzyme or substrate may be immobilized, can also aid in the detection of peptidase activity. In many such assays, activity can be detected in crude samples. These methods can be electrophoretic, with immobilized enzymes or immobilized substrates, or both. Plate assays, where enzymes diffuse into a gel matrix containing hydrolysable substrate were among the first assays devised for proteases; assays with substrate immobilized on porous supports, or assays using substrate bound to plastic supports are now more common.

4.1 Gel electrophoretic methods

Assays of this type fall into two main categories: those in which enzyme activity is determined *in situ* after electrophoresis, and those in which enzyme activity is detected by replica blotting of the electrophoretic gel on to a substrate-containing gel.

These assays are all based on incorporating a substrate (normally a protein) in a gel matrix. Protease-containing zones will appear as areas where the protein substrate is lost from the gel. The reader is referred to an excellent review (49) of the many zymographic techniques that have been used to study microbial proteases; this review provides several methods that can be readily used in most laboratories.

However, since electrophoretic methods are so commonly utilized, we provide two methods (*Protocols 12* and *13*) for protease detection following electrophoresis; other methods have also been published (37,49).

Protocol 12

Peptidase assay for electrophoretically separated enzymes using synthetic substrates.

Equipment and reagents

- Substrate: 0. 1–1.0 mg/ml aminoacyl- or *N*-blocked amino acyl-NHNap in the buffer of choice
- Staining solution: Fast Garnet GBC[a] (2 mg/ml in water) Prepared fresh just prior to use
- PAGE apparatus
- Assay buffer: the choice of buffer will depend on the enzyme being studied; examples are phosphate, Tris-HCl, Tricine and HEPES. Buffers should be at least 50 mM

Method

1 Fractionate enzyme samples by polyacrylamide gel electrophoresis under either native or denaturing conditions.[b]

2 Remove the gel from glass plates or tubes, and transfer it to the assay buffer. Wear latex gloves to avoid contaminating the gel.[c] Incubate the gel in the assay buffer for 10 min. Change the buffer once, then replace it with at least two times the volume of gel of fresh buffer, at the desired temperature, containing 0. 1–1.0 mg/ml 2-naphthylamide substrate.[d] Incubate the gel in the substrate solution for 10–60 min.[e]

3 Add enough freshly prepared Fast Garnet GBC[f] to yield a final concentration of 0.5

Protocol 12 continued

mg/ml of in the assay mixture.

4 Remove the gel from the solutions containing the substrate and diazonium salt when enzyme bands are well defined. The reaction should be terminated when the stained bands appear red against a pale yellow background, and have enough contrast for photography.

5 Wash the gel two or three times in water. The reaction product is insoluble in water, so the gels can be stored indefinitely in it.

[a] Fast Garnet GBC is a possible carcinogen and should be handled with caution.

[b] Consult (37) for information on gel electrophoretic conditions. If SDS is used during electrophoresis, it should be removed by washing the gels extensively (3–5 gel volumes) in a suitable buffer containing 1–2% Triton X-100. Also, note that not all enzymes will remain active after native gel electrophoresis and many will be irreversibly denatured by SDS.

[c] Acrylamide is a neurotoxin, and 2-naphthylamine is a carcinogen. Avoid contact with skin.

[d] Select a 2-naphthylamide substrate for which the electrophoretically fractionated enzymes show high specificity. This obviates the need for prolonged incubation times and will also result in sharper, more intense bands. See *Table 5* for solubility properties of 2-naphthylamide substrates. Alternatively, NHMec substrates can be used and zones of activity captured by UV-transillumination. This is a preferred method if zones of enzymatic activity are to be further manipulated after gel electrophoresis (e.g., 2-dimensional PAGE, recovery of activity, etc.)

[e] Increasing times of incubation will generally result in broad diffuse bands, especially if the incubation temperatures are around 37 °C. This normally does not interfere with interpretation of the results, provided that the enzymes fractionated in the gel have different mobilities. However, when isozymes of similar mobilities are being fractionated, overlap of activity stained bands is common. In such cases, gels of higher acrylamide concentration, or isoelectric focusing procedures are recommended to resolve enzymes.

[f] The reaction between 2-naphthylamine and Fast Garnet GBC is inhibited in the presence of sulphydryl groups; thus, if reducing agents are present in the assay buffer, it is necessary to remove these compounds by washing, or preferably by oxidation, prior to staining with the diazonium salt (12). Peptidases are also inhibited to varying degrees by stable diazonium salts, which can be a problem when these salts are present in the assay solution with a substrate. When enzyme activity is high, the addition of Fast Garnet GBC solutions of 0.25–0.5 mg/ml directly to the assay buffer can enhance the clarity of staining.

4.1.1 *In situ* gel procedures

In situ gel procedures either use histochemical methods (*Table 15*) or require the impregnation of the gel with a proteinaceous substrate (*Table 16*). Electrophoresis is commonly performed in polyacrylamide gels, although both starch and agarose gels have been used (37,49).

4.1.2 Replica blotting of gel fractionated enzymes

These methods rely on the ability of electrophoretically separated enzymes to hydrolyse substrates incorporated into a second gel or on strips of absorbent material such as cellulose acetate, yielding a 'zymogram' (49). Alternatively, fractionated enzymes can be transferred to nitrocellulose or PVDF sheets, and then assayed for activity (37,49).

Protocol 13

Assays with protein incorporated into polyacrylamide during polymerization[a,b]

Equipment and reagents

- Protein substrate: gelatin, haemoglobin, casein, azocasein or BSA are good initial choices
- PAGE apparatus.

Method

1. Dissolve a suitable protein substrate (0.5–10 mg/ml) in the separating gel solution. Initiate polymerization and pour the gel.
2. Layer the stacking gel, commonly prepared without substrate, over the separating gel and allow it to polymerize.
3. Load the enzyme samples and perform electrophoresis until the dye front is at or near the bottom of the separating gel.
4. Remove the gel and incubate it in a suitable buffer. The choice of the buffer and the conditions of incubations will vary depending on the activity of the enzymes being studied. The buffer should be 50–100 mM and the pH should be chosen for optimal enzyme activity. Some peptidases can be renatured in gels containing SDS by repeatedly washing the gel in buffer, or buffer containing 1% Triton X-100 (43).
5. Stop the reactions by placing the gel in a staining solution containing 0.1% Coomassie Brilliant Blue in 40% (v/v) methanol, 10% (v/v) acetic acid.[c]
6. Destain the gel. Areas of enzyme activity should appear as clear or light blue zones against a dark blue background.

[a] Adapted from different sources (46–49).

[b] Consult (48) for the use of gels containing radioactive protein substrate. If azocasein is used as a substrate, the gel can be soaked in 0. 1 M NaOH solution at the end of the incubation period. Areas containing peptidases should appear as clear zones against an orange background. Streaking of samples may occur, especially at the top of the separating gel, when electrophoresis is performed under non-denaturing conditions. Streaking can be minimized by performing electrophoresis at 4 °C under pH conditions that are not optimal for enzyme activity (46,49).

Protocol 14

Detection of proteases using unprocessed X-ray films[a]

Equipment and reagents

- X-Omat XAR 5 X-ray film from Kodak Corporation (several sizes are available)
- Controlled-temperature incubator

Protocol 14 continued

Method

1. Take a sheet of unprocessed X-ray film. Add 50 μl drops of different dilutions of a protease in an appropriate buffer,[b] and incubate the film at 42 °C for 1 h.
2. Wash film gently under running tap water for 1 min.
3. Dry and visually determine clearing of film.

[a] Adapted from (53).

[b] Film is not compatible with buffers containing guanidine hydrochloride (3 M) or potassium thiocyanate (1 M)

4.2 Plate assays

Plate assays rely on the radial diffusion of proteolytic enzymes into agar containing a suitable substrate. The method and numerous protocols for analyses of different proteases have been extensively described (50). These assays can be relatively non-specific and are helpful if suitable chromogenic substrates are unavailable or not known for the enzyme being assayed. However, these assays are prone to interference from other compounds present in whole-tissue homogenates (51). An improvement of plate assays has been reported recently by Hagen *et al.* (52).

Proteinases have been detected by incubation on the surface of Kodak Corp, X-Omat XAR films (53) (*Protocol 14*). This procedure utilizes the gelatin coating of the unprocessed films as substrate. Proteolytic degradation of gelatin on the surface results in a clear area against a black background. The utility of this simple assay has been extended to study the pH optima of uncharacterized fungal proteinases (54).

4.3 Miscellaneous solid-phase assays

Large proteins such as proteoglycans and gelatin can be immobilized by coupling to polyacrylamide or Sepharose beads according to standard procedures (37). Proteolytic enzymes are incubated with the substrate, and the peptides released by enzymatic hydrolysis are quantified. Protein substrates may be radioactively labelled, tagged with luminogenic, fluorogenic, chromogenic or other reporter molecules such as biotin, denatured or used in their native states. Incubation temperatures and times have to be determined by experimentation. When radioactively labelled proteins are used, the radioactive peptides released are measured by scintillation counting.

If native proteoglycan immobilized on polyacrylamide is used as a substrate, digestion products released can be measured by quantifying the amount of uronate or by dye binding (55). Dye-binding methods appear to have the same degree of sensitivity as radioactive protocols, at least for this substrate (55).

Yoshioka *et al.* (56) were among the first to exploit a ELISA for detecting mammalian collagenases (see *Protocol* in ref. 37). Several extensions of microtitre plate techniques have since been developed for the assay of proteases. Koritsas and Atkinson (57) have demonstrated the utility of using biotinylated-gelatin for the simultaneous assay of numerous samples in a microwell plate format (*Protocol 15*). This procedure was essentially duplicated using biotin-labelled casein (58).

4.4 Assays for histochemical studies

Many techniques used for histochemical localization have been reviewed by Smith *et al.* (59) and the reader is referred to this article for concise information. More recently, the use of 7-amino-3-trifluormethylcoumarin substrates for histochemical localization have been presented (60).

Protocol 15

Microtitre plate assay for proteases using biotinylated-protein substrates[a]

Equipment and reagents

- Microplate reader interfaced to a microcomputer
- Microwell plates
- Temperature controlled incubator
- Biotinylated-gelatin: dissolve gelatin at 1 mg/ml in 0.2 M Na_2HCO_3, pH 8.8, containing 0.15 M NaCl. To 5 ml of this solution add 100 µl of a 4 mg/ml solution of biotin–N-hydroxysuccinamide ester in dimethylformamide. Gently stir at room temperature for 15 min. Stop the reaction by adding 75 µl of 1 M NH_4Cl, pH 6.0. Dialyse exhaustively against water, aliquot and store frozen at −20 °C
- Substrate-coated microtitre plates: dilute stock biotinyl-gelatin to 25 ng/ml in coating buffer (15 mM Na_2CO_3 35 mM $NaHCO_3$, pH 9.6). Add 50 µl of the diluted substrate to each well and allow binding overnight at 4 °C. Wash wells with PBS. Incubate wells with 50 µl of 1% gelatin in PBS for 30 min at 37 °C to block all remaining sites. Wash three times with 200 µl PBS followed with three washes with 200 µl water. Plates can be stored dry or in 0.025% azide at 4 °C for over 6 months

Method

1 Incubate wells of substrate-coated plates with 50 µl of the appropriate buffers for 10 min at 37 °C.

2 Discard buffer and add 50 µl of a suitably-diluted proteinase solution and incubate for varying times (0–60 min) at 37 °C.

3 Stop reactions by washing wells with five 200-µl aliquots of PBS-Tween (normally 0.05% Tween-20 in PBS).

4 Add 50 µl of streptavidin-alkaline phosphatase conjugate (1: 2500 dilution) in water to each well and incubate for 15 min at 37 °C.

Protocol 15 continued

5 Rinse wells several times with PBS-Tween.

6 Add 200 μl of a 1 M diethanolamine-HCl buffer, pH 9.6, containing 5 mM $MgCl_2$ and 1 mg/ml *p*-nitrophenylphosphate[b] and incubate at 37 °C for 30 min.

7 Read absorbance at 405 nm using a plate reader.

9 Calculate protease activity using the absorbance of wells treated only with buffer to indicate amount of total available substrate.

[a] Adapted from (57). This protocol is amenable to almost any proteinaceous substrate. Potential limitations are for proteases that require significant folding in a substrate.

[b] Use of chemiluminescent substrates for alkaline phosphatase should greatly extend this useful protocol.

Acknowledgements

We thank our colleagues for their reading of this manuscript. Financial support from the Centre For Biotechnology, funded through the Nebraska Research Initiative is gratefully acknowledged (G. S.).

References

1. Bradford, M. M. (1976). *Anal. Biochem.* **72**, 248.
2. Buroker-Kilgore, M. and Wang, K. K.W. (1993). *Anal. Biochem.* **208**, 387.
3. Bickerstaff, G. F. and Zhou, H. (1993). *Anal. Biochem.* **210**, 155.
4. Jones, B. L., Fontanini, D., Jarvinen, M., and Pekkarinen, A. (1995). *Anal. Biochem.* **263**, 214.
5. Twinning, S. S. (1984). *Anal. Biochem.* **143**, 30.
6. Wilson, K. A. (1986). In *Plant Proteolytic Enzymes.* (ed. Dalling, M. J.), Vol. 2, p. 19. CRC Press, Boca Raton.
7. Liepnicks.J. J. and Light, A. (1979). *J. Biol. Chem.* **261**, 2610.
8. Dalbey, R. E. and von Heijne, G. (1992). *Trends. Biochem. Sci.* **17**, 474.
9. Smeekens, S. P. (1993). *Biotechnology* **11**, 182.
10. Pehrson, J. C., Weatherman, A., Markwell, J., Sarath, G., and Schwartzbach, S. D. (1999). *Biotechniques* **27**, 28.
11. Tsien, R. Y. (1998). *Annu. Rev. Biochem.* **67**, 509.
12. Mitra, R. D., Silva, C. M., and Youvan, D. C. (1996). *Gene* **173**, 13.
13. Matayoshi, E. D., Wang, G. T., Krafft, G. A., and Erikson, J. (1990). *Science* **247**, 954.
14. White, I. J., Lawson, J., Williams, C. H., and Hooper, N. M. (1999). *Anal. Biochem.* **268**, 245.
15. Wang, W., and Liang, T. C. (1994). *Biochem. Biophys. Res. Comm.* **201**, 835.
16. Bargetzi, J. P., Kamar, K. S.V., Cox, D. J., Walsh, K. A., and Neurath, H. (1963). *Biochemistry* **2**,1468.
17. Tuppy, H., Wiesbauer, U., and Wintersberger, E. (1962). *Z. Physiol. Chem.* **329**, 278.
18. Bratton, A. C. and Marshall Jr, E. K. (1939). *J. Biol. Chem.* **128**, 537.
19. Whittaker, R. G., Manthey, M. K., Le Brocque, D. S., and Hayes, P. J. (1994). *Anal. Biochem.* **220**, 238.

20. Cathers, B. E. and Schloss, J. V. (1996). *Anal. Biochem.* **241**, 1.
21. Lee, H. J., LaRue, J. N., and Wilson, J. (1971). *Anal. Biochem.* **41**, 397.
22. Goldbarg, J. A. and Rutenburg, A. M. (1958). *Cancer* **11**, 283.
23. Barrett, A. J. (1972). *Anal. Biochem.* **47**, 280.
24. Powers, J. C., and Kam, C. M. (1995). *Methods Enzymol.* **248**, 3.
25. Feder, J. (1968). *Biochem. Biophys. Res. Comm.* **32**, 326.
26. Wagner, F. W., Ray, L. E., Ajabnoor, M. A., Ziemba, P. E., and Hill, R. L. (1979). *Arch. Biochem. Biophys.* **197**, 63.
27. De Lumen, B. O. and Tappel, A. M. (1972). *Anal. Biochem.* **48**, 378.
28. Zimmerman, M., Yurewicz, E., and Patel, G. (1976). *Anal. Biochem.* **70**, 258.
29. Yaron, A., Carmel, A., and Katchalski-Katzir, E. (1979). *Anal. Biochem.* **95**, 228.
30. Bratovanova, E. K. and Petkov, D. D. (1987). *Anal. Biochem.* **162**, 213. 24.
31. Fitzgerald, M. C., Laco, G. S., Elder, J. H., and Kent, S. B.H. (1997). *Anal. Biochem.* **254**, 226.
32. Beebe, K. D. and Pei, D. (1998). *Anal. Biochem.* **263**, 51.
33. Lawler, J. F. and Snyder, S. H. (1999). *Anal. Biochem.* **269**, 133.
34. Maeda, H. (1979). *Anal. Biochem.* **92**, 222.
35. Schade, S. Z., Jolley, M. E., Sarauer, B. J., and Simonson, L. G. (1996). *Anal. Biochem.* **243**, 1.
36. Levine, L. M., Michener, M. I., Toth, M. V., and Holwerda, B. C. (1997). *Anal. Biochem.* **247**, 83.
37. Sarath, G., de la Motte, R. S., and Wagner, F. W. (1989). In *Proteolytic enzymes: a practical approach* (ed. Benyon, R. J. and Bond, J. S.), p. 25. IRL Press, Oxford.
38. Auld, D. S., Latt, S. A., and Vallee, B. L. (1972). *Biochemistry* **11**, 4994.
39. Auld, D. S. and Prescott, J. M. (1983). *Biochem. Biophys. Res. Comm.* **111**, 946.
40. Skidgel, R. A., Wickstrom, E., Kumamoto, K., and Erdos, E. G. (1981). *Anal. Biochem.* **118**, 113.
41. Rucklidge, G. J. and Milne, G. (1990). *Anal. Biochem.* **185**, 265.
42. Kuwada, M. and Katayama, K. (1984). *Anal. Biochem.* **139**, 438.
43. Green, J. D. (1986). *Anal. Biochem.* **152**, 83.
44. James, P. (1997). *Quart. Rev. Biophys.*, **30**, 279.
45. Chu, Q., O'Dwyer, M., and Zeece, M. G. (1997). *J. Capillary Electrophor.* **3**, 117.
46. Brown, T. L., Yet, M.-G., and Wold, F. (1982). *Anal. Biochem.* **122**, 164.
47. Lacks, S. A. and Springhorn, S. S. (1980). *J. Biol. Chem.* **255**, 7467.
48. Miskin, R. and Soreq, H. (1981). *Anal. Biochem.* **118**, 252.
49. Lantz, M. S. and Ciborowski, P. (1994). *Methods Enzymol.* **235**, 563.
50. Schumacher, G. F.B. and Schill, W. B. (1972). *Anal. Biochem.* **48**, 9.
51. Wickstrom, M., Elwing, H., and Linde, A. (1981). *Anal. Biochem.* **118**, 240. 49.
52. Hagen,H-E., Kläger, S. L., McKerrow, J. H., and Ham, P. J. (1997). *Anal.Biochem.* **251**., 121.
53. Cheung, A. L., Ying, P., and Fischetti, V. A. (1991). *Anal. Biochem.* **193**., 20.
54. Buckwold, V. E., Alvarado, M., Carraso, R. M., and Amils, R. (1999). *Anal. Biochem.* **267**, 420.
55. Nagase, H. and Woessner, J. F. (1980). *Anal. Biochem.* **107**, 385.
56. Yoshioka, H., Oyamada, I. and Usuku, G. (1987). *Anal. Biochem.* **166**, 172.
57. Koritsas, V. M. and Atkinson, H. J. (1995). *Anal. Biochem.* **227**, 22.
58. Gan, Z., Marquardt, R. R., and Xiao, H. (1999). *Anal. Biochem.* **268**, 151.
59. Smith, R. E., Reynolds, C. J., and Elder, R. A. (1992). *Histochem. J.* **24**, 637.
60. Lodja, Z. (1996). *Acta. Histochem.* **98**, 215.

Chapter 4
Determination of protease mechanism

Ben M. Dunn
Department of Biochemistry and Molecular Biology, University of Florida College of Medicine, Gainesville, FL 32610–0245, USA

1 Introduction—the importance of mechanistic classification

The opening chapter of this volume has outlined the four mechanistic classes: serine and cysteine proteases (those that form covalent enzyme complexes) and aspartic and metalloproteases (those that do not form covalent enzyme complexes). This distinction into two major mechanistic groups is of profound consequence since the strategy for inhibition is totally different for these two classes. Those enzymes of the first general class have strongly nucleophilic amino acids at their catalytic site. These are usually aligned with hydrogen bond acceptors to promote the dissociation of the nucleophile in the approach to the transition state and, thus, increase the fraction in the hyperreactive state. Design and synthesis of inhibitors of this broad class will therefore be concentrated on introduction of electrophilic groups (−C=C−, −C−C(=O)−Cl, etc.) that will chemically modify the nucleophile or general base. This will render the catalytic apparatus inactive and prevent the action of that protease. The second category of proteolytic enzymes includes two classes that catalyse the hydrolysis of peptide bonds without nucleophilic attack by a functional group of the enzyme. These enzymes rely more upon general acid/general base catalysis of the attack of a water molecule and therefore the catalytic residues lack the aggressive nucleophilicity of the serine or cysteine proteases. Inhibitor design for this second broad group relies on more subtle means of complexation, including a greater necessity for secondary binding interactions along the active-site cleft and transition state analogues. The metalloproteinases offer an additional handle for inhibitors, since functional groups such as -SH can be introduced at precise points to lead to nearly irreversible metal ion chelation.

In the following section the general mechanisms for the four classes of proteolytic enzymes will be briefly reviewed. In this analysis, it should be kept in mind that the overall process (peptide bond scission) is identical in all cases and that the differences between the catalytic mechanisms are rather subtle. Attack on a carbonyl group will require some nucleophile, either oxygen or sulphur, to approach the slightly electrophilic carbonyl carbon atom. This process will be assisted by general base catalysis to remove a proton from the attacking nucleophile and by some type of electrophilic influence on the carbonyl oxygen to increase the polarization of the −C−O bond. The exact groups that perform these three functions (nucleophilic attack, general base catalysis, and electrophilic assistance) are different in the four classes of peptidases, but the net result is the same. Also, the breakdown of the tetrahedral intermediate that is formed in the initial nucleophilic attack will require general acid catalysis to make departure of an amine easier. Again, different groups will play this role in each instance.

1.1 The serine peptidases

This is the most thoroughly studied class of enzymes in the protease field, and perhaps in all of enzymology. Our current understanding of the mechanism of serine peptidases is illustrative of the whole group of enzymes involved in fragment transfer (1), although in this case the transfer is to water. The intermediate transfer of the acyl portion of a substrate to form a covalent bond with a functional group of the enzyme is the common feature between the serine peptidases and other transferases in biology.

A scheme portraying the essential elements of the mechanism of the serine peptidase family of enzymes is provided in *Figure 1*. The important feature of this scheme is the formation of an ester between the oxygen of serine 195 (chymotrypsin numbering system) and the acyl portion of the substrate, with release of the 'amino' portion of the substrate as the first product. The ester thus formed will be the same for a series of substrates that differ in their leaving groups. This fact will be of consequence in the later discussion of the kinetics of this process. The other element of importance in the catalysis of peptide bond cleavage by the serine peptidases (2) is the presence of two backbone -NH- groups that are available for hydrogen bonding to the developing oxygen anion from the carbonyl group of the peptide bond undergoing attack. This effect is difficult to evaluate on a quantitative basis because the -NH- groups are essential for the structural integrity of the protein. Recently, Kent and colleagues have pioneered synthetic methods for chemical ligation of large peptides. This permits the incorporation of a thiol ester in place of an amide group at selected positions in proteins, thus allowing the removal of a potential hydrogen bonding group and the evaluation of the effects on catalytic activity (3,4). In subtilisin, the side-chain of Asn155 is believed to be one of the donors of a hydrogen bond forming the oxyanion hole (5). This has been tested in a site-directed mutant where Leu replaces Asn with a 200- to 300-fold decrease in catalytic efficiency (6).

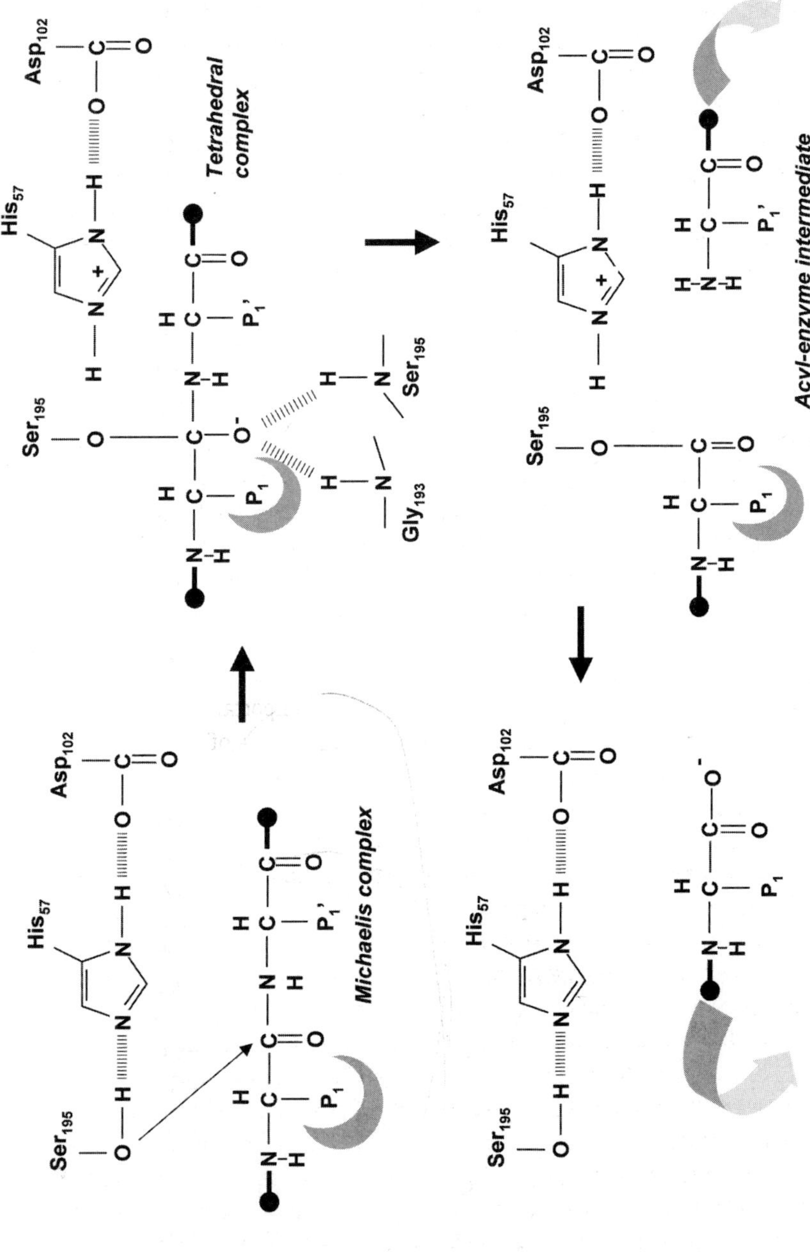

Figure 1 Schematic representation of the steps involved in catalysis by the serine peptidase type of enzyme. The reaction proceeds through formation of a tetrahedral intermediate followed by loss of the right-hand half of the substrate to give an acyl enzyme intermediate. Breakdown of that intermediate occurs by enzyme-catalysed attack of water to generate the product acid.

1.2 The cysteine peptidases

This family of peptidases bears great similarity to the serine peptidase group because, again, a covalent intermediate is formed (7–9). In this instance, the attacking nucleophile is the sulphur atom of a cysteine side chain (cysteine 25 in the papain numbering system). Also, a histidine side-chain is again involved in a hydrogen acceptor/shuttle role (His159). These discoveries followed the analysis of the serine peptidase mechanism so that the techniques developed there could be rapidly applied to this system. Of importance were the findings that chemical modification of a hyperreactive Cys residue destroyed activity (10), that the Cys

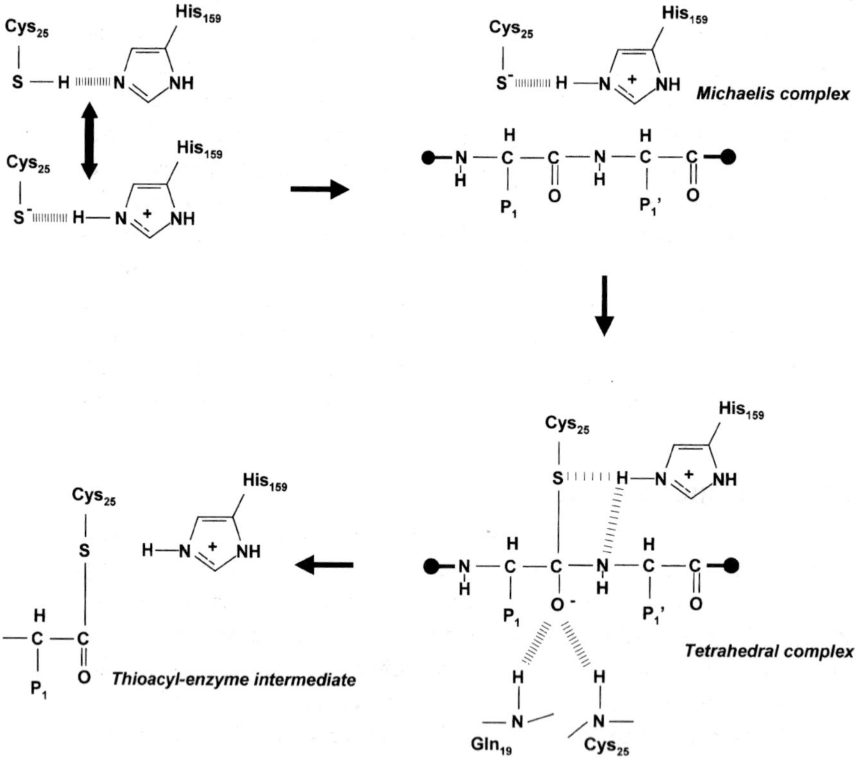

Figure 2 Schematic representation of the steps involved in catalysis by the cysteine peptidase type of enzyme. The catalytic Cys is involved in a tautomeric equilibrium between the neutral and zwitterionic forms. It is believed that the anionic sulphur is involved in direct nucleophilic attack on the substrate carbonyl. This scheme represents only the reaction through the acyl-enzyme intermediate. Breakdown of that again involves enzyme-catalysed attack of water.

was required to be in a free state, and that the structure of the active site includes a nearby His residue (11–13).

A simple mechanistic scheme to represent this family of enzymes is presented in *Figure 2*. This includes two -NH- groups, this time from residues Gln19 and Cys25, of the enzyme acting in the role of the 'oxyanion hole' in perfect analogy to the serine peptidase class.

1.3 The aspartic peptidases

This group of enzymes is believed to catalyse the cleavage of peptide bonds without the use of nucleophilic attack by a functional group of the enzyme (14). Therefore, there will be no covalent intermediate formed between the enzyme and a fragment of the substrate (*Figure 3A*). The catalytic apparatus of the aspartic peptidases consists of two aspartic acid side chains (residues 32 and 215 of the porcine pepsin numbering system). It is not surprising that nucleophilic attack is not involved since the carboxyl group is not noted as a potent nucleophile. The importance of these two carboxyl groups was inferred from chemical modification studies (15,16). In addition, the low pH optimum of this group of enzymes leads naturally to the belief that carboxyl groups were probably involved in the catalytic apparatus. Structural information from X-ray crystallographic work has confirmed the location of these two side-chains at the appropriate position to cause the cleavage of the substrate peptide bond (17,18). These analyses have also revealed that the two carboxyl groups are close enough to share a hydrogen bond between two of their oxygens (*Figure 3A*). In addition, the native enzyme contains a tightly bound water molecule hydrogen bonded to both of the active-site aspartic acids.

1.4 The metallopeptidases

This group of enzymes, which, like the aspartic peptidases, does not form covalent intermediates, has evolved with a more direct solution to the necessity for catalytic effect on the carbonyl group of the bond undergoing cleavage. Rather than rely on hydrogen bonding via an 'oxyanion hole', the metallopeptidases utilize coordination to a metal ion to exert this effect (19). This metal is usually zinc, although in some instances other transition metals can substitute. It is also possible that at some later time newer examples of this class will be purified and found to contain other metals at the active site.

The metal ion provides a strong electrophilic 'pull' to assist in attack by a water molecule. The native enzyme has a water molecule coordinated to the fourth tetrahedral site (the other ligands to the metal are two histidines and a glutamic acid in carboxypeptidase A and thermolysin). This water molecule may be displaced upon coordination of the substrate carbonyl to the metal atom, but it is believed to remain at the active site. It has been suggested that it may remain coordinated to the metal atom at least in a transition state (20). This water molecule is also hydrogen-bonded to a glutamic acid (residue 270 in the carboxypeptidase numbering scheme). That carboxyl group serves as a general base to remove a proton and assist the attack of the same water molecule on the

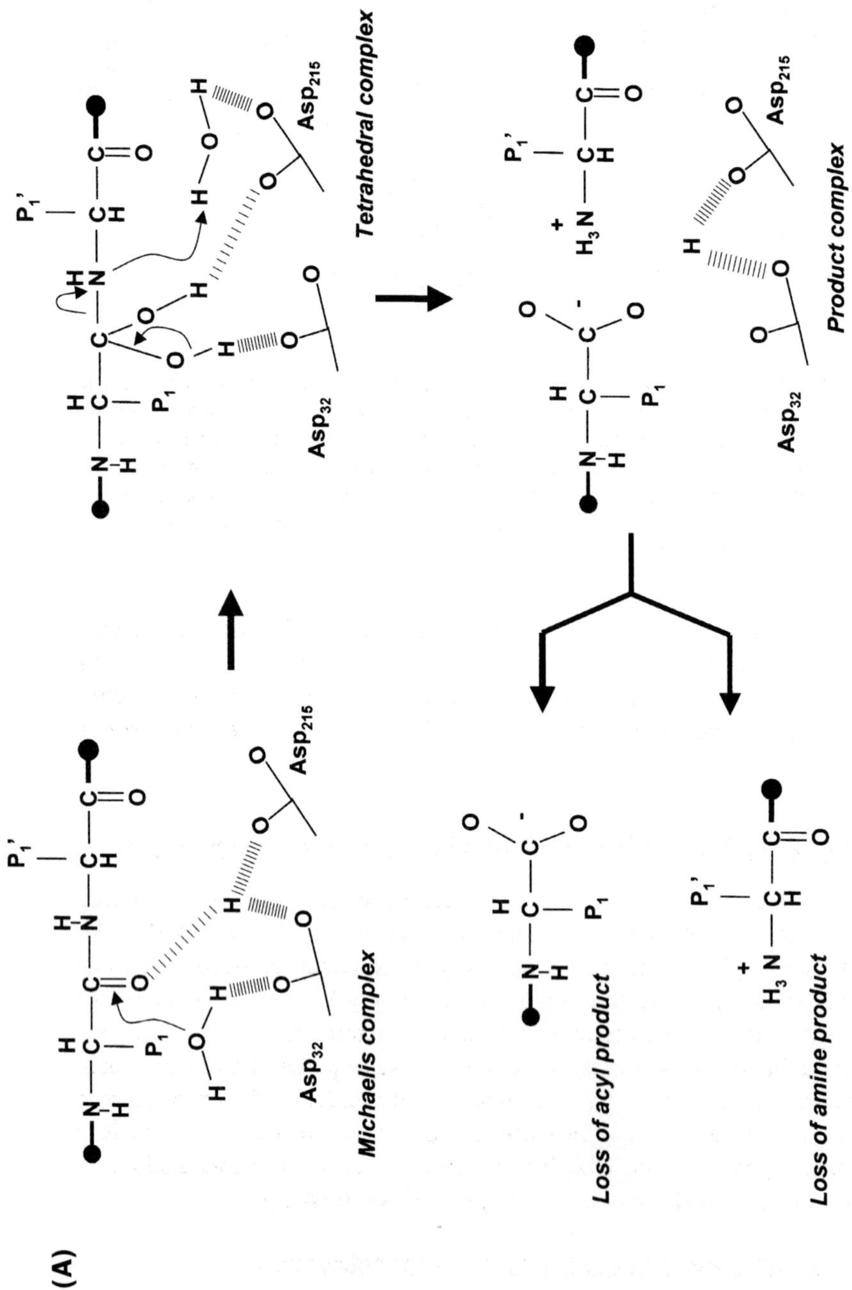
(A)
Michaelis complex
Tetrahedral complex
Product complex
Loss of acyl product
Loss of amine product
Asp32
Asp215

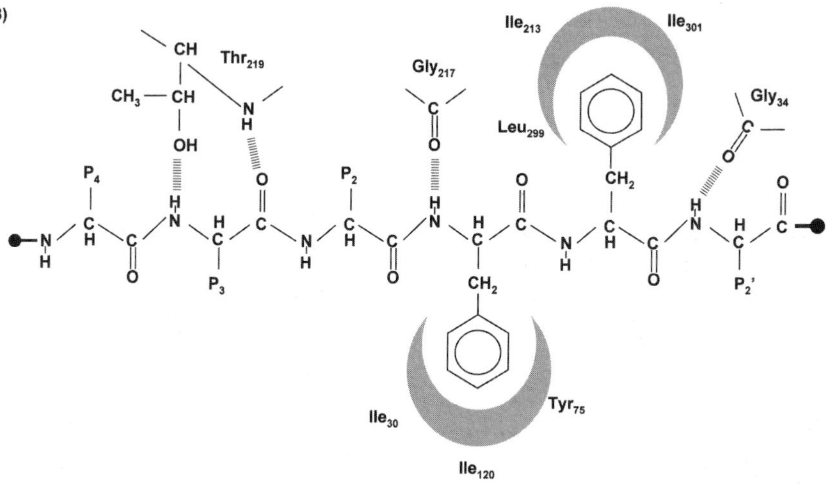

Figure 3 (A) Schematic representation of the general acid–general base catalytic mechanism of the aspartic peptidase type of enzyme. Breakdown of the tetrahedral intermediate gives a product complex containing both halves of the substrate, and this scheme indicates that dissociation of either can follow to give an acyl product complex or an amino product complex. (B) Schematic representation of the multitude of hydrogen -bonding interactions available in the active-site cleft of a typical aspartic peptidase. Some of the hydrogen bonds indicated may be critical to promotion of the catalytic events.

peptide carbonyl (*Figure 4*). Again, a proton must be transferred to the leaving nitrogen atom and this could be derived from the glutamic acid. Thus, the glutamic acid would be acting as a 'shuttle' in analogy to one of the catalytic groups in the aspartic peptidases and to the histidine in the serine and cysteine peptidases.

2 Methods for determining the mechanistic class

Describing the detailed mechanism of a protease can range from a crude classification based on susceptibility to a group of protease inhibitors (PMSF, E-64, pepstatin, and *o*-phenanthroline) to an elegant discussion of stereo-electronic factors based on high-resolution X-ray crystallography. Between these extremes there are numerous experimental protocols for obtaining data on the mechanistic grouping of a new activity. Among these are response to a broader range of inhibitors, chemical modification studies, pH dependence of activity, solvent deuterium isotope effects, and cleavage site specificity studies. The last topic is helpful in designing a more selective inhibitor and allows a further distinction into 'mechanistic' subclasses, such as 'trypsin-like serine protease'.

2.1 Classification based on 'standard' inhibitors

It is now possible to recommend a limited set of inhibitors to use for initial classification of a newly discovered protease. As indicated in *Protocol 1*, these

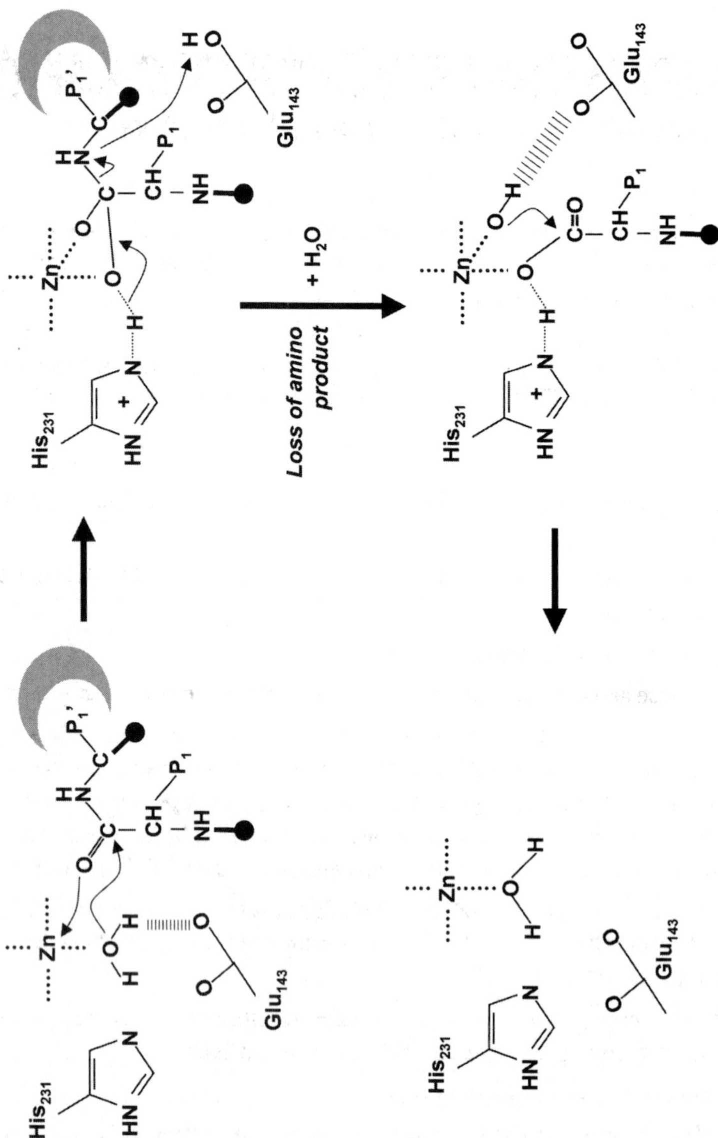

Figure 4 Schematic representation of the catalysis of peptide bond cleavage carried out by a member of the metallopeptidase class of enzyme.

should be applied individually with the effect on the activity of the enzyme carefully measured (see Chapter 5). Each inhibitor should be incubated with the enzymatic activity for up to 1 h at 20 °C, and the free enzyme should be incubated under the same conditions without the inhibitor to serve as the control. The value of this set of inhibitors is that their effects are exclusively on one class of peptidase.

Protocol 1

Determination of mechanistic class of a protease

Reagents[a]

- Serine protease inhibitor: 100 mM PMSF in DMSO (prepare fresh)
- Cysteine protease inhibitor: 10 mM E64 in water (stable at −20 °C)
- Aspartic protease inhibitor: 100 μg/ml pepstatin in DMSO (stable at −20 °C)
- Metalloprotease inhibitor: 100 mM 1,10-phenanthroline in DMSO (stable at −20 °C)
- Assay buffer[b]
- Stock solution of protease (approximately 20 μg/ml; store at 4 °C)

Method[c]

1. To six tubes, each containing 0.98 ml of buffer, add 10 μl protease, and incubate at 30 °C for 5 min.
2. To separate tubes, add 10 μl of each inhibitor (tubes 1–4), 10 μl of DMSO (tube 5) or 10 μl of buffer (tube 6).
3. Incubate all six tubes at 30 °C for 10–60 min.
4. Initiate the protease assay by addition of substrate.[b] Monitor reaction as appropriate.
5. Tube 6 should show the same activity as a sample of protease that is diluted and assayed immediately. If not, then the enzyme is inactivating during the inhibitor incubation. If the inactivation is significant (> 10% loss) then repeat the experiment under conditions in which the enzyme is more stable such as lower temperature or addition of stabilizing solutes to the buffer (glycerol, or less desirably, albumin).
6. If there is inactivation of the enzyme in tube 5, the enzyme is intolerant to DMSO. Repeat the experiment using a stronger inhibitor solution (and therefore less DMSO) or change to MeOH or PrOH.
7. If the enzyme is only partly inhibited, the experiments should be repeated for longer incubation times, or at higher inhibitor concentrations.

just a single protease here

[a] See Appendix III for further details of inhibitors.

[b] No specific details of buffer, substrate, etc., can be given. See Chapter 2 for several assay methods.

[c] Note the following complications. PMSF will inhibit cysteine proteases (also strongly inhibited by E64); this inhibition is readily reversible by addition of DTT to 10 mM 1,10-Phenanthroline will also inhibit metal-activated proteases (such as calcium-activated cysteine proteases). Examine 1,10-phenanthroline inhibition in the light of other inhibitor data.

Phenylmethanesulphonyl fluoride is listed in *Protocol 1* rather than the traditional Dip-F, because the former is safer to use, being less volatile. If only limited inhibition is observed after a 1-h incubation with PMSF, it is advisable to conduct a longer incubation, since PMSF is not as reactive as Dip-F. Both of these reagents will react with cysteine residues and will inactivate cysteine peptidases, but sulphydryl reagents such as DTT can readily reverse that effect.

Pepstatin is among the most specific inhibitors known in enzymology and is highly selective for the aspartic peptidases (21,22). The commercially available form is isovaleryl-Val-Val-statine-Ala-statine, where statine is the common name given to 4-amino-3-hydroxy-6-methylheptanoic acid. This compound has K_i values in the 10^{-8}–10^{-12} M range for the well-known aspartic peptidases. Pepstatin is only sparingly soluble in water and must be first dissolved in methanol or DMSO, followed by dilution (100× or greater) into aqueous buffer. The dilution is important, since alcohols inhibit many aspartic peptidases.

1,10-Phenanthroline acts by chelating the essential metal atom of the metallopeptidase class of enzyme. It is important to note that some members of the other classes may be activated or stabilized by metal ions. Therefore, the diagnosis of the metallopeptidase class may best be based on its insensitivity to the other three inhibitors plus the observation of inhibition by the chelating agent.

Another compound that is of considerable utility in classifying metallopeptidases is phosphoramidon, *N* (α-L-rhamnopyranosyloxyhydroxyphosphinyl)-L-Leu-L-Trp (23–25). It has been shown to be an excellent inhibitor of thermolysin and several other metallopeptidases. It does not, however, inhibit all metallopeptidases (26).

2.2 Chemical modification/identification

Because the serine and cysteine proteases both possess strongly nucleophilic catalytic residues, the covalent chemical modification of those enzymes using group-specific reagents is a reasonable way to establish the mechanistic class of the enzyme. The classic example is the use of Dip-F to chemically inactivate the serine proteases. This reaction relies on the enhanced reactivity of the catalytic serine residue in those enzymes. Chymotrypsin possesses 33 serine residues, but the active-site Ser at position 195 is the most reactive, and is modified by Dip-F most rapidly. The application of such reagents is dangerous, as it is possible that chemical modification of a non-catalytic residue may alter the conformation of the enzyme in such a way as to decrease activity in a linear fashion. This would look exactly like active-site modification with respect to stoichiometry as well as activity loss. The use of substrate or inhibitor protection experiments will be the best way to eliminate that possibility. The enzyme is incubated with a substrate or inhibitor at concentrations above the K_m or K_i value and then exposed to the modifying reagent. If the chemical modification occurs outside the active site, the presence of the substrate or inhibitor will not affect the chemical modification in a major way. If, on the other hand, the reagent does modify an active-site residue, then filling the active-site cleft with substrate or inhibitor will decrease accessibility of that amino acid. This should decrease the chemical modification and limit the effect on the enzyme activity.

In the ideal case, the chemical modification should lead to stoichiometric incorporation of reagent into the protein. Furthermore, the modification should result in a one-to-one decrease in activity with chemical modification, that is, a plot of mole per cent reagent incorporated versus per cent of decrease in activity should be linear and of slope 1.0.

Following chemical modification that results in loss of activity due to active site reaction, the enzyme can be denatured and digested with a small amount (1% w/w) of a standard protease such as trypsin to produce peptide fragments. If the reagent used for the modification is radioactively labelled (27) or contains a spectral probe, then the peptide fragment containing that group should be easy to isolate from the digest. Chemical analysis by gas-phase sequencing or mass spectrometry is now possible on amounts in the low picomole range. Obviously, the presence of a derivative on the side-chain will interfere with the identification by standard amino acid analysis of the amino acid modified. In that case, the same peptide from a fragmentation of the native protein with trypsin could be isolated and the modified amino acid identified by differential amino acid analysis, that is, if one amino acid is present in the analysis of the native peptide and missing in the analysis of the modified peptide.

As knowledge of selective binding to the active sites of various proteolytic enzymes has increased, it has been possible to incorporate design principles into the creation of selective covalent reagents. The four chapters listed in ref. (28) provide an exhaustive survey of reagents created to probe the active site of most known proteolytic enzymes. These efforts are essential in the identification of the second group of critical residues at the active site. Until now, this section has focused on the identification of a hyperreactive residue directly involved in nucleophilic catalysis. With the use of the more general 'active-site directed' or 'affinity' reagents, it is also possible to identify other amino acid residues that are within the active-site cleft (29–31). This is necessary in the description of possible mechanisms since it will help to catalogue the available residues that can serve as proton transfer agents.

2.3 Site-directed mutagenesis

With the advent of simple methods for the selective mutation of a single amino acid within a protein sequence, it is now possible to extend the concept of chemical modification to those enzymes that do not operate through a strongly nucleophilic residue. Essentially, mutagenesis is the alteration of one amino acid side-chain into a different one, accomplished by genetic means rather than harsh chemical reaction. Typically, this is achieved by replacement of a putative catalytic residue by an alanine. This substitution is believed to be the most conservative, in that the Ala side-chain consists of only a β-CH_3 group. This small functionality can fit into most spaces previously occupied by a larger amino acid, such as Glu or His. Alanine can also fit into most conformations, thus it should not disturb the local geometry.

Site-directed mutagenesis can be used in two ways: first, replacement of a suspected catalytic residue by an Ala would result in the loss of enzymatic activity if the residue replaced is critical. In the case of a loss of function, it is absolutely essential to confirm that the substitution has not disrupted the conformation of the protein in a significant way. This is usually evaluated by measurement of the circular dichroism spectra of a protein; however, this is a gross measure of protein conformation and may not reveal small alterations in local geometry. An alternative method is to evaluate the binding of a low-molecular weight inhibitor. The K_i value should not be altered significantly, as long as the binding to the enzyme does not depend completely on interaction with the catalytic residue in question.

With loss of function by mutagenesis without disturbing conformation, one can safely conclude that the residue replaced is essential for enzymatic activity. The report by Lin *et al.* (32) provides an example of such an analysis of a catalytic residue.

A second use of site-directed mutagenesis is more subtle than the first. Changes can be made in residues that are not absolutely essential for function, but which interact with the catalytic residues. An excellent example of this was provided by the work of Lin *et al.* (33), where the Ser in the sequence, Asp32-Thr-Gly-Ser35, of porcine pepsin was altered to Ala. In this enzyme, the Asp32 is one of two catalytically essential residues. The Ser at position 35 hydrogen bonds to the outer oxygen of the carboxyl group of Asp32 and influences the pK_a of the catalytic group. The mutation to Ala does not eliminate catalytic activity, but perturbs the catalytic pK_a values. A second example can be found in the work of McGrath *et al.* (34), where the environment of Asp102 of trypsin was altered by replacement of a conserved Ser at position 214. Asp102 is discussed in Section 1.1 (see *Figure 1*).

Site-directed mutagenesis can also be used in the case of proteases to probe the binding interactions within the often extended active site cleft. Lowther *et al.* (35) explored the contribution of Asp residues in the active site of *Rhizopus chinensis* aspartic proteinase, and determined that Asp77, located on a prominent β-hairpin loop that overhangs the active site cleft, is largely responsible for the ability of the fungal enzymes to cleave peptides following a Lys residue.

2.4 Mechanistic distinctions—intermediates

The original evidence for the presence of a covalent enzyme–substrate intermediate in the reaction catalysed by chymotrypsin was the observation of a ‘burst’ of product formation when a continuous assay was used. This results from the rapid reaction of one mole of substrate with each mole of enzyme to produce the first product and the substituted enzyme. Observation of this phenomenon requires that the portion of the substrate released serves as the spectroscopic handle for the reaction. For chymotrypsin, *p*-nitrophenol was released from the ester, *p*-nitrophenylacetate. The substituted enzyme was acetylated at the active-site serine. In this case, the hydrolysis of the acyl-enzyme intermediate was slower than its formation, so further product formation was held up by the necessity of turning over the enzyme intermediate. A stoichiometric 'burst' of the *p*-nitrophenol product was seen, followed by the slower release of additional product as the enzyme turned over and new substrate molecules reacted.

2.4.1 Methods for detection of intermediates

The preceding discussion reveals the value of using a continuous spectrophotometric assay for the enzyme under investigation. A single-point assay measuring the extent of cleavage of a protein or peptide substrate would not permit the observation of 'burst' kinetics. In general, it is usually sufficient to incorporate a spectroscopic probe into the 'amino' part of a substrate (also known as the 'prime' side or right-hand half (see *Figure* 5)

$$X_{aa}-Y_{aa}-Z_{aa}\overset{\downarrow}{-}X_{aa'}-Y_{aa'}-Z_{aa'}$$

Figure 5 Hypothetical chromophoric substrate for assay of a peptidase: arrow indicates point of cleavage, and one of the 'prime' residues contains a chromophoric group that changes its spectrum upon cleavage of the Z_{aa}–$X_{aa'}$ bond.

This is true since the clearly defined samples of covalent enzyme–substrate intermediates all involve acyl-enzyme structures (36). However, the observations of Hofmann and colleagues concerning the possibility of 'amino-enzyme' intermediates in the aspartic peptidases (37) mean that one should consider the use of substrates such as in *Figure* 5, where one of the non-prime amino acids contains such a chromophoric group. In the case of aspartic peptidase, the amino acid *p*-nitrophenylalanine has been useful, since it can be placed in the P_1 or the $P_{1'}$ position. Cleavage of the peptide bond involving that residue then leads to an increase in absorbance for cleavage following *p*-nitroPhe or a decrease in absorbance for cleavage preceding that residue. Thus, the possibility of formation of intermediates releasing either the tripeptide, X_{aa}-Y_{aa}-Z_{aa}, or the alternate tripeptide, $X_{aa'}$-$Y_{aa'}$-$Z_{aa'}$, as the first product can be readily tested.

The search for an intermediate species via kinetic analysis then simply requires the study of the reaction under a wide variety of conditions of substrate and enzyme concentration to adjust the time frame to a measurable period. The amount of enzyme present must be large enough so that release of a stoichiometric amount of the product with the chromophore can be accurately quantified, but small enough that the rate of appearance and turnover of the intermediate can be followed.

The most important aspect of the observation of a burst of product formation is the quantification of the amount of product released. This must correspond to the amount of enzyme used. A stoichiometry of greater than one-to-one would indicate that some or all of the product formation is due to acylation of non-active site residues on the enzyme surface. Upon completion of the modification of the enzyme, the resulting slower rate of turnover of substrate could represent the rate of reaction of the modified enzyme. This is one danger of using very reactive species such as *p*-nitrophenylacetate as an enzyme substrate. Fortunately, real peptide substrates are not particularly reactive species and intermediates observed from their reaction are most likely of mechanistic importance.

A second technique that has been applied in the search for covalent enzyme intermediates is the use of nucleophilic trapping. Alcohols, amines, hydroxylamines, hydrazines and other highly nucleophilic reagents can be added to an enzyme–substrate mixture in the hope that the covalent enzyme intermediate will be particularly reactive. Because the value of nucleophilic catalysis by the enzyme is to accelerate the cleavage of the normally stable peptide bond, this assumption of enhanced reactivity is reasonable. Reaction between the added reagent and the enzyme–substrate covalent intermediate will yield a modified product which should have altered chromatographic properties. The addition of such nucleophilic reagents can, in some instances, alter the kinetics of the enzymatic reaction as it should lead to faster turnover of the intermediate (38,39).

If the covalent intermediate is derived from a particularly unreactive group, as in the case of the $(CH_3)_3C{-}C{=}O$ group, then isolation is possible. This would permit analysis of the chemical composition, particularly if the acyl group is labelled with a radioactive or fluorescent moiety. In favourable cases, crystallographic analysis has been possible. Further analysis of this species will be detailed in the next section.

2.4.2 Criteria for importance of intermediates

The observation of burst kinetics for a newly discovered proteolytic enzyme would be good evidence for the nucleophilic mechanism of catalysis provided several criteria can be clearly established. These are listed in *Table 1*.

Table 1 Criteria for involvement of covalent intermediate in catalysis.

Observation	Method(s)	Conclusion
Isolation of substituted enzyme	Chromatography at low temperature; $(NH4)_2SO_4$ precipitation	Fragment is covalently attached
Substituted enzyme produces product	Return protein to assay conditions: isolation of product	Covalent attachment is temporary
Rate of turnover equals normal kinetics	Spectrophotometric observation of product release	Intermediate is kinetically competent
Common intermediate	Isolation of intermediate from different substrates with common fragments	Intermediate independent of substrate
Intermediate reacts with nucleophiles	Addition of strong nucleophiles, e.g. NH_2OH	Intermediate is reactive
Common rate-limiting step	Several substrates with good leaving groups	Common pathway

3 Kinetic studies to probe the mechanism in more detail

3.1 Notes on 'ideal' assays

Other chapters in this volume describe assays for the detection and analysis of proteolytic enzymes. Exacting studies of the mechanism of an enzyme require a well-characterized assay system. In particular, the substrate must be pure and stable (see *Table 2*), at least throughout the course of exposure to enzyme. The products of attack by the enzyme must also be fully characterized. Ideally, there should be one or more readily identifiable differences between the substrate and the cleavage products that will permit direct determination of their changing concentrations during enzymatic attack.

Reactions can be monitored by following the decrease in concentration of the substrate or the increase in concentration of the product, but only if the presence of one does not interfere with the determination of the other. Proteolytic enzymes are almost ideal in this respect, since their action results in the splitting of a peptide chain into two (or more) fragments of smaller size and, usually, different properties. Although there are numerous studies describing the activity of proteolytic enzymes due to their cleavage of protein substrates, and although these experiments are used in Section 4.4 in studies of specificity, the optimal analysis of the kinetics of a proteolytic enzyme uses small, well-defined oligopeptides. In such a system, only a single peptide bond will be cleaved and the observed kinetics will therefore be relevant to a single chemical step. Of course, there will be several microscopic steps including binding and conformational changes also involved. In some instances, these events can be analysed through conventional enzyme kinetic methods. The other problem with the use of protein substrates for the study of peptidase mechanisms is that, in many cases, the effect of alteration of pH is upon the structure of the substrate itself rather than upon the functional groups of the enzyme. The use of small peptide substrates eliminates this distinction since they usually have no stable, well-defined structure, but exist as random, extended chains. Spectrophotometric and spectrofluorimetric substrates are listed in *Table 3*.

Table 2 Desirable properties of substrates.

Property of substrate	Importance for assays
High purity	Simplification of analysis; absence of competition
Substantial stability	Changes in reaction rate can be ascribed to action of enzyme
Measurable difference from products; size, spectra, etc.	Permits monitoring of change from substrate to product
Contains single point for cleavage	Kinetics simplified; observed rate can be attributed to one chemical event
Conformational structure insensitive to conditions	Kinetics not influenced by conformational transition

Table 3 Spectrophotometric or spectrofluorometric substrates useful for mechanistic studies on representative proteases.

Enzyme	Substrate
Serine peptidases	
Chymotrypsin	Succinyl-Leu-Leu-Val-Tyr*MCA Succinyl-Ala-Ala-Pro-Phe**p*-NA
Elastase	Succinyl-Ala-Pro-Ala**p*-NA Succinyl-Ala-Ala-Ala*MCA
Trypsin	*o*-ABZ-Gly-Arg*Nph-NH_2 *N*-α-tBOC-Leu-Thr-Arg*MCA
Subtilisin	Z-Gly-Gly-Leu**p*-NA
Cysteine peptidases	
Cathepsin B	*N*-α-BZ-Phe-Val-Arg**p*-NA DNS-Phe-Arg*Nph-Leu Z-Ala-Arg-Arg*F_3MCA Z-Arg-Arg*MBNA
Aspartic peptidases	
Pepsin	Lys-Pro-Ile-Glu-Phe*Nph-Arg-Leu Lys-Pro-Ala-Lys-Phe*Nph-Arg-Leu
Chymosin	Leu-Ser-Nph*Nle-Ala-Leu-OCH_3
HIV-1 and -2	Tyr-Val-Ser-Gln-Asn-Phe*Pro-Ile-Val-Gln-Asn-Arg Lys-Ala-Arg-Ile-Nle-Nph-Glu-Ala-Nle-NH_2
Metalloendopeptidases	
Thermolysin	FA-Gly*Leu-NH_2 Succinyl-Ala-Ala-Phe*MCA

* Point of cleavage.

Abbreviations: MCA, 4-methylcoumaryl-7-amide; *p*-NA, p-nitroanilide; *o*-ABZ, *o*-aminobenzoyl; tBOC, *tert*-butyloxycarbonyl; DNS, dimethylaminonaphthalenesulphonyl; Nph, *p*-nitrophenylalanine; BZ, benzoyl; Z, carbobenzoxy; MBNA, 4-methoxy-α-naphthylamide; FA, furylacryloyl; F_3MCA, 4-tritluoromethylcoumaryl-7-amide. All amino acids are of the L-configuration. Most of the substrates listed here are commercially available from Serva Fine Biochemicals, the Peptide Institute, Chemical Dynamics, Peninsula Laboratories, Bachem Bioscience, and Enzyme Systems Products (for addresses see Core list of Suppliers).

3.2 Kinetic determination of K_m and k_{cat}

Assuming a simple peptide of the type A–B is available, that it is split to yield products A and B, that all three peptides are readily separable from one another, and that there is a readily observable change in some parameter to define the kinetics of cleavage, one can proceed with collection of data to define the catalytic reaction. Initially, a concentration of 100 μM substrate peptide should be used to scan the pH range to seek the 'pH optimum' for the reaction. This will not be a valid measure of the true pH optimum, since that will also vary with the concentration of substrate, but this will be useful in establishing initial conditions for kinetic analysis. The concentration of 100 μM should be adjusted higher or lower, depending on the size of the oligopeptide substrate. As a rule of thumb, dipeptide or tripeptide substrates should be assayed at concentrations closer to 1 mM, since their K_m, values are likely to be much higher than larger

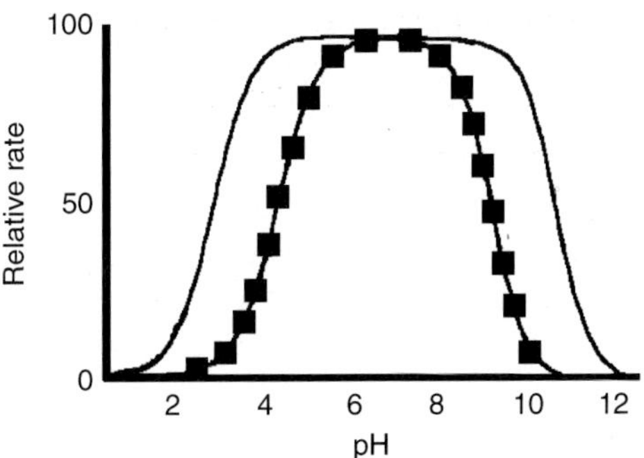

Figure 6 Plots of the activity vs. pH for an enzyme-catalysed reaction. The symbols (■) represent the activity obtained when the enzyme is assayed at that pH. The curve through those points is derived based on a relationship that includes a critical group that must be in the basic form for activity ($pK_a = 4.0$) and a group that must be in the acidic form for activity ($pK_a = 9.0$). The second curve, without points, represents the results of preincubation of enzyme at the indicated pH for 1 h, followed by the assay of that sample at the pH optimum, between pH 6 and pH 7. Since this curve represents the stability of the enzyme, this result indicates that it is safe to ascribe the observed pH activity profile to kinetically important ionizations, rather than to effects of pH on the enzyme structure.

oligopeptides. Substrates of about 6–10 residues in length can be assayed at the 100 μM level with some confidence (40), and oligopeptides of 12–20 residues in length could probably be assayed at 20–50 μM. This is unwise until the true K_m has been carefully defined, but in this initial phase, some assumptions are justified.

An important control is to incubate the enzyme at a range of pH values (see *Figure 6*) and then assay for residual activity at the optimal pH determined above. In the ideal case, the enzyme will be stable over a wider pH range than the pH range of activity. This condition will permit kinetic data to be obtained in detail over the whole range of interest. If incubation at one of the extremes of pH leads to loss of catalytic activity, this experiment will define the region where valid kinetic data can be obtained.

Next, in the pH range where the activity is highest, studies should be performed to measure K_m and k_{cat} by measuring initial velocity as a function of substrate concentration. Special care should be taken to ensure that the analytical method employed yields valid measurements of the concentration of substrate or product versus time. If the appearance of product is being measured, four or five time points should be taken between time zero and 20% of total cleavage. If a continuous spectroscopic assay is available, one can follow the curvature of the resulting line of concentration versus time. The line should be linear up to at least 20% of reaction, and the initial slope can be readily measured. Many new spectrophotometers are equipped with programs to permit this measurement automatically. If the progress is being followed by withdrawing samples followed by HPLC separation (41), it may be necessary to quench the reaction by a rapid drop in pH before injection. In any event, the HPLC system used should be optimized so that multiple analytical points can be obtained in the minimum time. The collection of sufficient data to define V_{max} and K_m could take as little as 1 h with a spectrophotometric assay or as long as 1 week with a labour-intensive HPLC analysis. In either instance, the spectrophotometric analysis should be followed up with an HPLC separation of the products to confirm that cleavage has occurred uniquely at a single bond in the substrate. Multiple cleavages would indicate that kinetic analysis will be complicated, if not impossible.

Assuming that a single cleavage has occurred and can be quantitatively followed, the initial velocity data can be plotted as a function of the concentration of substrate by any one of the usual linear transformations of the Michaelis–Menten equation or by non-linear curve fitting to permit calculation of K_m and V_{max} for the reaction. In many cases, it will be advisable to plot the initial velocity versus the concentration of substrate directly in order to ensure that curvature is seen (see *Figure 7*. This indicates that the higher substrate concentrations are in the range of K_m and therefore the data will yield a reliable K_m value.

The V_{max} value can be converted to a value for k_{cat} only when a strong, essentially irreversible active-site titrant is available to enable the active enzyme concentration to be accurately evaluated. In the absence of such data, a concentration of enzyme obtained from amino acid composition data and the assumption that the enzyme is 100% active could be used. In that case, the value of k_{cat} calculated will be an underestimate of the true k_{cat} value. In no instance should an enzyme concentration be based on the amount of protein estimated from dye binding in polyacrylamide gels, since the staining level is variable from protein to protein.

All this effort to obtain values for K_m and k_{cat} is essential, since the following series of experiments depend on that result. First, the resulting value of K_m,

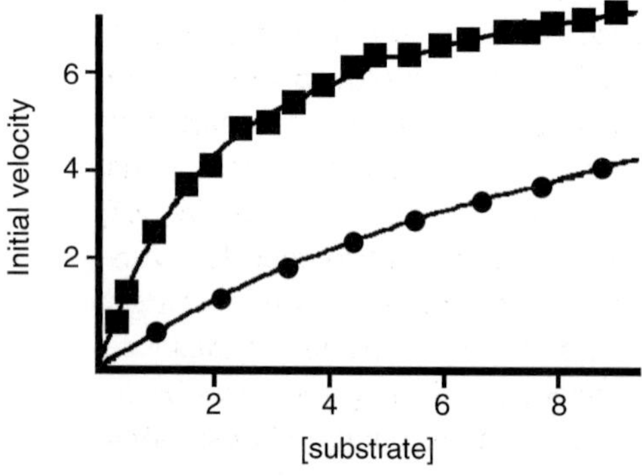

Figure 7 Plots of measured activity v. concentration of substrate utilized for two different systems. In the upper curve, K_m is approximately 2.0 and the line shows considerable curvature. These data can safely be employed to derive the Michaelis–Menten constant, K_M. The second curve illustrates a case in which the highest concentration of substrate does not exceed the value of K_m, which is approximately 10.0. This line exhibits only minimal curvature and these data are not adequate to define experimentally the K_m value with confidence.

should be compared with the concentration of substrate used in the initial pH activity study. If the initial substrate concentration used was well below the actual K_m at the optimal pH, then the series should be repeated at higher initial substrate concentration. This initial screening will help to define the activity to be expected at the various pH values used in the more precise study of the pH dependence to follow. For pH values where the activity is more than 10-fold lower, a larger aliquot of enzyme can be used to boost the activity to a readily measurable level.

3.3 pH dependence of the kinetic parameters

The next step will then be to determine K_m and k_{cat} values over the whole range of pH where activity can be observed and where the enzyme is stable. Construction of buffered solutions for this purpose should be done according to the suggestions of Ellis and Morrison (42). Their recommendations are designed to avoid the large changes in ionic strength normally encountered when buffers of different pH are prepared using a single buffer species. Following collection of sufficient data at each pH to allow dissection of K_m from k_{cat}, plots can be constructed of k_{cat}/K_m, k_{cat} or K_m vs. pH or log k_{cat}/K_m, log k_{cat} or pK_m (negative logarithm) vs. pH. The first plotting technique will yield sigmoidal curves for kinetically significant dissociation processes in either the free enzyme or free substrate (k_{cat}/K_m vs. pH), the enzyme-substrate Michaelis complex k_{cat} vs. pH), or in both (K_m vs. pH). The midpoints of the sigmoidal curves define pK_a values for the kinetically significant dissociation events. The second plotting technique, employing the log/log plots, will yield linear sections with slopes that should have unitary values (+1.0, 0.0, −1.0). The intersections of these straight lines should define pK_a values for the kinetically significant events (43–46).

The importance of these studies, which are expensive in terms of enzyme, substrate and manpower, is that the observation of kinetically significant pK_a values can aid in definition of essential catalytic residues. If the functional groups present on the substrate can be analysed in the absence of the enzyme by titrimetric methods, or if the substrate can be prepared in a form that will not have any significant dissociations in the pH range of interest, the observed pK_a values can be ascribed to the enzyme itself. This is a point to consider in the design of a synthetic peptide substrate. If one wishes to add a positively charged amino acid to increase the aqueous solubility of the substrate, arginine is a better choice than lysine, since the arginine side-chain pK_a (13.5) will be far outside the range of most enzymatic studies and therefore the acid dissociation of the side-chain will not affect the kinetics. One can also prepare the *N*-terminal acetylated and *C*-terminal aminated version of the substrate to eliminate those charges. Although the microenvironments present in the active sites of most enzymes can shift the actual pK_a values over a considerable range (see ref. 36, p. 154), it is still generally valid to assume that an observed pK_a value of between 2 and 5 is probably due to a carboxyl group (Asp or Glu); values in the range 5–7 are probably due to imidazole groups (His); values in the range 7–8 are probably due to sulphydryl groups (Cys); and values in the range 8–11 are probably due to hydroxyl groups (Ser or Tyr) or amino groups (Lys). These tentative assignments will require confirmation through chemical modification and/or structural analysis (47).

Defining the pH dependence of the catalytic process is also useful in that it is frequently possible to find a region of the profile where the enzyme is operating at optimal efficiency and there are no significant dissociation events. This pH region will be optimal for studies of the solvent deuterium isotope effect or of the effect of transition state analogue inhibitors.

3.4 Solvent deuterium isotope effects

Solvent deuterium isotope effects can be used in two ways. First, the catalytic activity should be measured in pure D_2O to determine if there is an effect of the substitution of deuterium for hydrogen on the catalytic process. If protons are transferred in the rate-limiting step of the mechanism, the rate will be reduced by a factor of two to three. This observation is diagnostic of a general acid- or general base-catalysed process and is expected for enzymes of the second, non-nucleophilic category discussed above. However, all proteolysis reactions require proton transfer at some point to aid in departure of the leaving amino group. Therefore, solvent deuterium isotope effects can also be observed for enzymes in the first category. The presence or absence of the effect will depend on the relative rates of the two steps in the catalytic process: nucleophilic attack and breakdown of the resulting tetrahedral intermediate.

For studies of the solvent deuterium isotope effect, the substrate and the enzyme must be dissolved in solvent of $>$ 99.5% deuterium, and lyophilized three times. This process will remove most of the hydrogen atoms originally present in exchangeable positions. The actual kinetic measurements should then be done in buffers of 99.9% deuterium oxide. The pH can be adjusted with DCl or NaOD solutions. The kinetic measurements are conducted in exactly the same fashion as the original experiments in pure water.

It is important to mention that the D_2O solvent will perturb the pK_a of most groups by 0.4 pH unit. Therefore, it is not wise to carry out the comparison of the kinetic parameters measured in H_2O and D_2O at pH values close to a known break point in the pH profile in water. It is essential that these studies be done in a pH-insensitive region, or that complete pH profiles be done in both solvents so that a fair comparison can be made.

Next, if an effect is seen when the substitution is made from water to deuterium oxide, the technique of 'proton inventory' can be applied to add to the analysis of the mechanism (48). This method requires that kinetic data be obtained at a variety of ratios of H_2O and D_2O. The kinetic parameter is then plotted against the mole ratio of deuterium oxide. For reactions in which a single-proton transfer step is occurring in the slowest rate step, the resulting plot will be linear. Where two or more protons are being transferred at the same time, the plot will be curved. The degree of curvature can be studied by curve fitting and the number of protons involved can then be determined.

3.5 Transition state analogues and substrate alteration

Given the chemical changes that must occur during attack on a carbonyl group, it was recognized early that compounds that resemble the transition state of the reactions of peptidases might be excellent inhibitors (49,50). Accordingly, peptide derivatives of boronic acids (51), reduced peptide isosteres (52), hydroxyethylene analogues (53–55), ketomethylene analogues (56), amino alcohols (57) and phosphorus-containing compounds (58) have been used to probe binding at the active sites of a wide variety of these enzymes. When these groups are incorporated into a peptide sequence that matches the specificity of a particular enzyme (see below), it is possible to construct a tight-binding inhibitor that can be very useful in X-ray crystallography (59,60). The resulting structural information can provide important clues to possible mechanistic pathways, since this is as close as one can get to actually observing the catalytic transition state. It should be stated here that simply constructing a putative transition state analogue and demonstrating that it inhibits a new protease will be of little value in defining the mechanistic class, since all peptidases must proceed through a tetrahedral transition state at one or more points along the reaction pathway. However, the demonstration of a very low value of K_i will give some confidence that structural data from crystallography will be a valid picture of the actual transition state of the reaction. Bartlett and Marlowe (58, and references therein) have discussed the use of a correlation of log K_i with log K_M/k_{cat} for two series of homologous inhibitors and substrates.

In addition to replacement of the peptide bond by tetrahedral surrogates, some mechanistic information can be provided by replacement of the carbonyl oxygen with sulphur (61,62). Storer *et al.* (53) have employed resonance Raman spectroscopic studies to provide information on transient acyl enzyme species and have correlated this with kinetic studies by stopped-flow and steady-state methods. Their extensive structure–reactivity correlations have provided insights into the geometry of the intermediates at the active site and raised new questions about the fundamental mechanisms for catalysis by papain.

Extensive work has also been carried out on the replacement of the oxygen atom of the reactive Ser in the serine peptidases. These studies are summarized in an excellent review by Phillip and Bender (63).

4 Determination of primary specificity of a protease

There are two general ways to explore the specificity of a new enzyme. The first is to examine the cleavage pattern of the activity on a standard substrate such as the insulin B-chain. This sequence contains a useful variety of bonds available for cleavage. The products are small peptides that can readily be separated by HPLC, and a considerable literature exists documenting the properties of the products of digestion by the common proteases. Incubation of insulin B-chain with the newly discovered activity, followed by HPLC and comparison with the standard peptide maps will reveal the cleavage specificity. There also exists a growing list of new oligopeptide substrates designed to test the specificity of new enzymes and to characterize the activity and specificity of known enzymes. This has been necessary because the insulin B-chain, and other naturally occurring peptides, contain only a subset of all possible cleavage dipeptides. Furthermore, in some cases, the secondary specificity of the protease may be as important as the primary specificity. By this we mean that the amino acids in the P_2 to P_n positions on the left of the cleavage site and in the P_2' to P_n' positions on the right of the cleavage site may provide enhanced binding to the active-site cleft of the protease. This can lead to changes in the K_m of orders of magnitude, thus increasing the ratio of k_{cat}/K_m. This will in turn lead to more efficient cleavage in the limited time chosen for the specificity test.

4.1 Degradation of standard proteins and peptides

The oxidized B-chain of insulin has served for over 30 years as the substrate of choice for the initial screening of the specificity of a newly discovered protease. The sequence, given in *Table 4*, contains a selection of peptide bonds 'on offer' to a proteolytic enzyme, and almost certainly has an unfolded structure so that all peptide bonds are freely accessible.

Table 4 Insulin B-chain, oxidized: cleavage points with several proteases

Sequence	F		V		N		Q		H		L		C*		G		S		H		L		V		E		A		L		Y		L		V		C*		G		E		R		G		F		F		Y		T		P		K		A
Porcine pepsin																						↕						↕				↕																↕		↕									
Trypsin																																												↕														↕	
Elastase																												↕								↕										↕													
Papain						↕								↕												↕						↕																		↕		↕							
Cathepsin B						↕				↕				↕		↕				↕						↕						↕								↕										↕		↕							
Subtilisin								↕										↕												↕		↕																		↕									
Meprin										↕		↕				↕				↕								↕		↕				↕						↕								↕		↕									

References: pepsin: (64–66), trypsin, (67), elastase, (68), papain, (69), cathepsin B, (70,71), subtilisin, (72), meprin, (73)

Protocol 2

Procedure for analysis of cleavage of insulin B-chain

Equipment and reagents

- Insulin B-chain
- HPLC apparatus (C_{18} reversed-phase)
- Solvent A: 0.1% TFA
- Solvent B: CH_3CN containing 0.1% TFA

Method

1 Prepare an incubation mixture of 150 μg of insulin B-chain, oxidized (50 nmol), with a measured amount of the new protease activity. Choose the buffer for this reaction on the basis of previous studies of the pH optimum for kinetics and stability.

2 Incubate for 1 h.

Protocol 2 continued

3 Withdraw an aliquot of 20% of the sample and fractionate by HPLC (C_{18} reversed phase, 0.1% TFA as solvent A and CH_3CN containing 0.1% TFA as solvent B, 10-100% B linear gradient at 1 ml/min over 45 min).

4 If sufficient cleavage has not occurred in that interval, incubate the remaining sample for a longer period, up to 24 h, followed by separation.

5 Collect the resulting peptide fragments, hydrolyse, and analyse for amino acid content or sequence to determine the points of cleavage, since the sequence is known and this is a linear peptide.

6 Alternatively, if MALDI-TOF or ESI-MS is available, it may be sufficient to acquire the accurate masses of the products, or in complex digestion mixtures, to collect peaks from reversed phase HPLC and then measure peptide masses individually. In this instance, it is essential to remove the TFA from the sample, as this ion-pairing reagent has a strong ion suppression effect.

The products of cleavage of this peptide (*Protocol 2*) are of a size that can be conveniently separated by HPLC (see *Figure 8*), and the presence of aromatic groups throughout the sequence provides a convenient handle for detection by conventional absorbance monitors.

4.2 Cleavage of homologous synthetic peptides

The previous section indicated that some of the specificity of cleavage of proteins by these enzymes can be explained by their preference for a particular amino acid on the acyl side of the peptide bond that is cleaved. This is known as the primary specificity of the enzyme. As data were obtained from a wider variety of protein cleavages, it was frequently observed that some 'anomalous' cleavages were occurring. In some instances, those events may have occurred because of the impurity of the initial preparations. As more information became available about the three-dimensional structure of the proteolytic enzymes, in particular the gastrointestinal digestive enzymes, it became obvious that the active sites are elongated trenches, designed to bind to extended polypeptide chains. In mammalian digestion, dietary proteins come into contact with the acidic conditions of the stomach and are usually denatured. This permits attack by the gastric enzyme pepsin and leads to the generation of multiple fragments. Even if this does not lead to total digestion of the protein substrate, the fragments probably lack any organized structure. Thus, further digestion by trypsin and chymotrypsin also involves protease attack on extended polypeptide chains. Thus, it is clear why these peptidases are designed with long active-site crevices.

In addition to the structural data available from X-ray crystallography, several studies have provided evidence for the role of 'secondary enzyme–substrate' interactions. Fruton (74) has listed several instances where elongation of a peptide substrate increases the value of k_{cat}/K_m, dramatically. Striking examples

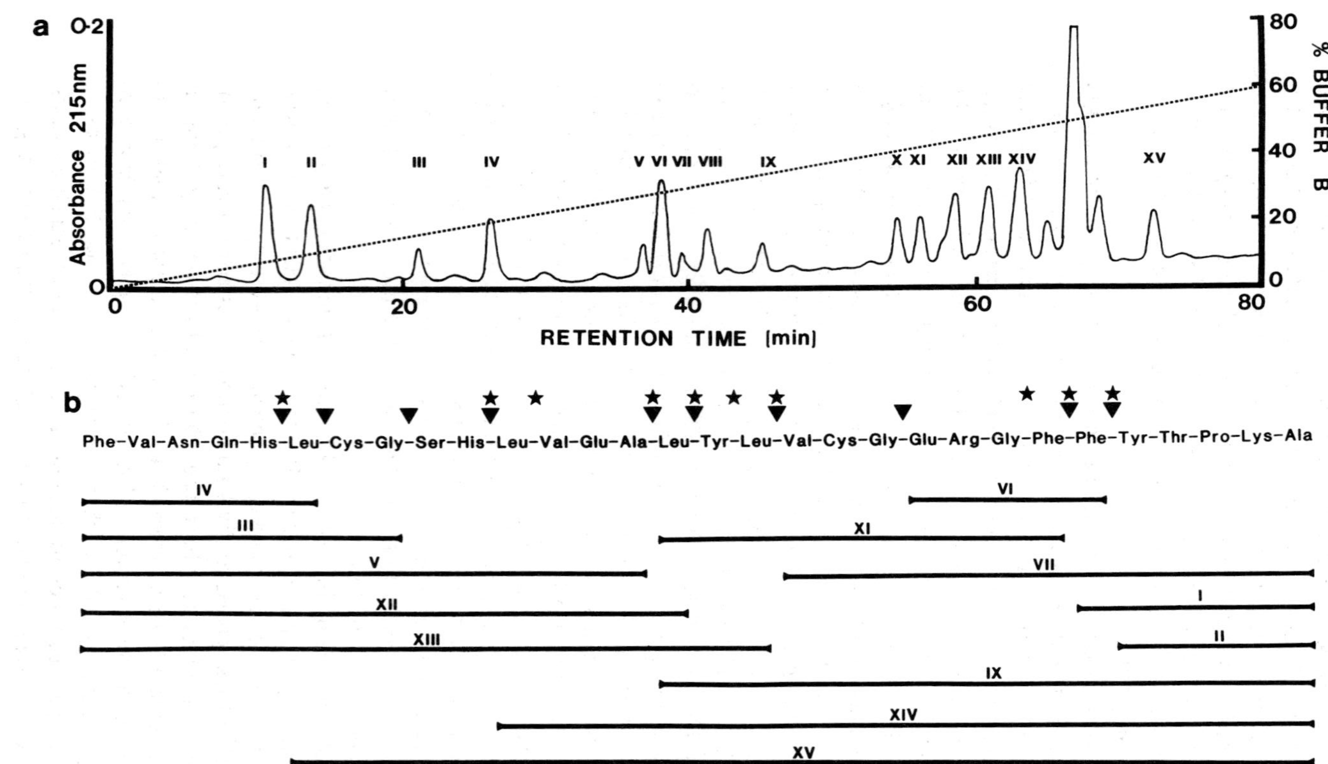

Figure 8 Illustration of the separation of the fragments generated by digestion of oxidized insulin B-chain by HPLC. Peptide peaks indicated by Roman numerals were collected and analysed for amino acid content and the *N*-terminal amino acid residue. The information permits the identification of the amino acid sequence of each fragments, based on the known structure of the insulin B chain (From ref. 73, with permission of the authors).

include comparison of the subtilisin or elastase cleavage of the ester group of Ac-Ala-OCH_3 with the cleavage of the same ester in the substrate Ac-$(Ala)_2$-Ala-OCH_3; the ratio k_{cat}/K_m is 2300 times or 3900 times higher for the longer substrate when the enzyme is, respectively, subtilisin (75) or elastase (76); and comparison of the cleavage of the Phe-Phe bond in the substrates Z-Phe-Phe-O-4-pyridylpropyl and Z-$(Ala)_2$-Phe-Phe-O-4-pyridylpropyl by porcine pepsin, where the ratio k_{cat}/K_m is 1900 times higher for the longer substrate (77). These examples point out the effect, described in Section 1.3 for the aspartic peptidase mechanism, of adding hydrogen bonds along the active-site cleft between substrate backbone groups and enzyme groups.

Thompson and co-workers (78,79) have demonstrated that significant changes in both binding and catalytic efficiency result from the occupancy of the S_4 and S_5 subsites of elastase, chymotrypsin, and *Streptomyces griseus* protease 3. Schechter and Berger (Appendix I) have described the preference of papain for a large, hydrophobic amino acid in the P_2 position of both substrates and inhibitors. The structural basis for this is the existence of a hydrophobic pocket comprised of the side-chains of Pro68, Ala160, Val133, Val157, Tyr67, Trp69 and Phe207 (80).

Studies by Dunn and co-workers (81–83) have revealed specific interactions in the active site of the aspartic peptidases at subsites remote from the scissile bond. These effects were observed by comparing a series of homologous substrates of the type Lys-X_{aa}-Y_{aa}-Z_{aa}-Phe-Nph-Arg-$X_{aa'}$-$Y_{aa'}$-$Z_{aa'}$, where X_{aa}, etc., are variable amino acids, Nph is the abbreviation for *p*-nitrophenylalanine and cleavage is at the Phe-Nph peptide bond. Since porcine pepsin has a strong primary preference for aromatic amino acids on both sides of the bond being cleaved, it is possible to use those interactions to anchor the substrate while varying the flanking residues. These studies underlined the importance of examining the kinetic parameters over a variety of pH values, as electrostatic interactions will be of different value as the charges on the substrates or enzymes are altered.

The preceding studies point out that the mechanisms of these enzymes involved a subtle blend of both binding and catalytic effects. A complete understanding of the mechanism of any one class of peptidases will require a total description of all the forces that are involved with binding as well as proton transfer and nucleophilic catalysis.

Finally, it is obvious from the results with the well-characterized digestive enzymes that binding of extended peptide substrates is preferred. However, interest has now shifted to the study of proteolytic cleavage by less common enzymes, including those coded for by viral genomes. It is becoming apparent that some of those enzymes may have been designed to cleave projections on the surface of their substrate proteins that will exist in a native, folded conformation. For example, the specificity of the poliovirus 3C protease is known to involve -Gln-Gly- bonds, and the AMV protease as well as the HIV protease apparently cleave their polyprotein substrates at -Tyr-Pro- or -Phe-Pro- bonds. Both of these types of sequences would have a high probability of occurring in a β-bend structure. It may well be that those enzymes have an active-site cleft that is more like a deep hole than a shallow trench.

References

1. Walsh, C. (1979). *Enzymatic reaction mechanisms*, p. 49. Freeman, San Francisco.
2. Fersht, A. R., Blow, D. M., and Fastrez, J. (1973). *Biochemistry* **12**, 2035.
3. Baca, M. and Kent, S. B. (1993). *Proc. Natl. Acad. Sci. USA* **90**, 11638.
4. Lu, W., Qasim, M. A., Laskowski Jr, M., and Kent, S. B. (1997). *Biochemistry* **36**, 673.
5. Robertus, J. D., Kraut, J., Alden, R. A., and Birktoft, J. J. (1972). *Biochemistry* **11**, 4293.
6. Bryan P., Pantoliano, M. W., Quill, S. G., Hsiao, H. Y., and Poulos, T. (1986). *Proc. Natl. Acad. Sci. USA* **83**, 3743.
7. Smith, E. L. (1958). *J. Biol. Chem.* **233**, 1392.
8. Bender, M. L. and Brubacher, L. J. (1964). *J. Am. Chem. Soc.* **86**, 5333.
9. Lowe, G. (1976). *Tetrahedron* **32**, 291.
10. Finkle, B. J. and Smith, E. L. (1958). *J. Biol. Chem.* **230**, 669.
11. Drenth, J., Jansonius, J. N., Koekoek, R., and Wolthers, B. G. (1971). In *The enzymes* (ed. Boyer, P. D.), Vol. 3, p. 485. Academic Press, New York.
12. Drenth, J., Kalk, K. H., and Swen, H. M. (1976). *Biochemistry* **15**, 3731.
13. Glazer, A. N. and Smith, E. L. (1971). In *The enzymes* (ed. Boyer, P. D.), Vol. 3, p. 501. Academic Press, New York.
14. Hofmann, T., Dunn, B. M., and Fink, A. L. (1984). *Biochemistry* **23**, 5241.
15. Delpierre, G. R. and Fruton, J. S. (1965). *Proc. Natl. Acad. Sci. USA* **54**, 1161.
16. Hartsuck, J. A. and Tang, J. (1972). *J. Biol. Chem.* **247**, 2575.
17. Pearl, L. H. (1987). *FEBS Lett.* **214**, 8.
18. Antonov, V. K. (1979). In *Protein: structure, function, and industrial applications* (ed. Hofmann, E., Pfeil, W., and Aurich, H.), p. 95. Pergamon Press, Oxford.
19. Powers, J. C. and Harper, J. W. (1986). In *Proteinase inhibitors.* (ed. Barrett, A. J. and Salvesen, G.), p. 219. Elsevier, Amsterdam.
20. Hangauer, D. G., Monzingo, A. F., and Matthews, B. W. (1984). *Biochemistry* **23**, 5730.
21. Aoyagi, T. and Umezawa, H. (1975). In *Proteases and biological control* (ed. Reich, E., Ritkin, D. B., and Shaw, E.), p. 429. Cold Spring Harbor Laboratory Press, New York.
22. Umezawa, H. and Aoyagi T. (1983). In *Proteinase inhibitors: medical and biological aspects* (ed. Katunuma, N., Umezawa, H., and Holzer, H.), p. 3. Springer-Verlag, Berlin.
23. Suda, H., Aoyagi, T., Takeuchi, I., and Umezawa, H. (1973). *J. Anribiot.* **26**, 621.
24. Umezawa, S., Tatsuta, K., Izawa, O., Tsochiya, T., and Umezawa, H. (1972). *Tetrahedron Lett.* **1**, 97.
25. Komiyama, T., Suda, H., Aoyagi, T. Takeuchi, T., Umezawa, H., Fujimoto, K., and Umezawa, H. (1975). *Arch. Biochem. Biophys.* **171,** 727.
26. Bond, J. S. and Butler, P. E. (1987). *Annu. Rev. Biochem.*, **56**, 333.
27. Rauber, P., Wekstrom, P. and Shaw, E. (1988). *Anal. Biochem.* **168**, 259.
28. Barrett, A. J. and Salvesen, G. (eds). (1986). *Proteinase inhibitors, Section B.* Elsevier, Amsterdam.
29. Shaw, E. (1975). In *Proteases and biological control.* (ed. Reich, E., Ritkin, D. B., and Shaw, E.), p. 455. Cold Spring Harbor Laboratory Press, New York.
30. Bazaes, S. (1987). In *Chemical modification of enzymes: active site studies.* (ed. Eyzaguirre, J.), p. 35. Halsted Press, New York.
31. Plapp, B. V. (1982). In *Methods in enzymology* (ed. Colowick, S. P. and Kaplan, N. O.), Vol. 87, p. 469. Academic Press, New York.
32. Lin, X-L., Wong, R. N. S., and Tang, J. (1989). *J. Biol. Chem.* **264**, 4482.
33. Lin, Y., Fusek, M., Lin, X., Hartsuck, J.A., Kezdy, F.J., and Tang, J. (1992). *J. Biol. Chem.* **267**, 18413.
34. McGrath, M. E., Vasquez, J. R., Craik, C. S., Yang, A. S., Honig, B., and Fletterick, R. J. (1992). *Biochemistry* **31**, 3059.

35. Lowther, W.T., Majer, P., and Dunn, B.M. (1995). *Protein Sci.* **4**, 689.
36. Fersht A. (1977). *Enzyme Structure and Mechanism*, p. 303. Freeman, San Francisco.
37. Blum, M., Cunningham, A., Bendiner, M., and Hofmann, T. (1985). *Biochem. Soc. Trans.* **13**, 1027.
38. Bauer, C. A., Lofqvist, B., and Pettersson G. (1974). *Eur. J. Biochem.* **41**, 45.
39. Houmard, J. (1976). *Eur. J. Biochem.* **68**, 621.
40. Martin, P. (1984). *Biochim. Biophys. Acta* **791**, 28.
42. Rossomando, E. F. (1987). *High performance liquid chromatography in enzymatic analysis.* John Wiley, New York.
43. Ellis, K. J. and Morrison, J. F. (1982). In *Methods in Enzymology.* (ed. Colowick, S. P. and Kaplan, N. O.), Vol. **87**, p. 405. Academic Press, New York.
44. Dixon, M. (1953). *Biochem. J.* **55,** 161.
45. Tipton, K. F. and Dixon, H. B.F. (1979). In *Methods in enzymology* (ed. Purich, D. L.), Vol. 63 p. 183. Academic Press, New York.
46. Cleland, W. W. (1982). In *Methods in enzymology* (ed. Purich, D. L.), Vol. 63 p. 390. Academic Press, New York.
47. Wilkes, S. H. Bayliss, M. E., and Prescott J. M. (1988). *J. Biol. Chem.* **263**, 1821.
48. Schowen, K. B. and Schowen, R. L. (1982). In *Methods in enzymology* (ed. Purich, D. L.), Vol. 63 p. 551. Academic Press, New York.
49. Wolfenden, R. (1972). *Acc. Chem. Res.*, **5**, 10.
50. Lienhard, G. E. (1973). *Science* **180**, 149; and references therein.
51. Bone R. Shenvi A. B., Kettner C. A., and Agard, D. A. (1987). *Biochemistry* **26,** 7609.
52. Szelke, M., Leckie, B., Hallett, A., Jones, D. M., Sueiras, J., Atrash, B., and Lever, A. F. (1982). *Nature* **299**, 555.
53. Szelke, M. (1985). In *Aspartic proteinases and their lnhibirors.* (ed. Kostka, V.), p. 421. de Gruyter, Berlin.
54. Kati W. M., Pals, D. T., and Thaisrivongs, S. (1987). *Biochemistry* **26**, 7621.
55. Rich, D. H., Bernatowicz, M. S., Agarwal, N. S., Kawai, M., Salituro, F. G., and Schmidt, P. G. (1985). *Biochemistry* **24**, 3165.
56. Shoham, G., Christianson, D. W., and Oren, D. A. (1988). *Proc. Natl. Acad. Sci. USA* **85**, 684.
57. Gordon, E. M., Godfrey, J. D., Pluscec, J., von Langen, D., and Natarajan, S. (1986). *Biochem. Biophys. Res. Commun.* **126**, 419.
58. Bartlett, P. A. and Marlowe, C. K. (1987). *Biochemistry* 26, 8553.
59. Foundling, S. I., Cooper J., Watson F. E. Cleasby A., Pearl, L. H., Sibanda, B. L. *et al.* (1987). Nature **327**, 349.
60. Suguna, K., Padlan, E.A., Smith, C. W., Carlson, W. D., and Davies, D. R. (1987). *Proc. Natl. Acad. Sci. USA* **84**, 7009.
61. Storer, A. C., Angus, R. H., and Carey, P. R. (1988). *Biochemistry* **27**, 264; and earlier references cited therein.
62. Campbell, P., Nashed, N. T., Lapinskas B. A., and Gurrieri, J. (1983). *J. Biol. Chem.* **258**, 59.
63. Philipp, M. and Bender, M. L. (1983). *Mol. Cell Biochem.* **51**, 5.
64. Sanger F. and Tuppy, H. (1951). *Biochem. J.*, **49**, 481.
65. Ryle, A. P. Leclerc, J., and Falla F. (1968). *Biochem. J.*, **110**, 4P.
66. Powers, J. C., Harley, A. D., and Myers, D. V. (1977). In *Acid proteinases – structure, function, and biology*. (ed. Tang, J.), p. 141. Plenum Press New York.
67. Wang, S.-S. and Carpenter F. H. (1967). *Biochemistry* **6**, 215.
68. Sampath Narayanan A. and Anwar R. A. (1969). *Biochem. J.* **114**, 11.
69. Johansen, J. T. and Ottesen, M. (1968). *Compt. Rend. Trav. Lab. Carlsberg* **36**, 265.
70. Otto, K. and Bhakadi, S. (1969). *Hoppe-Seyler's Z. Physiol. Chem.* **350**, 1577.

71. Keilova, H. (1971). In *Tissue proteinases* (ed. Barrett, A. J. and Dingle, J. T.), p. 45. North Holland, Amsterdam.
72. Morihara, K. and Tsuzuki, H. (1969). *Arch. Biochem. Biophys.* **129**, 620.
73. Butler, P. E., McKay, M. J., and Bond, J. S. (1987). *Biochem. J.* **241**, 229.
74. Fruton, J. S. (1975). In *Proteases and biological control.* (ed. Reich, E., Ritkin, D. B., and Shaw, E.), p. 33. Cold Spring Harbor Laboratory Press, New York.
75. Morihara, K. and Oka, T. (1973). *FEBS Lett.* **33**, 54.
76. Gertler, A. and Hofmann, T. (1970). *Can. J. Biochem.* **48**, 384.
77. Sachdev, G. P. and Fruton, J. S. (1970). *Biochemistry* **9**, 4465.
78. Thompson, R. C. and Blount, E. R. (1973). *Biochemistry* **12**, 66.
79. Bauer, C.-A., Thompson, R. C., and Blount, E. R. (1976). *Biochemistry* **15**, 1291
80. Baker, E. N. and Drenth, J. (1987). In *Biological Macromolecules and Assemblies. Vol. 3. Active sites of enzymes.* (ed. Jurnak, F. A. and McPherson, A.), Ch. **6**, p. 313. John Wiley, New York.
81. Dunn, B. M., Jimenez, M., Parten, B. F., Valler, M. J., Rolph, C. E., and Kay, J. (1986). *Biochem. J.* **237**, 899.
82. Dunn, B. M., Valler, M. J., Rolph, C. E., Foundling, S. I., Jimenez, M., and Kay, J. (1987). *Biochem. Biophys. Acta* **913**, 122.
83. Pohl, J. and Dunn, B. M. (1988). *Biochemistry* **27**, 4827.

Chapter 5
Inhibition of proteolytic enzymes

Guy S. Salvesen
The Burnham Institute, 10901 North Torrey Pines Road, La Jolla, CA 92037, USA

Hideaki Nagase
Kennedy Institute of Rheumatology Division, Imperial College School of Medicine, Hammersmith, 1 Aspenlea Road, London, W6 8LH, UK

1 Introduction: which inhibitors?

The two most frequent questions asked by those not working on proteases every day are: 'How do I prevent proteolysis in my particular system?' and 'How do I find out which protease is cutting my protein?' There are no 'bullet-proof' recipes to help with these questions, possibly because of the many varied or even overlapping specificities and catalytic strategies adopted by proteases. Nevertheless, there are a number of general issues and specific techniques that we and others use to address these and other related questions. This usually involves employing natural and synthetic protease inhibitors in complex biological environments such as tissue fluids or cells in culture. Many of the methods rely on the theoretical backgrounds illustrated in Chapter 6.

In keeping with the scope of this book, we concentrate primarily on synthetic inhibitors in the characterization and prevention of proteolysis. With a few notable exceptions, the synthetic inhibitors generally afford more practical uses in experimental biology. Readers interested in the protein inhibitors of proteases are referred to Chapter 6, and reviews in the literature (1–4).

Inhibitors of proteases come in many different forms and are often grouped on the basis of their reaction mechanism, origin or structural similarity. We divide them here into specificity groups on the assumption that readers will be more interested in how to inhibit given proteases than how to classify inhibitors. Three classes are designated: firstly, those that react with more than one class of protease; secondly, those that are ostensibly specific for one of the classes; and finally, those that show high selectivity for a single protease.

2 Principles for using irreversible and reversible inhibitors

2.1 General structure of synthetic inhibitors

Proteases bind their substrates in a cleft on the enzyme surface, with the scissile bond positioned over the catalytic machinery, and with side-chains of the substrate-filling specific pockets in the cleft. Most design strategies for synthetic protease inhibitors have used this property to attach a group ('warhead') reactive with components of the catalytic machinery to a short stretch of peptide chain, as shown in *Figure 1*. The peptide side-chains (P) deliver a certain degree of specificity, and the N-blocking group is frequently added to prevent digestion by aminopeptidases and increase binding since many endopeptidases will not accept a free α-amine in their specificity pockets. Usually, the peptide is designed to fit the unprimed (S) protease subsites, but sometimes it fits the primed (S′) ones, as in the case of several effective hydroxamate peptidyl inhibitors of metalloproteases.

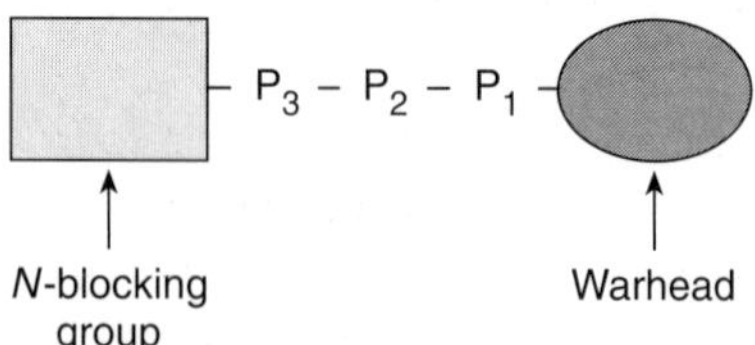

Figure 1 General structure of a tripeptidyl protease inhibitor. Sometimes a single residue (P_1) is sufficient for directing inhibition, but extending the peptide chain satisfies additional extended subsite preferences and imparts selectivity to the inhibitor. The 'warhead' may be one of a number of moieties that are reactive with specific protease catalytic components.

2.2 How to live with crossreactivity

A new inhibitor is usually intended to be directed against a specific target enzyme; it is then tested against a number of other related and unrelated proteases to gauge the specificity and selectivity of the compound. The inhibitors in this chapter have been invaluable in characterizing proteases, but ultimately there is often a later study that demonstrates unexpected crossreactivity with other proteases, and sometimes even other non-proteolytic proteins. Crossreactivity is particularly noticeable between closely related proteases, but occasionally occurs even between distinct catalytic classes of serine, cysteine and threonine (proteasome). Members of these classes tend to stabilize the transition state by similar mechanisms, and so transition-state inhibitors such as aldehydes can inhibit all three of these classes. All contain powerful catalytic nucleophiles that participate in covalent catalysis, so good electrophilic warheads such as chloromethyl ketones can demonstrate cross-mechanism reactivity.

These issues do not pose a problem if all that is needed is to abolish general proteolytic activity, but if specificity is desirable they must be considered. Use of a single inhibitor to demonstrate a single protease involvement in a biological process can, therefore, lead to misinterpretations. Happily, there are solutions to this problem. So, whenever using synthetic inhibitors to characterize or discover proteases or biological functions, it is always best to consider cross-reactivity and use a concerted approach of several of the inhibitors available. Most inhibitors in this chapter are commercially available, and those that are not can sometimes be obtained from the original source.

Surveys of the literature and experience have led the authors to recommend the procedures listed here and, where space permits, reasons are given for particular choices made. Where it was felt necessary to point out errors resulting from popular (but often inappropriate) methods, we have tried to replace these with more valid procedures.

3 Practical use of inhibition constants

The importance of reaction kinetics as a basis for using inhibitors to investigate proteolysis is detailed in the Chapter 6. If one wishes to suppress the activity of a given protease and knows the kinetic constants for the reaction of this protease with an inhibitor, one can determine how much inhibitor to add and how much time to allow for inhibition. Although more complex reaction pathways can be written for all protease–inhibitor interactions, for most practical applications the following simplifications apply, considered in the generic case of a two-step irreversible inhibitor with the following definitive kinetic quantities. If the inhibitor is reversible, one can simply truncate the mechanism by omitting the second step that is, by omitting the rate governing the formation of the EI* complex (k_2).

Reaction pathways $$E + I \underset{k_{-1}}{\overset{k_1}{\rightleftharpoons}} EI \xrightarrow{k_2} EI^*$$

Definitions $$k_i = \frac{k_{-1}}{k_1} \qquad k_{ass} = \frac{k_2}{K_i}$$

Figure 2 Protease inhibitor reaction pathway and definitions. K_i and k_{ass} are constant for a particular enzyme–inhibitor reaction under specified environmental conditions. To make relevant comparisons of enzymes or inhibitors, and derive useful predictions, relationships between the constants and the total inhibitor concentration are used as described in the text.

3.1 Irreversible inhibitors

If an inhibitor operates by an essentially irreversible mechanism, both the rate of complex formation (the second order reaction rate, k_2/K_i) and the equilibrium concentration of the first step (K_i) are important in principle. For most practical purposes k_{ass} takes into account both steps in the reaction and so the magnitude of the K_i step can be ignored. This is because most good inhibitors have a K_i well below the common working concentration of the inhibitor, and the first step in the reaction pathway (see above) is therefore mostly saturated. Examples of such inhibitors are α_1-protease inhibitor, TLCK, PMSF, and Z-VAD-FMK, which form covalent bonds with the enzyme that often remain when the enzyme is denatured.

The most important quantity to consider when using irreversible protease inhibitors is the time required for the free enzyme concentration to decrease by 50% (half life, $t_{1/2}$). During the reaction of a protease with an irreversible inhibitor is then given by:

$$t_{1/2} = \ln 2 / k_{ass}[I] \quad (1)$$

One may use this relationship to compare the inactivation of an enzyme by a variety of inhibitors (2,5,6). It is used later in this chapter to compare the efficacy of different inhibitors for a variety of proteases. The relationship, as discussed by Bieth (7), can be very powerful in revealing the potential role of protease inhibitors in physiological processes, and is widely applicable. It is also extremely useful in deciding how long a synthetic inhibitor should be allowed to incubate with a protease solution to achieve adequate inhibition. The time required for an irreversible inhibitor to diminish the activity of a protease by 99.9% (the delay time) is 10 half-lives of the reaction.

3.2 Reversible inhibitors

An equilibrium between the complexed form of an enzyme and free enzyme characterizes reversible inhibition. Sometimes, a reversible inhibitor may bind so tightly that it is not possible to measure a binding constant. Under this condition the inhibitor is sometimes classified as 'pseudo-irreversible' and it is then appropriate to analyse it as an irreversible inhibitor.

For freely reversible inhibitors, provided that $[I] > [E] \times 10$, the fraction of free enzyme existing in equilibrium with a reversible inhibitor is given by the relationship.

$$[EI]/[E] = [I]/K_i \quad (2)$$

If $[I] > 1000K_i$, and if $[I] > 10[E]$, then less than 0.1% of the total enzyme is free at any time. Only when [I] approaches K_i should one anticipate significant substrate hydrolysis. Therefore the ratio $[I]/K_i$ describes the efficiency of a reversible inhibitor. The concentration of a reversible inhibitor required to diminish the activity of a protease by 99.9% (the stability concentration) is $K_i \times 1000$. Thus the stability concentration is an extremely useful quantity to use when adequate inhibition of a protease by an reversible inhibitor is required (see Box 1 for examples).

The effect of substrate on the delay time or stability concentration (7) is not considered, as this is irrelevant to comparisons of inhibitor efficacy *in vitro* and *in vivo*.

Box 1

Guide for effective inhibition

According to the relationships above, one can calculate how long (for an irreversible inhibitor, or at what concentration a reversible inhibitor should be incubated to decrease enzyme activity by 99.9% (essentially complete inhibition). This depends on the respective k_{ass} or K_i values and the concentration of added inhibitor. Below are recommended incubation times for the addition of 1.0 μM of an irreversible inhibitor and recommended concentrations for a reversible one.

Irreversible (see text for definition of delay time)

if $k_{ass} = 5 \times 10^3$/M.s (relatively slow) delay time = 23 min

if $k_{ass} = 5 \times 10^5$/M.s (relatively fast) delay time = 14 s

Reversible (see text for definition of stability concentration)

if K_i = 100 nM (relatively weak) stability concentration = 100 μM

if K_i = 1 nM (relatively strong) stability concentration = 1 μM

4 Non-specific inhibitors

4.1 α-Macroglobulins

The most noteworthy example of an inhibitor that fails to discriminate between protease classes is the human protein α_2M, a member of a group of high molecular weight proteins known collectively as the α-macroglobulins that bind and inhibit most endopeptidases. The inhibitory capacity of α_2M results from a conformational change following endopeptidase reaction that appears physically to entrap the reacting molecule(s) (8). Thus, α_2M is mechanistically different from the active site-directed protein inhibitors of proteases. Binding of a protease to α_2M is irreversible; however, the complexed enzyme retains substantial activity against small peptides, but is almost completely inactive against larger ones (proteins larger that 25 kDa). This unusual reactivity is a result of steric hindrance placed on the protease by enclosure within the large α_2M molecule (8). Frequently, inhibited endopeptidases may also become covalently attached via displacement of an internal α_2M thiol-ester (9).

The covalent attachment phenomenon can be extremely useful in determining whether a protein of interest is an endopeptidase. The first step in the reaction of an endopeptidase with α_2M is cleavage of the 'bait region', a stretch of about 20 residues that is the most sensitive and non-specific of all known protein substrates. After bait region cleavage, the protease is frequently covalently attached to α_2M, in a reaction that depends partly on the lysine content of the protease (10). Thus, if the protein of interest (purified or recombinant) is a protease, there is a very good chance that it will become covalently linked when added to α_2M. This can be monitored by a 90–200 kDa shift in the apparent size of the catalytic chain of the putative protease upon SDS-PAGE. This reaction will only be seen with endopeptidases, since exopeptidases do not cleave the bait region. Not all endopeptidases will cleave the bait region, so absence of a gel-shift does not mean the candidate is not an active endopeptidase. However, an obvious gel shift is very characteristic and will almost certainly mean endopeptidase activity. Caution must be exercised to ensure that the samples for SDS-PAGE are boiled for 5 min in 1% SDS and at least 10 mM DTT to overcome artefactual, non-specific binding.

The lack of specificity or selectivity of α_2M usually argues against its use in most instances, since it binds almost all endopeptidases. Moreover, because proteases in complex with α_2M may retain significant activity against protein substrates, especially disordered ones such as casein, the use of α_2M as an additive to prevent proteolysis in sensitive samples is discouraged.

4.2 Peptide aldehydes

- Advantages: relatively stable in aqueous solution; can be cell-penetrating
- Disadvantages: lack of specificity; recognize serine, threonine and cysteine proteases

Peptide aldehydes, such as leupeptin and antipain, are di-, tri- or tetra-peptides containing an aldehyde moiety in place of the usual α-carboxylate. They are reversible inhibitors and are often referred to as transition-state analogues because they act by mimicking the tetrahedral intermediate formed during peptide bond hydrolysis. Leupeptin and antipain in particular are sometimes injudiciously used to classify the catalytic mechanism of a newly identified protease. Such use is inadvisable as these react equally well with serine and cysteine proteases; for example, leupeptin reacts equally well with trypsin and cathepsin B, and antipain equally well with trypsin and calpain I (11,12). Similarly, some aldehydes that are used to inhibit calpain also inhibit the proteasome, for example Ac-Leu-Leu-nVal-CHO.

Many peptide aldehyde derivatives have been synthesized in attempts to improve specificity (13) and, largely because of their low toxicity, they are used frequently in animal and cell culture models of proteolysis. However, two major drawbacks limit their usefulness as selective or specific inhibitors. Firstly, the lack of selectivity allows for the inhibition of proteases other than the one(s) under investigation. This is especially true in cell culture and animal experiments. Secondly, in aqueous solution most aldehydes exist primarily as inactive hydrates, and in the case of arginals as cyclic carbinolamines. Accordingly, the rate-limiting step for inhibition can be formation of the aldehyde rather than encounter with a suitable target enzyme (14).

4.3 Peptide chloromethyl ketones

- Advantages: covalent reacting; good for affinity labelling; more stable than other covalent inhibitors
- Disadvantages: lack of specificity; recognize serine, threonine and cysteine proteases.

These irreversible inhibitors represent one of the first systematic approaches in the study of active-site-directed protease-modifying reagents (15). The two best known members of the group are TLCK and TPCK. The mechanism of inactivation of serine proteases has been studied extensively, and it is clear that inhibition results from the protease binding the inhibitor in a substrate-like manner followed by alkylation of the active-site histidine by the chloromethyl moiety. Chloromethyl ketones also inactivate cysteine proteases and the mechanism of inactivation is thought to be similar to that for serine proteases, although this remains unproven. The design of selective inhibitors has reached a high art in these compounds. However, given the poor specificity of some of them (16), we caution against their use in analysing catalytic mechanisms. Nevertheless, peptide chloromethyl ketones are considered by many to be invaluable in preventing proteolysis in well-defined systems. Their use in cell culture and animal studies is limited by toxicity resulting from alkylation of cellular components and interaction with intracellular glutathione (13).

4.4 Metal chelators

- Advantages: reversible; some selectivity can be achieved for zinc peptidases
- Disadvantages: can inactivate many other enzymes and proteins.

Metalloproteases contain a zinc ion that participates in catalysis by polarizing a water molecule to attack the substrate–peptide bond. This differentiates them from proteases such as the calpains and some trypsin-like serine proteases, whose activities are stabilized by, but not necessarily dependent on, the presence of calcium. Both the zinc-dependent metalloproteases and some calcium-stabilized proteases from other classes can be inactivated by chelating agents such as EDTA and EGTA. Care should therefore be exercised when using these to characterize new proteases. 1,10-Phenanthroline is preferred as an inactivator of metalloproteases, as it has a much higher stability constant for zinc (2.5×10^6/M) than for calcium (3.2/M). Thus, 1 mM 1,10-phenanthroline will inactivate a metalloprotease even in the presence of 10 mM calcium, whereas this concentration of 1,10-phenanthroline will not remove calcium from a calcium-binding protein. Therefore, this chelator is usually diagnostic for a zinc metalloprotease. The related, non-chelating compound 1,7-phenanthroline is generally used as a control to rule out non-specific effects of the relatively hydrophobic heterocyclic frame of the chelator. Confirmation of a protease as metallo-enzyme is accomplished by removal of zinc (see *Protocol 1*).

Protocol 1

Spin-column method for rapid exchange of metal ions

Equipment and reagents

- Enzyme solution
- Buffer used for enzyme solution
- Sephadex G-10 (Pharmacia)
- 15 ml conical polystyrene tube (Corning)
- Bench-top clinical centrifuge (e.g. International Equipment Co.)

Protocol 1 continued

Method

1. Plug the bottom of a 1-ml plastic tuberculin syringe (column) with small pieces of polymer wool and fill the syringe with Sephadex G-10, equilibrated with buffer, to the top of the syringe. The sample will exchange into this buffer.
2. Place the syringe in a 15-ml conical polystyrene tube and spin in a bench-top clinical centrifuge at 1000 r.p.m. for 1 min. Do not worry about the appearance of the column.
3. Transfer the syringe to a fresh 15 ml tube and apply 100 μl of enzyme solution,[a] and centrifuge exactly as in step 2. About 90 μl of effluent is usually collected, and the recovery of the protein after the spin column is about 90%.[b]

[a] The enzyme solution should be treated with a buffer containing 5 mM EDTA to remove zinc from the active site of the enzyme.

[b] If the recovery of the apoenzyme is poor, add a non-ionic detergent (e.g. 0.05% Brij 35, Sigma) to the Sephadex gel.

This is followed by titration with zinc (0.1–100 μM) to restore full activity. Removal of zinc, followed by re-addition, if desired, can also be used to quantify the contribution of metalloproteases to proteolysis in a complex mixture. High concentrations of zinc (in the mM range) often inhibit metalloproteases. Inhibition by zinc results from the formation of zinc monohydroxide which bridges the catalytic zinc ion to a side-chain in the active site of the enzyme (17). This inhibition is therefore competitive with substrate. The non-competitive inhibition by other heavy metal ions is attributed to binding of the ion to a site distinct from the active site (18).

5 Class-specific inhibitors

5.1 Serine proteases

5.1.1 Organophosphates

Diisopropylphosphofluoridate (DFP), historically one of the most widely used protease inhibitors, is a member of the organophosphorus compounds that irreversibly inactivate serine proteases. DFP is a toxic compound by virtue of its inactivation of acetylcholine esterase, and has lost favour as a general inhibitor of serine proteases. On the whole, DFP and its relatives react less rapidly with trypsin-like serine proteases than with chymotrypsin-like ones. The toxicity of DFP, its relatively low reaction rates (see *Table 1*) and its instability in aqueous solutions limit its practical use as a general inhibitor of serine proteases. Radioactive derivatives of DFP are, however, of considerable use as active site labels for serine proteases and are therefore suitable for autoradiography.

5.1.2 Sulphonyl fluorides

Phenylmethylsulphonyl fluoride (PMSF), another well-known irreversible inhibitor of serine proteases, reacts relatively slowly with common proteases (*Table 1*), although it also reacts reversibly with some cysteine proteases (19), so care should be exercised in interpreting results. Attempts to design faster-acting, more selective derivatives have met with some success, but the derivatives tend not to discriminate very well between closely related proteases. Thus, 2-(benzyloxycarbonylaminoethyl carbonyl)-benzene sulphonyl fluoride reacts about 100-fold faster with chymotrypsin-like enzymes than does PMSF, but more slowly with elastases (20). Similarly, APMSF reacts much more rapidly with trypsin-like proteases than does PMSF, but very slowly with chymotrypsin (21). These data point to the potential use of sulphonyl fluorides as inhibitors of general types of serine proteases. PMSF, the least selective and therefore most useful as a general inhibitor of serine proteases, decays relatively rapidly in aqueous solutions ($t_{1/2}$ 25 °C, pH 7.5, 55 min) (22) and this must be taken into account when using this particular inhibitor. Spontaneous hydrolysis in aqueous solutions is a characteristic of the sulphonyl fluorides. A recently popular alternative to PMSF is AEBSF (frequently known by its trade name Pefabloc), which shares a broad range of serine protease targets, but is reportedly more soluble and stable in aqueous solutions. AEBSF also is reported to inhibit NADH oxidase activation through a non-proteolytic route (23), so clearly some non-protease targets must be expected. In all instances, one should assume that the sulphonyl fluorides are unstable in aqueous solutions, and repeated additions are required for each fractionation step during protein purification.

5.1.3 Coumarins

Research into the design of inhibitors of serine proteases based on coumarin and related heterocyclic compounds has been reviewed before (13). Interest in the design of heterocyclic protease inhibitors takes advantage of their capacity to act as mechanism-based inactivators, and the more pragmatic fact that most successful drugs are heterocyclics. Considerable current synthetic inhibitor re-

Table 1 Relative efficacies of general serine protease inhibitors[a]

	Half-life (s) for inhibition at 1 mM inhibitor				
	Trypsin	**Chymotrypsin**	**Neutrophil elastase**	**Cathepsin G**	***S. aureus* protease**[b]
DFP	43	2.8	—	—	>300
PMSF	237	2.8	33	49	NI[c]
3,4-DCI	3.5	1.2	0.08	25	0.25

[a]Data taken from refs. (24–26). Data is expressed as half-time for inhibition, therefore a lower number means faster inhibition. Note that 3,4-DCI is by far the most effective inhibitor for all proteases listed.

[b]*Staphylococcus aureus* serine protease is sometimes called *S. aureus* V8 protease and is frequently used for protein digestion. Its only other known inhibitor is α_2M (J. Potempa, personal communication).

[c]NI, not inhibited.

search is geared towards the production of potentially therapeutic compounds to inhibit proteolysis *in vivo,* and the coumarins show promise in this area. Only one coumarin-based inhibitor 3,4-DCI, is commercially available and this compound shows good reactivity with a large number of serine proteases (24). It is relatively non-toxic, does not react at appreciable rates with acetylcholine esterase, papain, leucine aminopeptidase or β-lactamase and is faster-reacting with serine proteases than DFP or PMSF. It does, however, inactivate glycogen phosphorylase b (27), so clearly some non-protease targets must be expected. 3,4-DCI is not diagnostic for serine proteases since it also inhibits caspase-1 (J. Powers, personal communication) and the proteasome. Nevertheless, its use as a general inhibitor of serine proteases is strongly recommended. Although kinetically irreversible, activity of some proteases acylated by 3,4-DCI can be recovered following treatment with hydroxylamine (24), a property that could be of use in mechanism studies of protease–inhibitor interactions.

5.1.4 Peptide boronic acids

The extended boronic acid peptides shown in *Table 2* are among the most potent reversible protease inhibitors yet developed. The boronic acids are similar to the peptide aldehydes in their apparent function as transition state analogues, but they appear to be much more selective for serine proteases. Relatively few studies on design of peptide boronic acids for protease selectivity have been reported but it is clear from the current data that these derivatives have a future. The compounds show little toxicity at millimolar concentration in cell-culture experiments, although it is not known whether they penetrate cells (C. Kettner, personal communication).

5.1.5 Protein inhibitors of serine proteases

The protein inhibitors of serine proteases fall into several groups according to various schemes. The avian ovomucoids, the Kunitz-type trypsin inhibitors (such as aprotinin) and soybean trypsin inhibitor all form very tight, but reversible, complexes with a number of proteases. The high-molecular-weight inhibitors of serine proteases from the plasma of mammals, the serpins (serine protease inhibitors), form such tight complexes that reversibility is difficult to demonstrate.

In most instances, the protein inhibitors show little selectivity for proteases *in vitro,* although they probably show effective selectivity *in vivo* (2,7). The most extensively studied group of protein inhibitors of proteases, the avian ovomucoids, demonstrate hypervariability of the amino acid residues that are in contact with proteases (28) and this has probably arisen from attempts to fit the specificity of, as yet undefined, proteases during the evolution of avian species. The serpins, although less well conserved in primary structure than the ovomucoids, also show hypervariability of their presumed reactive-site loops (29). Given these points, it is surprising that the members of these two groups have not achieved higher degrees of selectivity, although this may be due to the relative invariability of extended substrate binding sites of proteases, or because our understanding of the requirements for protease selectivity are far from complete (28). Attempts to alter selectivity of inhibitors using site-directed mutation and recombinant DNA strategies have met with partial success, but it has seldom been possible to achieve the hoped for degree of selectivity. Indeed, these approaches have unfortunately raised more questions about inhibitory selectivity than they have answered. This may change as our understanding improves.

5.1.6 Inhibitor selectivity for serine proteases

The main goal of inhibitor design is to maximize selectivity for closely related proteases. Thus, it should be possible to suppress the activity of a single protease in a mixture and determine the result of this on any given biological phenomenon. Such a selective inhibitor should have a K_i for the protease of 1000-fold less than for any other protease in the mixture, or a k_{ass} of 1000-fold greater than for any other protease. This ideal has been very difficult to achieve and one can only speculate about the types of inhibitor that will provide the required degree of selectivity. Valuable lessons have, however, been learned from the chloromethyl ketones.

Most of the work on chloromethyl ketones has focused on exploring selectivity for plasma serine proteases, and on discriminating between the proteases from neutrophils. Examples of these successes are shown in *Table 2*. Because chloromethyl ketones react well with cysteine proteases, one should be wary of inferring too much regarding the role of a protease in a very complicated mixture (such as a cell or tissue extract or culture medium) from inhibitory data such as shown in *Table 2*. It is also likely that other, as yet unidentified, proteases reacting well with supposedly selective inhibitors are at work. Nevertheless, with the judicious manipulation of experimental conditions and use of cysteine protease inhibitors such as E-64 (for papain family proteases) or fluoromethyl ketones and acyloxymethyl ketones (see below), one can track down the likely causes of specific proteolysis.

Despite the selectivity engineered into chloromethyl ketones, it may transpire that the peptide boronic acids will ultimately be of greater use in cell culture experiments. They are generally deemed less toxic than chloromethyl ketones. Further, the principle of using a reversible inhibitor in a relatively long-term experiment is sounder than that of using an irreversible one. This is because an irreversible inhibitor will eventually inhibit proteases that it reacts with very slowly, whereas a reversible one will never decrease the activity of an untargeted protease when that inhibitor is at a concentration above K_i for the reaction. For example, for the inhibitors shown in *Table 2*, 10 μM Glu-Phe-Arg-CH_2Cl will inhibit 99.9% of free plasmin in 30 s and 99.9% of free t-PA in just under 3 h. This is an excellent degree of selectivity, but consider that at 20 μM MeO-Suc-Ala-Ala-Pro-Boro-Phe, cathepsin G will be 99.9% inhibited but elastase will never be more than 25% inhibited, even after an incubation lasting several hours. These points should be taken into account when choosing inhibitors for long-term experiments.

Table 2 Selectivity in serine protease inhibitors[a]

	Half-life (s) for inhibition at 10 μM inhibitor[b]				
	t-PA	**uPA**	**Plasmin**	**Neutrophil elastase**	**Cathepsin G**
D-Val-Gly-Arg-CH_2Cl	13	140	260	–	–
Glu-Gly-Arg-CH_2Cl	172	9	653	–	–
Glu-Phe-Arg-CH_2Cl	1000	3800	3	–	–
Z-Gly-Leu-Phe-CH_2Cl	–	–	–	N.I.	1360
MeO-Suc-Ala-Pro-Val-CH_2Cl	–	–	–	44	N.I.
K_i (nM)					
Meo-Suc-Ala-Ala-Pro-Boro-Val				0.57	359
Meo-Suc-Ala-Ala-Pro-Boro-Phe				74 000	21

[a] Data from refs. (19, 30–32). Reaction rates for the irreversible chloromethyl ketones are expressed as half-lives in the presence of 10 μM inhibitor.

[b] t-PA, tissue plasminogen activator; u-PA, urinary plasminogen activator; NI, not inhibited.

5.2 Cysteine proteases

Popular class-specific inhibitors of cysteine proteases fall into five categories: the synthetic peptide diazomethanes, the peptide epoxides, the fluoromethyl ketones, the acyloxymethyl ketones and the natural protein inhibitors known as the cystatins.

5.2.1 Peptide diazomethanes

These are a group of oligopeptide derivatives in which the OH of the terminal carboxylate is replaced with a diazomethyl moiety, CH-N = N. Inhibition is irreversible and is thought to proceed, via a thiohemiketal intermediate, to the formation of an alkylated active site cysteine existing as a thio-ether to the methyl group of the inhibitor. These inhibitors are almost completely specific for cysteine proteases (see *Table 3*), although one report indicates the Z-Phe-Arg-CHN_2 can also inhibit the serine protease plasma kallikrein (33). Although most strategies have focused on inhibition of endopeptidases, the compound Gly-Phe-CHN_2 is a useful inhibitor of dipeptidyl peptidase I, the activator of several white blood cell associated serine proteases (34).

5.2.2 Peptide epoxides

The peptide epoxides are a group of irreversible inhibitors based on compound E-64 isolated by Hanada *et al.* (35) from an extract of *Aspergillus japonicus*. Several derivatives of E-64 have been synthesized in attempts to explore this group as class-specific inhibitors of cysteine proteases. Inhibition of papain family proteases by E-64 and its relatives results from occupancy of the enzyme subsites followed by alkylation of the catalytic cysteine by the *trans* epoxide group. Interestingly, most structures show the inhibitor peptide backbone binding the unprimed subsites in the opposite direction to substrate, with the exception of

Table 3 Selectivity in papain family protease inhibitors[a]

	Half-life (s) for inhibition at 10 μM inhibitor[b]				
	Papain	**Cathepsin B**	**Cathepsin H**	**Cathepsin L**	**Calpain**
E-64(L)	0.1	0.8	17	0.7	9.2
Ep-479	0.08	0.2	33	*0.5*	14
Ep-460	0.15	0.4	89	0.4	3
CA074	> 5000	0.6	> 5000	> 5000	–
Z-Phe-Phe-CHN_2	23	374	–	0.4	3
Z-Phe-Ala-CHN_2	2	57	–	1.0	–
Z-Phe-Tyr(*O*-but)-CHN_2	–	7000	–	0.3	–
Z-(iodo-Tyr)-Ala-CHN_2	–	2.5	–	0.06	> 7000
Iodoacetic acid	30	156	592	–	–
***K_i* (nM)**					
Human cystatin A	0.019	8.2	0.31	1.3	NI
Human cystatin B	0.12	73	0.58	0.23	NI
Human cystatin C	0.005	0.25	0.28	0.005	NI
Human kininogen	0.015	600	1.2	0.017	1.0
Chicken ovocystatin	0.005	1.7	0.06	0.019	NI

[a] Data from refs. (3, 39–44).
[b] NI, not inhibited; –, no data; *O*-but, orthobutyl.

CA074 which binds in the same direction as substrate, but occupies the primes subsites (36). The inhibition of several papain family proteases by E-64 and its relatives is quantified in *Table 3*. The degree of selectivity is minimal, with the exception of CA074 , which shows remarkable specificity for cathepsin B compared with other members of the family (37). The compounds are relatively non-toxic and appear to be completely specific for papain family proteases under commonly used conditions. At very high concentration (> 0.1 mM) E-64 may inhibit trypsin (38) or the cysteine protease R-gingipain, although binding is weak and reversible (J. Potempa, personal communication). Since their rate of reaction with protein or low molecular weight thiols other than the active cysteine of a protease is negligible, peptide epoxides are excellent inhibitors in the presence of reducing agents used to activate cysteine proteases. This is an additional advantage over sulphydryl alkylating agents such as iodoacetate.

5.2.3 Fluoromethyl ketones

The fluoromethyl ketones (CH_2F, frequently abbreviated FMK) are useful specific cysteine protease inhibitors, and should not be confused with serine protease specific trifluoromethyl ketones (CF_3). Originally Z-Phe-Ala-CH_2F was used to target the lysosomal cysteine protease cathepsin B (45). CH_2F compounds have the advantage of being less reactive with serine proteases and far more stable in the presence of thiols than CH_2Cl equivalent compounds. They are cardiotoxic, limiting their usefulness *in vivo*, but experiments in cell culture and *in vitro* have

Table 4 Selectivity in caspase inhibitors

	Half-life (s) for inhibition at 1 μM inhibitor								
Inhibitor[a]	**Casp-1**	**Casp-2**	**Casp-3**	**Casp-4**	**Casp-5**	**Casp-6**	**Casp-7**	**Casp-8**	**Casp-9**
Z-VAD-FMK	2.5	2400	43	130	5.3	98	39	2.5	3.9
K_i (nM)									
Ac-DEVD-CHO	18	1710	0.23	132	205	31	1.6	0.92	60
Ac-IETD-CHO	<6	9400	195	400	223	5.6	3280	1.05	108
Ac-YVAD-CHO	0.76	NI	NI	362	163	NI	NI	352	970

Data from ref. (47).

[a]VAD, Val-Ala-Asp; DEVD, Asp-Glu-Val-Asp; IETD, Ile-Glu-Thr-Asp; YVAD, Tyr-Val-Ala-Asp. NI, $K_i > 10\,000$ nM.

been extremely informative. The good specificity of CH_2F compounds has been exploited particularly well with another family of cysteine proteases, completely unrelated to the cathepsins, namely the caspases. The caspase-specific inhibitor Z-Val-Ala-Asp-CH_2F has been used in many studies to establish the role of these enzymes in apoptosis (46). Unfortunately, this compound does not discriminate well between caspases, and, as demonstrated in *Table* 4, extreme caution must be exercised in interpreting results because of the non-discriminatory nature of this frequently used compound.

5.2.4 Acyloxymethyl ketones

The AOMK unit ($-COCH_2OCOR$), containing a weak leaving group, functions as an irreversible affinity label. Inhibitors with this unit are remarkably selective for cysteine proteases. Extensive work on a number of leaving groups within the frame Z-Phe-Ala-CH_2OCOR demonstrate that R = 2,6-$(CF_3)_2$-Ph provides an extremely powerful and rapid cathepsin B inhibitor, primarily because this unit is such a good leaving group (48). As with the fluoromethyl ketones (see above) the AOMKs react with members of the papain family and the caspases, depending on specificity dictates in the substrate subsites, and so do not distinguish between these separate families. The general order of effectiveness is similar to the FMKs, and from a practical viewpoint there is presently little to choose between them.

5.2.5 Cystatins

The cystatin superfamily of cysteine protease inhibitors has been divided into three families based on primary amino acid sequence homology and domain structure (49); the cystatins are tight-binding reversible inhibitors of papain family cysteine proteases and are not known to inhibit other families of cysteine proteases. The structure of cystatin B in complex with papain reveals three non-contiguous loops of the inhibitor, creating a three dimensional wedge that blocks the substrate cleft of the enzyme (50), and provide a pleasing explanation of why cystatins do not inhibit other protease families.

5.2.6 Inhibitor selectivity for cysteine proteases

Neither the cystatins nor the peptide epoxides show a significant degree of selectivity for individual proteases (see *Table 3*), with the notable exception of the cathepsin B-inhibitor CA074, and the possible exception of human plasma kininogen. This protein contains three cystatin-like tandem repeats of which the third one, which can be released by limited digestion with trypsin, has a very high affinity for cathepsin L and a low affinity for cathepsin B but does not inhibit calpains (43).

Papain family proteases tend to lack a well-defined primary specificity. This has hampered attempts to design inhibitors with selectivity for proteases, with the result that random approaches have proved more effective than directed ones. Several of these proteases seem to have a favourable interaction with hydrophobic amino acids in the P_2 position, a finding that aided Shaw and co-workers (33) in their design of peptide diazomethanes. The most selective diazomethane inhibitors favour cathepsin L (see *Table 3*). The problem of differentiating the activities of cathepsin B and L in cell culture may be overcome with the aid of compounds such as Z-(iodo-Tyr)-Ala-CHN_2, which shows promising selectivity for cathepsin L (51).

Occupancy of the S_1 pocket of caspases by Asp is definitive of this family and in stark contrast to the papain family. This is exploited by using peptides containing Asp adjacent to the warhead, and the currently most useful reagent is Z-Val-Ala-Asp-FMK, which inhibits most caspases (see *Table 6*). The Asp-OMe derivative is reportedly cell-penetrating, and has been used in a large number of studies to ablate caspase activity in cell culture (46). However, caution should be exercised in attributing caspase activity solely on the basis of the use of Z-Val-Ala-Asp-FMK, since this compound can inhibit the proteasome, at least in vitro (G. Salvesen, unpublished data). Perhaps more useful from the objective of distinguishing between caspases are some of the aldehydes described in *Table 4*. Here, it is fairly easy to distinguish the pro-inflammatory caspase 1 from the apoptotic ones (caspases 3, 6, 7, 8, and 9) by using Ac-Tyr-Val-Ala-Asp-CHO since there is over a 100-fold difference in K_i. However, the frequently used Ac-Asp-Glu-Val-Asp-CHO only demonstrates less than 10-fold selectivity between caspases 3, 7, and 8, and is therefore of dubious value in dissecting apoptotic pathways. Presumably the same would be true for outer warheads placed on the -Asp-Glu-Val-Asp- frame.

5.3 Proteasome

The proteasome presents a particular problem in inhibitor design owing to its ability to react with several warheads, including aldehydes, chloromethyl ketones, isocoumarins and even diazomethanes (52) which were originally designed for other proteases. Many investigators use tripeptide aldehydes such as Ac-Leu-Leu-nLeu-CHO or Ac-Leu-Leu-Val-CHO in cell culture to demonstrate an involvement of the proteasome, however, these may crossreact with calpain or lysosomal papain family members. Therefore, the use of lactacystin is recommended. This reagent, a microbial natural product has become widely used. Lactacystin binds some, but not all, of the catalytic subunits of the 20S proteasome and ablates proteolysis. The mechanism is complex, and varies with individual catalytic subunits, but effectively terminates all proteolytic activity (53). It is a cell-penetrant compound and is currently thought to be proteasome-specific.

Another inhibitor warhead based on peptidyl vinyl sulphones (54) has recently shown promise in proteasome investigations. Although also reactive with members of the papain family of cysteine proteases (55), biotinylated tripeptide or tetrapeptide vinyl sulphones may be used as affinity labels for individual proteasome subunits (54).

5.4 Metalloproteases

5.4.1 Peptidic inhibitors

The design of class-specific inhibitors of metalloproteases has focused on attempts to chelate the catalytic zinc atom. Synthetic inhibitors, therefore, commonly contain a chelating moiety, such as a carboxyl, a thiol, a phosphorous or a hydroxamic acid group, to which is attached a series of other groups designated to fit the specificity pocket of a particular metalloproteases. Examples are shown in *Table 5*. Some naturally occurring low molecular weight inhibitors such as actinomin, which is an aminopeptidase M inhibitor found in actinomycetes, also contain a hydroxamate moiety.

Many hydroxamate inhibitors originally targeted towards matrix metalloproteases (MMPs or matrixins) have been designed. Although selective inhibitors towards particular MMPs have been successfully generated (56,57) many also inhibit metalloproteases other than MMPs. For example, tumour necrosis factor α-converting enzyme (a member of the ADAM/reprolysin family) is inhibited by hydroxamate inhibitors originally designed for MMPs (58–60). Useful distinctions between MMPs and other family members of metalloproteases can be made by the use of TIMPs, which tend to be specific for MMPs.

5.4.2 Protein inhibitors

The most widely used protein inhibitors of metalloproteases are the TIMPs of which four (TIMPs 1–4) have been described. TIMPs inhibit MMPs, but not metalloproteases belonging to other families and are therefore useful reagents to distinguish MMPs from other families. TIMPs and MMPs form a tight 1:1 complex with subnanomolar K_i values (61). The crystal structure of the complex of TIMP-1 and the catalytic domain of MMP-3 revealed that the α-aminoacyl group of Cys1 bidentately chelates the catalytic zinc and the side-chain of Thr2 extends into the Large S_1' pocket of MMP-3 (62). Because of their slow dissociation rates, they are suitable for active site titration of MMPs. However, TIMP-1 fails to inhibit the membrane-bound MT1-MMP (MMP-14) whereas TIMP-2 and TIMP-3 inhibit MT1-MMP.

Other protein inhibitors of metalloproteases are not so widely used as TIMPs,

Table 5 Potentially useful synthetic metalloprotease inhibitors

Compound	Target enzyme	K_i (nM)	IC_{50} (nM)
(i) Carboxylates			
N-[1-(S)-carboxy-3-phenylpropyl]-Leu-Trp	Thermolysin		300
N-[1-(S)-carboxy-3-phenylpropyl]-Ala-Pro (enalapril)	ACE	0.05	
(ii) Thiols			
D-HS-CH_2CH(Me)CO-L-Pro (captopril)	ACE	1.3	
HS-CH_2-CH(CH_2ph)-CO-Gly	ACE	140	
(Thiorphan)	Neprilysin	4	
Ac-Trp-Leu-SH	MMP-1		2000
(iii) Phosphorus-containing inhibitors			
Rhamnosyl-PO_2-Leu-Trp (Phosphoramidon)	Neprilysin Thermolysin Pseudolysin	3 28 250	
$C_6H_5CHCH_2$-PO_2-Ala-Pro	ACE	0.5	
Phthaloyl-N-$(CH_2)_4$-PO_2-Ile-(β-naphthyl)-Ala-NH-CH_3	MMP-3	7	
Pro-L-Phe (PO_2CH_2)Gly-Pro	Neurolysin Thimet oligopeptidase	4 8100	
(iv) Hydroxamates			
HONHCOCH$_2$CH(iBu)CO-Trp-NHMe	MMP1	0.4	
(GM6001)	Thermolysin Pseudolysin	20 200	
Batimastat™	MMP-1 MMP-2 MMP-3 MMP-9 Atrolysin C		3 4 20 1 6
Actinonin	Meprin A Astacin	135 130 000	

[a] ph, phenyl. ACE, angiotensin converting enzyme.

but originate from organisms throughout the biotic kingdoms. For example, oprin is a single-chain glycoprotein of 52 kDa isolated from opossum (63), which inhibits snake venom metalloprotease atrolysin AB (formally known as Ht-b) from *Crotalus atrox*. The serum of Japanese Habu snake *Trimerasurus flavoviridis* contains a 323-amino acid glycoprotein that inhibits several metalloproteases in Habu venom. The molecule contains two copies of a cystatin-like domain (64), but it does not inhibit cysteine proteases. A 12-kDa inhibitor from *Streptomyces nigrescens* TK-23 inhibits thermolysin and *Pseudomonas aeruginosa* metalloproteases (65). Microbial proteinaceous inhibitors of metalloproteases are also found in *Erwinia crysanthemi* (66), in *Serratia marscescens* (67) and in *P. aeruginosa* (68).

Unlike its counterpart in mammals, the ovomacroglobulin from chicken eggs (ovostatin) inhibits only metalloproteases well. Ovostatins from other species show broader reactivity with proteases (69).

5.5 Aspartic proteases

The best characterized aspartic proteases from mammals (pepsin, chymosin, cathepsin D and renin) are all inhibited by pepstatin A, a pentapeptide-like compound secreted by *Streptomyces* species, which contains two residues of the unusual amino acid statine [(3S,4S)-4-amino-3-hydroxy-6-methylheptanoic acid]. K_i values range from 4.5×10^{-11} M for pepsin to 5×10^{-7} M for renin, and the original pepstatin structure has been modified in many ways to increase selectivity for different aspartic proteases (11). Renin is an extremely selective enzyme (its only known protein substrate is angiotensinogen) and the other mammalian aspartic proteases encountered during protein isolation are inactive above pH 6. This class of enzymes is unlikely to cause problems during protein isolations. A tremendous amount of work has gone into designing potent, selective, and bioavailable inhibitors of HIV protease, and the topic will not be covered in this chapter.

6 Inhibitors as active-site titrants

Often, one must determine the concentration of active enzyme in a crude, or pure but partially inactive, preparation of a protease. This is most accurately accomplished using specific 'burst titrants' (substrates that are turned over only once). However, these titrants are available for only certain serine proteases and require relatively large quantities of protein (about 1–10 μM enzyme per assay). If material is precious, or burst titrants are unavailable, then titrations using protease inhibitors may be employed. To successfully determine the active protease concentration one should use an active-site-directed inhibitor that: (i) has defined purity, (ii) is relatively stable in aqueous solution, (iii) shows no turnover during inhibition, (iv) reacts quickly and tightly enough to allow essentially complete inhibition in a short time. The most reliable inhibitors for performing active-site titrations are, in principle, irreversible ones.

Not all appropriate inhibitor titrants react sufficiently quickly to be useful in determining small concentrations of proteases. (Refer to Box 5.1 to calculate the minimum reliable concentration, assuming the association rate is known.) Affinity titrants can only rarely be used in crude samples because other proteins may sequester the titrant to give an erroneously high estimate of the target protease concentration. For crude samples, consider using a radioactive or biotinylated inhibitor affinity label followed by SDS-PAGE and quantification of bound label.

6.1 Cysteine proteases

Given their broad specificity for papain family members and fast-reacting mechanism the *trans*-epoxysuccinyl peptides are almost ideal active-site titrants. Compound E-64 is used to titrate these cysteine proteases primarily because of its ready availability. The procedure, taken from Barrett *et al.* (44), is applicable to the lysosomal proteases cathepsins B, H, L and S, and the plant proteases papain, chymopapain, ficin and bromelain, but not clostripain; it is carried out as described in Protocol 2. Caspases are not inhibited by E-64, and the authors recommend determining the active concentration of this family by using Z-Val-Ala-Asp-CH_2F, which is the broadest currently available caspase inhibitor. See *Table 4* for details.

Protocol 2

Active-site titration of cysteine proteases (papain family or caspases)

Equipment and reagents

Papain family (C1)

- Inhibitor titrant, E-64 (the unmethylated derivative, also known as E-64C). E-64D is the methylated derivative and is only appropriate for cell culture experiments
- Substrate, as appropriate for the protease being titrated, and depending on the available method of detection
- Assay buffer, as appropriate, noting that many papain family enzymes are unstable at neutral pH. Many assay buffers call for detergents such as Brij 35 or Triton X-100 in the activation buffer; the authors recommend peroxide-free grades of these as many commercial grades contain oxidants which inactivate cysteine proteases. Use freshly dissolved DTT.

Caspase family (C14)

- Inhibitor titrant, Z-Val-Ala-Asp-FMK (the unmethylated derivative)
- Substrate, Ac-Asp-Glu-Val-Asp-R (for caspases 2, 3, 6, 7, 8, 9 or 10), or Ac-Tyr-Val-Ala-Asp-R (for caspases 1, 4 or 5), where R = pNA, AMC or AFC, depending on the method of detection available
- Caspase buffer, 20 mM PIPES, 100 mM NaCl, 10% sucrose, 0.1% Chaps, 1 mM EDTA, 10 mM DTT,[a] pH 7.2 (use freshly dissolved DTT).

The assays are carried out in a total volume of 100 μl for detection in a 96-well plate reader, preferably operating in the kinetic mode. For analysis requiring larger volumes, increase proportionally.

Method

1. Prepare working solutions of inhibitor titrant at 1–10 μM in H_2O, as required, from a 1 mM stock. The E-64 stock is held in H_2O at −20 °C and is stable for at least 1 year, the Z-Val-Ala-Asp-FMK stock is in dry DMSO and is stable at −20 °C for at least 3 months. Discard the Z-Val-Ala-Asp-FMK stock no more than 12 h after thawing.
2. Activate the enzyme to be measured in the usual way in assay buffer (10 min at 37 °C is usually sufficient).
3. Add 20 μl of each working solution of inhibitor titrant to 30 μl of enzyme solution (≈ 3 μM) and incubate for 30 min at 37 °C.
4. Assay samples from each reaction mixture using 50 μl of an appropriate substrate in assay buffer to determine the residual activity. Plot residual activity as a function of inhibitor concentration.

Protocol 2 continued

5 The plot should be linear with an intercept on the *x*-axis equal to the concentration of active enzyme. Curvature of the plot as it approaches this axis indicates that the reaction has not gone to completion. This can be overcome by longer incubation or by increasing the concentrations of enzyme and inhibitor. Alternatively, the concentrations of the enzyme can be extrapolated by a straight line drawn from partial inhibition.

[a] May be substituted by 20 mM β-mercaptoethanol.

6.2 Serine proteases

The general serine protease-selective inhibitors DFP, PMSF and 3,4-DCI are too unstable in aqueous solutions to be used as active-site titrants. The protein inhibitors can, however, be useful for this purpose. A preparation of inhibitor is standardized using trypsin with known activity, by using the burst-titrant procedure of Chase and Shaw (70). Trypsin acts as the primary standard upon which other inhibitors may be standardized for secondary titrations. The standardized inhibitor is then used to titrate activity of proteases known to react well with it (*Table 2*).

Protocol 3

Active-site titration of serine protease using α_1-protease inhibitor.

Equipment and reagents

- Inhibitor titrant, α_1-protease inhibitor
- Substrate, depending on the protease being titrated
- Assay buffer, 50 mM Tris-HCl buffer, pH 8.0. containing 0.2 M NaCl

The assay is carried out in a total volume of 100 μl for detection in a 96-well plate reader, preferably operating in the kinetic mode. For analysis requiring larger volumes, increase proportionally.

Method

1 Working solutions of α_1-protease inhibitor (10–300 μg/ml) are prepared in assay buffer.

2 Add 25 μl of each inhibitor solution to 75 μl of a freshly prepared trypsin solution (40 μM in the assay buffer) and incubate at 37 °C for 15 min.

3 Assay 50 μl samples from each reaction mixture by adding 50 μl of an appropriate substrate in assay buffer, and plot residual activity against volume of inhibitor. The intercept on the *x*-axis will give the volume of the inhibitor solution that contains 4 nmol of active α_1-protease inhibitor.

Protocol 3 continued

4 The lack of selectivity of α_1-protease inhibitor allows titration of a large number of serine proteases in a similar manner to the cysteine protease titration above. The authors routinely use α_1-protease inhibitor to titrate chymotrypsin, neutrophil elastase and cathepsin G.

5 This method of titration can be applied, in principle, to many inhibitors provided that they form tight complexes with the target proteases. We recommend against the use of α_2M as a general protease titrant, as its ratio of binding varies between 1 and 2 mol of protease per mole of α_2M, depending both on the protease and on the experimental conditions.

7 Protease inhibitors in cell culture

Many of the forgoing procedures can be adapted from purified systems to cell culture (*Table 6*). The key is to use inhibitors that are cell-penetrating and stable to culture conditions. Often, these inhibitors are methylated to block acidic moieties and improve cell penetration and, whenever possible, it is preferable to use the methylated derivative. The methylated compounds are frequently inactive *in vitro*, and it is assumed that cellular esterases demethylate them to the active inhibitor. They are thus technically considered pro-drugs. Success has been reported with most types of proteases, and some examples are included below. Since the extent of cell penetrance by these compounds is not known, it is best to conduct several experiments at different inhibitor concentrations to determine the minimal realistic concentration required to achieve a measurable result.

Table 6 Inhibitors for cell culture

Proteases targeted	Inhibitor	Comments
Papain family	E-64D (methylated derivative of E-64) CA074-methyl ester Gly-Phe-CHN_2	Broadly reactive with members of the papain family Selective cathepsin B pro-inhibitor Selective for dipeptidyl peptidases over endopeptidases
Caspase family	Z-Val-Ala-(OMe)Asp-FMK	Non-selective, inhibits apoptosis and IL1β processing
Proteasome	Lactacystin	Reputedly, very selective
Calpains	Ac-Leu-Leu-nLeu-CHO[a] Ac-Leu-Leu-Met-CHO[b]	Possible cross-reactivity with proteasome.
Serine proteases	CH_2Cl derivatives	Depending on sequence, can be selective, but watch for cross-reaction with cysteine proteases and proteasome
Metallo-proteases	Batimastat	Broadly reactive with MMPs Also inhibits ADAMS

[a] Also known as calpain inhibitor 1.

[b] Also known as calpain inhibitor 2.

8 Suppression of proteolysis

There are many instances in which proteolysis has been identified as a cause of artefacts. The problem usually arises from the disruption of cells or tissues during analytical or purification procedures. Certain procedures are highly prone to adventitious proteolysis, including the isolation or identification of structural proteins and intracellular protein precursors following cell lysis, preparations of immunoprecipitates (IgG preparations are often contaminated by plasma kallikrein) from cell extracts and preparation of protein samples for SDS-PAGE. Denatured proteins are much more sensitive to proteolysis than native ones, and several proteases are active in 1% SDS.

Classically, biologists have included protease inhibitors to overcome some of these problems, and it is probably a good idea to incorporate this as part of a standard procedure when proteolytic artefacts are suspected. Therefore, the compounds in *Table 7* are recommended as general protease inhibitors for suppress-

Table 7 Inhibitors for prevention of adventitious proteolysis

	Inhibitor solutions		
Compound	**Stock solution**	**Working concentration**	**Catalytic class**
EDTA	0.5 M in H_2O, pH 7	5.0 mM	Metallo-
1,10-Phenanthroline	0. 1 M in DMSO	1.0 mM	Metallo-
3,4-DCI	5.0 mM in DMSO	0.1 mM	Serine/threonine
PMSF	0.1 M in dry PrOH	1.0 mM	Serine
E-64	1.0 mM in H_2O	0.02 mM	Cysteine
Iodoacetic acid	1.0 M in H_2O (fresh)	10 mM	Cysteine

EDTA can be omitted if the use of a calcium chelator presents a problem. 3,4-DCI is a much faster inhibitor of serine proteases than PMSF, but not as widely tested. Iodoacetic acid modifies cysteine residues on many other proteins. Therefore, these three compounds are used only if failure of the following cocktail is indicated.

General inhibitor cocktail solution from mixture of above stocks:

10 mM 1,10-phenanthroline

2 mM 3,4-DCI

0.4 mM E-64

Make up in H_2O (except for 1,10-phenanthroline) and use within 1 h. Dilute 20-fold into sample. Make up 1,10-phenanthroline in DMSO, as it is not soluble in aqueous solvents above 1 mM.

If samples are prepared from animal cell lysates, caspases may be a problem. Since E-64 does not inhibit caspases, lysates are treated with 20 μM (final concentration) of Z-Val-Ala-Asp-FMK (unmethylated derivative). This only applies when caspases are suspected to be a problem, and when iodoacetic acid cannot be used.

If checking for proteolytic artefacts arising from SDS-reducing gel electrophoresis, first add reducing agent in presence of cocktail and sample. Incubate 10 min, then add SDS. This will eliminate activation of cysteine proteases during simultaneous addition of SDS and reducing agent.

Boehringer supply a general inhibitor reagent which is a mixture of some different compounds. Although we have not compared its effectiveness with the above reagents, it should also prove acceptable when used according to the manufacturers recommendations.

ing proteolysis. They are mostly irreversible general inhibitors of low molecular weight that can readily be removed without regeneration of proteolytic activity. Occasionally, protein inhibitors such as aprotinin or soybean trypsin inhibitor may be added, but these are far less broad in their specificity than 3,4-DCI or PMSF. Some soybean inhibitor preparations are contaminated with metalloprotease activity, and the addition of the protein inhibitors is discouraged unless one wishes to inhibit a specific protease with which they are known to react well. Benzamidine (an inhibitor of trypsin-like serine proteases) is not recommended as its K_i values (0.01–1 mM) with various enzymes require it to be used at 10–1000 mM to be an effective inhibitor. The peptide aldehydes, such as leupeptin or antipain, should be used in the 1–100 mM range for complete inhibition but, since they are reversible inhibitors, proteolytic activity may reappear upon dilution or dialysis. Aspartic proteases do not present a problem as they are inactive at neutral pH, with the exception of renin which, however, is not a problem because of its highly restricted specificity. Their zymogens can, however, be stable at this pH and can be activated by a decrease in pH to below 5. If a procedure calls for decreasing the pH, and if aspartic proteases may be present, one should include 10 μg/ml pepstatin.

9 Therapeutic value of protease inhibitors

A major incentive in inhibitor research is that control of proteolysis is a valid pharmacological principle. Based on successes (and failures) of inhibitor therapy, Schnebli and Braun (71) noted common properties of pharmacologically effective protease inhibitors. First, the target enzyme for therapy must be precisely known. Second, the target enzyme must be in a compartment (plasma, lung lumen or peritoneum) easily accessible by common drug delivery routes. Third, the interaction between inhibitor and protease must be either irreversible or have $K_i < 10^{-10}$ M in order to maintain inhibitory activity following administration. Inhibitors have indeed proved useful in controlling pathogenesis in many animal models of proteolysis, and their application to human counterparts of these models shows promise. For example, captopril, a potent inhibitor of angiotensin I-converting enzyme is in therapeutic use for hypertension. A recent advancement is the design of a meta-substituted benzofused macrocyclic lactam that inhibits both ACE and neprilysin with IC_{50} values of 4 nM and 8 nM, respectively (72). The design of such dual inhibitors is considered to be a potent antihypertensives resulting from an increase of the circulating levels of atrial natriuretic peptide due to the inhibition of neprilysin and a decrease of angiotensin II due to the inhibition of ACE. Clinical trials with several synthetic inhibitors of coagulation and fibrinolytic enzymes, and those for MMPs in cancer metastasis and arthritis, are in progress. Successes with AIDS therapy by using inhibitors of HIV protease has led to searches for therapy of other viral diseases by selective targeting of viral processing enzymes (73). Significantly, the success with HIV protease inhibitors indicates that intracellular compartments are no longer a practical barrier to inhibitor therapy.

References

1. Laskowski, M. and Kato, I. (1980). *Annu. Rev. Biochem.* **49**, 593.
2. Travis, J. and Salvesen, G. S. (1983). *Annu. Rev. Biochem.* **52**, 655.
3. Barrett, A. J. and Salvesen, G. (1986). *Proteinase inhibitors*. Elsevier Science Publ., Amsterdam.
4. Bode, W. and Huber, R. (1992). *Eur. J. Biochem.* **204**, 433.
5. Bieth, J. (1974). *Bayer-symposium V 'proteinase inhibitors'* (ed. Fritz, H., Tschesche, H., Greene, L. J. and Truscheit, E.) p. 463. Springer-Verlag, Berlin.
6. Abrahamson, M., Barrett, A. J., Salvesen, G., and Grubb, A. (1986). *J. Biol. Chem.* **261**, 11282.
7. Bieth, J. G. (1984). *Biochem. Med.* **32**, 387.
8. Barrett, A. J. and Starkey, P. M. (1973). *Biochem. J.* **133**, 709.
9. Sottrup-Jensen, L. (1987). In *Plasma proteins* (ed. Putnam, F. W.), p. 191. Academic Press, London.
10. Salvesen, G. S., Sayers, C. A., and Barrett, A. J. (1981). *Biochem. J.* **195**, 453.
11. Rich, D. H. (1986). In *Proteinase inhibitors* (ed. Barrett, A. J. and Salvesen, G.), p. 153. Elsevier Science Publ., Amsterdam.
12. Umezawa, H. (1982). *Annu. Rev. Microbiol.* **36**, 75.
13. Powers, J. C. and Harper, J. W. (1986). In *Proteinase inhibitors* (ed. Barrett, A. J. and Salvesen, G.), p. 55. Elsevier Science Publ., Amsterdam,.
14. Schultz, R. M., Varma-Nelson, P., Ortiz, R., Kozlowski, K. A., Orawski, A. T., Pagast, P., and Frankfater, A. (1989). *J. Biol. Chem.* **264**, 1497.
15. Petra, P. H., Cohen, W., and Shaw, E. N. (1965). *Biochem. Biophys. Res. Commun.* **21**, 612.
16. Kettner, C. and Shaw, E. (1981). *Methods Enzymol.* **80**, 826.
17. Larsen, K. S. and Auld, D. S. (1991). *Biochemistry* **30**, 2613.
18. Mallya, S. K. and van Wart, H. E. (1989). *J. Biol. Chem.* **264**, 1594.
19. Beynon, R. J. (1987) *Methods in molecular biology*, 1–1. Humana Press, Clifton.
20. Powers, J. C., Tanaka, T., Harper, J. W., Minematsu, Y., Barker, L., Linclon, D., and Crumley, K. V. (1985). *Biochemistry* **24**, 2048.
21. Laura, R., Robison, D. J., and Bing, D. H. (1980). *Biochemistry* **19**, 4859.
22. James, G. T. (1978). *Anal Biochem* **86**, 574.
23. Diatchuk, V., Lotan, O., Koshkin, V., Wikstroem, P., and Pick, E. (1997) *J. Biol. Chem.* **272**, 13292.
24. Harper, J. W., Hemmi, K. and Powers, J. C. (1985). *Biochemistry* **24**, 1831.
25. Lively, M. O. and Powers, J. C. (1978). *Biochim. Biophys. Acta* **525**, 171.
26. Cohen, J. A., Oosterbaan, R. D., and Berends, F. (1967) *Methods Enzymol.* **11**, 686.
27. Rusbridge, N. M. and Beynon, R., J. (1990). *FEBS Lett.* **268**, 133.
28. Laskowski, M., Kato, I., Ardelt, W., Cook, J., Denton, A., Empie, M. W. *et al.* (1987). *Biochemistry* **26**, 202.
29. Huber, R. and Carrell, R. W. (1989). *Biochemistry* **28**, 8966.
30. Lijnen, H. R., Uytterhoeven, M., and Collen, D. (1984). *Thromb Res.* **34**, 431.
31. Powers, J. C. (1977). *Chemistry and biochemistry of amino acids, peptides and proteins*, pp. 1–65. Marcel Dekker, New York.
32. Kettner, C. A. and Shenvi, A. B. (1984). *J Biol. Chem.* **259**, 15106.
33. Zumbrunn, A., Stone, S. and Shaw, E. (1988). *Biochem. J.* **250**, 621.
34. McGuire, M. J., Lipsky, P. E., and Thiele, D. L. (1993). *J. Biol. Chem.* **268**, 2458.
35. Hanada, K., Tamai, M., Yamagishi, M., Ohmura, S., Sawada, J., and Tanaka, I. (1978). *Agric. Biol. Chem.* **42**, 523.
36. Yamamoto, A., Hara, T., Tomoo, K., Ishida, T., Fujii, T., Hata, Y. *et al.* (1997). *J Biochem* **121**, 974.

37. Katunuma, N. and Kominami, E. (1995). *Methods Enzymol* **251**, 382.
38. Sreedharan, S. K., Verma, C., Caves, L. S., Brocklehurst, S. M., Gharbia, S. E., Shah, H. N., and Brocklehurst, K. (1996). *Biochem. J.* **316**, 777.
39. Parkes, C., Kembhavi, A. A., and Barrett, A. J. (1985). *Biochem. J.* **230**, 509.
40. Crawford, C., Mason, R. W., Wikstrom, P., and Shaw, E. (1988). *Biochem. J.* **253**, 751.
41. Kirschke, H., Wikstrom, P., and Shaw, E. (1988). *FEBS Lett.* **228**, 128.
42. Towatari, T., Nikawa, T., Murata, M., Yokoo, C., Tamai, M., Hanada, K., and Katunuma, N. (1991). *FEBS Lett.* **280**, 311.
43. Salvesen, G., Parkes, C., Abrahamson, M., Grubb, A., and Barrett, A. J. (1986). *Biochem. J.* **234**, 429.
44. Barrett, A. J., Kembhavi, A. A., Brown, M. A., Kirschke, H., Knight, C. G., Tamai, M., and Hanada, K. (1982). *Biochem. J.* **201**, 189.
45. Rasnick, D. (1985). *Anal. Biochem.* **149**, 461.
46. Cohen, G. M. (1997). *Biochem. J.* **326**, 1.
47. Garcia-Calvo, M., Peterson, E. P., Leiting, B., Ruel, R., Nicholson, D. W., and Thornberry, N. A. (1998). *J. Biol. Chem.* **273**, 32608.
48. Krantz, A., Copp, L. J., Coles, P. J., Smith, R. A. and Heard, S. B. (1991). *Biochemistry* **30**, 4678.
49. Barrett, A. J., Rawlings, N. D., Davies, M. E., Machleidt, W., Salvesen, G., and Turk, V. (1986). In *Proteinase inhibitors* (ed. Barrett, A. J. and Salvesen, G.), p. 515. Elsevier Science Publ., Amsterdam.
50. Stubbs, M. T., Laber, B., Bode, W., Huber, R., Jerala, R., Lenarcic, B., and Turk, V. (1990). *EMBO J.* **9**, 1939.
51. Mason, R. W., Wilcox, D., Wiksrtom, P., and Shaw, E. N. (1989). *Biochem. J.* **257**, 125.
52. Reidlinger, J., Pike, A. M., Savory, P. J., Murray, R. Z., and Rivett, A. J. *J. Biol. Chem.* **272**, 24899.
53. Fenteany, G. and Schreiber, S. L. (1998). *J. Biol. Chem.* **273**, 8545.
54. Bogyo, M., Shin, S., McMaster, J. S., and Ploegh, H. L. (1997). *Chem. Biol.* **5**, 307.
55. Bromme, D., Klaus, J. L., Okamoto, K., Rasnick, D., and Palmer, J. T. (1996). *Biochem. J.* **315**, 85.
56. Bottomley, K. M., Borkakoti, N., Bradshaw, D., Brown, P. A., Broadhurst, M. J., Budd, J. M. *et al.* (1997). *Biochem. J.* **323**, 483.
57. Chander, S. K., Antoniw, P., Beeley, N. R., Boyce, B., Crabbe, T., Docherty, A. J. *et al.* (1995). *J. Pharm. Sci.* **84**, 404.
58. Mohler, K. M., Sleath, P. R., Fitzner, J. N., Cerretti, D. P., Alderson, M. *et al.* (1994). *Nature* **370**, 218.
59. Gearing, A., Beckett, P., Christodoulou, M., Churchill, M., Clements, J., Davidson, A. H. *et al.* (1994). *Nature* **370**, 555.
60. McGeehan, G. M., Becherer, J. D., Bast, R. C.J., Boyer, C. M., Champion, B., Connolly, K. M. *et al.* (1994). *Nature* **370**, 558.
61. Murphy, G. and Willenbrock, F. (1995). *Methods Enzymol.* **248**, 496.
62. Gomis-Rüth, F. X., Maskos, K., Betz, M., Bergner, A., Huber, R., Suzuki, K. *et al.* (1997). *Nature* **389**, 77.
63. Catanese, J. J. and Kress, L. F. (1992). *Biochemistry* **31**, 410.
64. Yamakawa, Y. and Omori-Satoh, T. (1992). *J. Biochem.* **112**, 583.
65. Seeram, S. S., Hiraga, K., Saji, A., Tashiro, M., and Oda, K. (1997). *J. Biochem.* **121**, 1088.
66. Letoffe, S., Delepelaire, P., and Wandersman, C. (1989). *Mol. Microbiol.* **3**, 79.
67. Kim, K. S., Kim, T. U., Kim, I. J., Byun, S. M., and Shin, Y. C. (1995). *Appl. Env. Microbiol.* **61**, 3035.
68. Duong, F., Lazdunski, A., Cami, B., and Murgier, M. (1992). *Gene* **121**, 47.
69. Nagase, H., Harris Jr, E. D., and Brew, K. (1986). *J Biol Chem* **261**, 1421.

70. Chase, T. J. and Shaw, E. (1967). *Biochem. Biophys. Res. Commun.* **29**, 508.
71. Schnebli, H. P. and Braun, N. J. (1986). In *Proteinase inhibitors* (ed. Barrett, A. J. and Salvesen, G.), p. 613. Elsevier, Amsterdam.
72. Ksander, G. M., de Jesus, R., Yuan, A., Ghai, R. D., McMartin, C., and Bohacek, R. (1997). *J. Med. Chem.* **40**, 506.
73. Babe, L. M. and Craik, C. S. (1997). *Cell* **91**, 427.

Chapter 6
Finding, purification and characterization of natural protease inhibitors

Hideaki Nagase
Kennedy Institute of Rheumatology Division, Imperial College School of Medicine, Hammersmith, 1 Aspenlea Road, London W6 8LH, UK

Guy S. Salvesen
The Burnham Institute, 10901 North Torrey Pines Road, La Jolla CA 92037, USA

1 Introduction

The activity of a proteolytic enzyme in a living organism is precisely regulated by synthesis and secretion of the enzyme, by zymogen activation and frequently by inhibition. Although some proteinase activities may be limited by very restricted substrate selectivity, or by specific cellular location (e.g. on the cell surface), most proteases are regulated by endogenous inhibitors. Thus, to understand the biological regulation of proteolysis, one must understand the role of endogenous inhibitors. In this chapter, we describe general approaches of how to discover, purify, and characterize endogenous protease inhibitors.

1.1 The meaning of inhibition

Any compound that decreases the measured rate of enzyme-catalysed hydrolysis of a given substrate is, in principle, an enzyme inhibitor. This definition applies to competing substrates in a reaction, for if two substrates are incubated with a protease, one can decrease the apparent rate of hydrolysis of the other and therefore appear to inhibit the enzyme. This is not a trivial point, as protease substrates have occasionally been misclassified as inhibitors. It is not our intention to make rigid definitions of what is and is not a protease inhibitor, as this has been explained before (1,2). Instead, we aim to give methods and examples for finding relatively tight binding and/or rapid, naturally occurring inhibitors of proteases. For readers who wish to assess the physiological function of protease inhibitors, ref. (1) is highly recommended.

2 Finding protease inhibitors

2.1 Screening of inhibitors from natural sources

In the initial search for proteinase inhibitors in tissue extracts or body fluids two points should be considered: (i) the concentration of the protease used for the assay, and (ii) apparent non-specific inhibition of the enzymatic activity by proteins in the extract.

When the enzyme concentration is much lower than the equilibrium dissociation constant K_i of inhibition, the apparent decrease in enzymatic activity by crude extract does not follow stoichiometric inhibition (see *Figure 3*). This may lead the investigator to ignore existence of a potential inhibitor. When the concentration of the enzyme in the assay is close to or higher than the K_i value, dose dependent inhibition is observed. On the other hand, low inhibitor concentrations in the crude tissue extract, especially if far below the test protease concentration, make it difficult to detect inhibition. Unfortunately, during screening of new inhibitors for a particular proteinase, K_i values are not known. It is generally found, however, that physiologically relevant inhibitors have K_i values of at most 10 nM. To detect low concentrations of an inhibitor it is, therefore, recommended that a low enzyme concentration is used, preferably less than 10 nM for initial screening.

Finding the richest source of an inhibitor simplifies purification. This can be simply tested by looking for percentage inhibition of the target protease in crude samples. Several dilutions of the sample should be tested to get a better idea of which starting material contains a larger amount of the inhibitor, and it is always best to dilute samples so that the testing material gives less than 75% inhibition. Specific inhibitory units (IU/mg protein) (see *Protocol 1*) are also a criterion to be considered. For a large number of samples, assay systems using microtitre plates facilitate the screening process. When the enzyme inhibition assay uses a protein substrate such as casein or gelatin, a possible inhibitory effect of contaminating α_2M must be taken into consideration when the sample is from an animal tissue. α_2M can be inactivated by treating with 0.2 M methylamine at pH 8.5 for 4 h at 23 °C.

Once inhibitory activity is detected in a crude sample, it is important to perform an initial simple fractionation of the inhibitor activity: e.g. ammonium sulphate, organic solvent, polyethylene glycol fractionation, gel filtration or ion exchange chromatography. This is to determine whether the specific activity of the inhibitor increases after such fractionation. A failure to increase the specific activity suggests that the initial apparent inhibition was due to endogenous substrates competing with the test substrate. An alternative test is to examine the inhibitory activity after reduction and alkylation of the sample, or after heat denaturation, which will destroy the folded structure of most protein inhibitors. However, some protein inhibitors of proteases are stable even after heat treatment at 95 °C for 15 min, e.g. TIMPs (3) and cystatins (4).

Protocol 1

Determination of specific inhibitory activity

1 Define one unit of the protease activity. In this example, one unit of the enzyme digests 1 μg of casein in 1 min at 37 °C.

2 Define one unit of inhibitory activity (IU), which is the amount of inhibitor sample to give 50% inhibition of two units of the protease. Note: since the linearity of enzyme inhibition depends on the enzyme concentration (see 4.4.1), the inhibition assay with the identical enzyme concentration must be used.

3 Dilute the inhibitor samples so that the enzymic activity is partially (25–75%) inhibited.

4 Calculate IU of the diluted sample in the assay from the percentage inhibition as follows:

$$\text{IU} = \text{units of enzyme} \times \frac{\text{A} - \text{B}}{\text{A}}$$

where A is the enzymic activity detected without inhibitor, and B is the enzymic activity with inhibitor.

5 Calculate the specific activity of the sample (IU/mg) as follows:

IU/the total protein amount (mg) of the sample used for inhibition assay.

2.2 Finding inhibitors by reverse zymography

Some proteinase inhibitors retain inhibitory activity after SDS-PAGE without reduction. Using this principle, Herron *et al.* (5) developed a reverse zymography technique to detect tissue inhibitors of metalloproteinase (TIMPs). In this technique the sample to be detected is run on SDS-polyacrylamide gel containing a protein substrate copolymerized into the gel. The gel is then incubated with the solution containing matrix metalloproteinases. The procedure was improved by Staskus *et al.* (6) by copolymerizing the substrate and the target enzyme in SDS-polyacrylamide gel. Since both the enzyme and the substrate are copolymerized in the acrylamide gel, they do not migrate significantly during electrophoresis of the sample. After electrophoresis, washing and incubating the gel with an appropriate buffer at 37 °C allows the protein substrate to be digested by the enzyme. Where inhibitor molecules are present, the protein substrate remains undigested and can be detected by staining for protein. *Protocol 2* describes the procedures for gelatin reverse zymography. Examples are shown in *Figure 1*.

This technique reveals not only how many species of inhibitors are present in the sample, but also molecular masses of the inhibitors and a semiquantitative description of their relative amounts. This is particularly useful for inhibitors that tightly bind to insoluble tissue matrices, because they can be extracted with SDS. TIMP-3 was discovered using this technique (7). However, this simple assay method requires that both enzyme and the inhibitor retain their activities after removal of SDS from the gel.

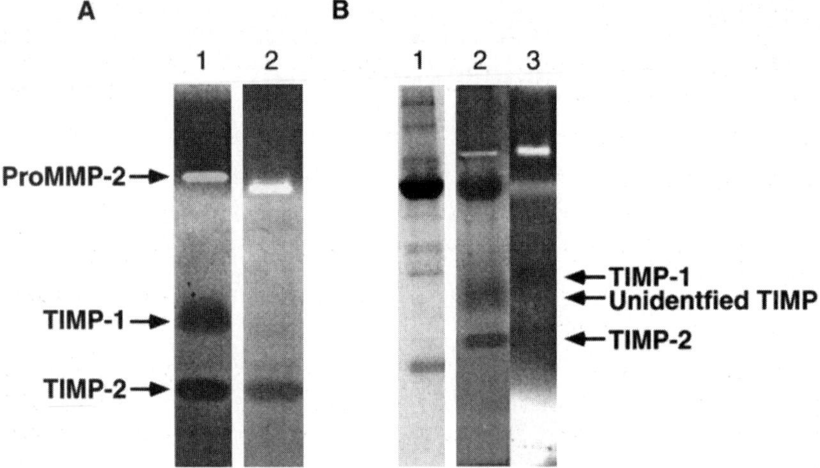

Figure 1 Reverse zymography. (A) Lane 1, crude culture medium (2.5 μl) of the phorbol ester-stimulated HT1080 fibroscarcoma cells, which contains TIMP-1 and TIMP-2; lane 3, purified proMMP-2-TIMP-2 complex. The complex is dissociated by treatment with SDS. Both TIMP-1 and TIMP-2 are detected as zones of purple-blue staining. (B) Identification of uncharacterized TIMP in human vitreous. Lane 1, 2.5 μl of vitreous, sample subjected to SDS-PAGE and stained with Coomassie Brillant Blue R250; lane 2, 2.5 μl of vitreous subjected to gelatin-reverse zymography; lane 3, the conditioned medium of human fibrosarcoma HT-1080 cells. The human vitreous sample (lane 2) exhibits three potential MMP-2 inhibition bands, but the top band corresponds to the major protein in vitreous in lane 1 and it is therefore unlikely to be an inhibitor of MMP.

Protocol 2

Reverse zymography for detection of matrix metalloproteinase inhibitors

Reagents

- Washing buffer: 50 mM Tris-HC1 (pH 7.5), 5 mM CaC_2, 5 μM ZnC_2, 0.02% NaN_3, 2.5% Triton X-100
- Incubation buffer: 50 mM Tris-HC1 (pH 7.5), 5 mM CaC_2, 5 μM ZnC_2, 0.02% NaN_3

Method

1. Make a 10% SDS-PAGE gel containing 0.8 mg/ml of gelatin (final conc.) and 0.5 μg/ml (final) of purified proMMP-2 (progelatinase A) or 6 U/ml (final activity in the gel; 1 U digests 1 μg gelatin at 37 °C in 1 min) of gelatinase A or B solution (e.g. conditioned-medium from cultured fibroblasts). It is not necessary to preactivate progelatinases, since SDS activates these zymogens.
2. Mix the sample with an equal volume of non-reducing loading buffer containing SDS. Do not boil the sample.

Protocol 2 continued

3 As standards, apply 5 μl of purified TIMP-1 and/or TIMP-2 (0.5 μg/ml) per lane or conditioned medium known to contain TIMPs.
4 Apply samples to the prepared gel, and run SDS-PAGE.
5 Wash the gels with washing buffer on a shaker for 15 min at room temperature four times.
6 Incubate the gels in the incubation buffer at 37 °C for 2–3 h without shaking.
7 Stain the gels with Coomassie Brilliant Blue R-250. Inhibitors are located as purple-blue bands with clear background.

Note: gelatin is stained in purple-blue. When dark-blue protein bands are detected, they are likely to be proteins in the sample rather than MMP inhibitors. In this case, run regular SDS-PAGE without reduction and compare the patterns of reverse zymography and Coomassie Blue-stained.

2.3 Finding inhibitors from DNA sequences

Recent advances in genome and EST sequencing have revealed a host of new proteins, and BLAST analysis of these and other databases are frequently used to identify new homologues of known proteins. This can obviously apply to protease inhibitors. To test for inhibition, one would need to pick a likely target enzyme, express the putative inhibitor, and analyse for inhibition. Rather than construct novel heterologous protein expression systems for production of proteins to be tested, we recommend a rapid screening system for protease inhibitors that relies on simple transcription/translation followed by analysis of binding of putative target proteases (*Protocol 3*). *Figure 2* shows the example of identification of caspace binding protein. After identifying binding partners, one can then move to more conventional expression strategies to understand the relevance of the binding phenomena.

Protocol 3

In vitro expression of protease inhibitors

1 Insert full-length cDNA into transcription/translation vector. T7-driven prokaryotic (e.g. pET23) or eukaryotic (e.g. pcDNA3) vectors are optimal since they allow use of T7-polymerase driven transcription.
2 Prepare plasmid mini-preps suitable for transcription.
3 Transcribe and translate the construct using a coupled transcription/translation system[a] employing a suitable radiolabelled amino acid.
4 Check for translation of the expected product by SDS-PAGE and autoradiography.
5 Immobilize equivalent amounts (in the range 5–50 ng) of potential target enzymes

Protocol 3 continued

on PVDF membrane, in a strip of closely spaced wells from a slot-blot apparatus. Block the membrane with an appropriate agent.

6 Incubate the blotted strip with the translation product for 45 min at 23–37 °C with gentle rocking. Drain, rinse, and autoradiograph the blot strip.

[a] Because of different transcription/translation regulators between prokaryotes and eukaryotes, it is important to match the vector to the transcription/translation system. Prokaryotic vectors are translated using the *Escherichia coli* T7S30 extract (Promega) or similar, and eukaryotic vectors are translated with the TNT T7 rabbit reticulocyte lysate (Promega) or similar.

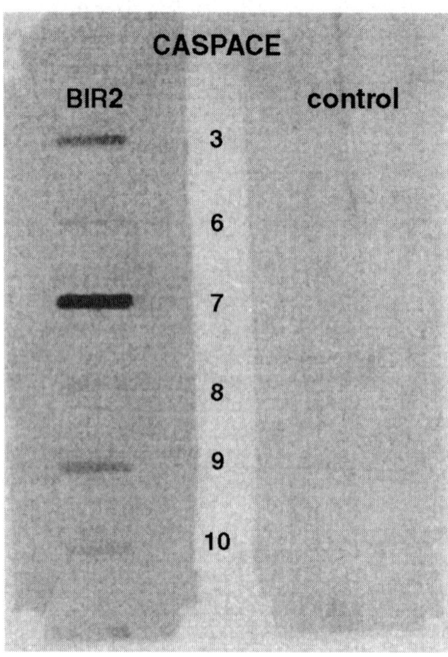

Figure 2 Detection of the binding of a specific caspase inhibitor. The second baculovirus inhibitor of apoptosis protein repeat (BIR) domain of a human inhibitor of apoptosis protein (XIAP) was ligated into pET23b, transcribed/translated using the *Escherichia coli* T7 S30 system in the presence of [^{35}S]Met, and used to probe a slot-blot containing 15 ng of each of the indicated purified caspases. The control strip contained another translated protein without caspase-binding activity. Both strips were autoradiographed using a phosphorimage scanner and binding specificity clearly demonstrated for caspases 3, 7 and 9.

This system detects protease-binding proteins, but it does not necessarily demonstrate inhibition. To demonstrate specifically that the binding protein is an inhibitor, it is necessary to express the protein. The constructs are usually cloned into a versatile *E. coli* or mammalian expression vector to simplify steps to express the protein. Ligation into vectors encoding an in-frame *N*-terminal or *C*-terminal His-6 tag is suitable for rapid purification of expressed protein. Since the His-tag can interfere with activity, both constructs are used to maximize success in the initial approach. In its simplest form the system allows detection, and even demonstrates the specificity, of inhibitors that do not require post-translational modifications. Therefore, it is useful for serpins and inhibitor of apoptosis proteins. However, it would not allow detection of proteins that require disulphide or other modifications (e.g. TIMPs, kazals and kunins). In this case, the use of coupled translation–translocation systems using dog microsomes may overcome the post-translational problem.

2.4 Combinatorial protease inhibitors

Although not directed towards finding new inhibitors, some recent approaches to protease and protease inhibitor specificity have focused on the identification of consensus sequences from combinatorial chemistry libraries or phage display. From the natural inhibitor viewpoint, success in altering specificity has been achieved using the frames of kunins (8) and cystatins (9) displayed as fusions on filamentous bacteriophage coat proteins. The technique of phage display is particularly attractive, because the good fit between protease specificity sites and complementary inhibitor side-chains in principle allows great selectivity to be achieved. It is most useful, not in obtaining extremely tight binding, but in obtaining exquisite selectivity of the type that may be difficult to achieve by synthetic inhibitors. Thus, positive selection and subtraction strategies have led to kunin derivatives remarkably selective for fibrinolytic proteases (10). Sometimes the technology has given surprising results: in kunin frames selected for chymotrypsin, the optimal P_1 residues were His and Asn, rather than the expected large hydrophobic types (11). For further exploration of the combinatorial methods, readers are directed to reviews that specifically address inhibitor phage (12,13).

3 Purification of natural protease inhibitors

3.1 Use of the target enzyme as an affinity ligand

Recombinant proteases mutated at the active site, or natural ones with a covalent modification to the catalytic nucleophile, can serve as ideal ligands for purification of inhibitors. In this case, a crude culture media or tissue extract can be applied directly to an inactive proteinase coupled to an insoluble support (e.g. Affi-Gel10, Bio-Rad labs), and the bound material can be eluted by a buffer at low or high pH, or by a chaotropic agent. SDS-PAGE analysis of the eluted material will often reveal how many species of endogenous inhibitor are present in the crude tissue extracts. This procedure was used to purify cystatins (14).

3.2 Conventional purification

In the absence of affinity purification, conventional purification procedures are used. To establish effective purification steps of an inhibitor it is important to determine the specific activity of the inhibitor sample after each purification step. When the inhibitor concentration is very high, almost complete inhibition of the enzyme activity is detected. In this case, the sample must be diluted to give only 25–75% inhibition of the enzymatic activity to determine the specific activity (see *Protocol 1*). This allows one to quantify the amount of inhibitor and provides a guidance for purity of the inhibitor. Specific inhibitory activity should increase after an effective purification step.

Many natural inhibitors of proteinases are proteins with molecular sizes ranging from 3000–800 000 Da. Microbial culture broth contains lower molecular weight inhibitors, and they are often peptidic in nature. Thus, isolation steps are similar to those of conventional protein purification: ammonium sulphate, organic solvent and polyethylene fractionations, gel filtration, ion exchange, affinity chromatographic, isoelectric focusing, preparative gel electrophoresis, etc. When a new inhibitor activity is found, it is advisable to test its stability by exposure to acidic and alkaline pH, higher temperature (60–100 °C) and treatment with urea or SDS. Intracellular proteinase inhibitors may be sensitive to oxidation. In this case, inclusion of 0.1–1 mM 2-mercaptoethanol, cysteine, or DTT may prevent oxidation. Knowing the stability of the inhibitor will provide purification strategies suitable for a specific inhibitor.

3.3 Reverse zymography

This technique allows localization of the inhibitor activity after SDS-PAGE. Elution from the gel after SDS-PAGE may be used to purify the inhibitor. However, the concentration of the inhibitor used for analytical reverse zymography is very low, and the inhibitor protein may not be detected by Coomassie Brilliant Blue staining (see *Protocol 2*). Thus, both preparative and analytical electrophoresis must be conducted in parallel. A good example is purification of TIMP-3 (6).

4 Characterization of inhibitors: inhibition kinetics

4.1 The importance of kinetics

The importance of a protease inhibitor is directly related to the efficiency with which it inhibits a protease. A given inhibitor may inhibit several proteases, and the relative efficiency is described by the kinetics of the interaction. Analysis of the kinetics of the reaction delineates the likely control point in complex biological media and gives crucial insights into the mechanism by which the inhibitor operates. The use of appropriate methods is essential, since inappropriate ones can lead to serious misconceptions.

4.2 IC_{50} and percentage inhibition.

The potency of an inhibitor is often described as IC_{50} (the molar concentration of the inhibitor that gives 50% inhibition of the target enzymatic activity) or simply the percentage inhibition of the enzymatic activity of a fixed concentration of the inhibitor. Such values are useful in following inhibitors during purification, giving qualitative analysis of the inhibitor under certain assay conditions. However, for an understanding of potency, mechanism and usefulness of the inhibitor, quantitative analyses is required. For example, the IC_{50} values are only useful when working at an enzyme concentration below the K_i and therefore usually only useful for poor inhibitors. Most good inhibitors have very low K_is and a simple IC_{50} value may represent just half the amount of the enzyme in the assay. In this case, the potency of the inhibitor cannot be evaluated, and equilibrium and/or rate constants are required. The two key kinetic constants that define protease–inhibitor interactions are the speed (usually described as association rate constant, k_{ass}) and strength (usually described as binding constant, K_i) of the interaction. In principle a good, or physiological inhibitor should react rapidly with its target enzyme to form a tight complex, with k_{ass} greater than 10^5/M.s, or K_i below 10^{-9} M. This section deals with simple and reliable methods to determine these kinetic parameters.

4.3 Practical inhibitor kinetics

Most protein inhibitors of proteinases are 'active-site-directed', i.e., they combine with the catalytic and substrate-binding sites of the proteinase to form a tight and stable complex (1,2,15). Typically, these natural or synthetic inhibitors mimic a substrate. However, good inhibitors have kinetic specificity constants (K_i) and turnover numbers (k_2) several orders of magnitude lower than the equivalent constants (K_m, k_{cat}) for comparable substrates (15). This means that proteinase inhibitors compete with substrate, and many are extremely tight-binding. Accordingly, tight-binding inhibitors cannot be evaluated by popular graphical treatments which are based on the general Michaelis–Menten equation (16,17). The overriding assumption in these treatments is that the total inhibitor concentration is not significantly decreased by binding to the enzyme, an assumption that is invalidated by the strong interaction between proteinases and inhibitors. Indeed, many proteinase inhibitors form essentially irreversible complexes with proteinases. Irreversible or tight-binding inhibitors decrease the number of active sites available for substrate. They therefore appear to act as non-competitive inhibitors when the reaction is analysed by the above treatments, despite the fact that their mechanism is strictly competitive with substrate. It is therefore necessary to analyse inhibition kinetics by using pre-steady-state treatments.

The key equation for proteinase inhibition, in simplified form, describes a bimolecular reaction between proteinase (E) and inhibitor (I), which results in the formation of a stable complex (EI) via an intermediate analogous to substrate binding (EI*) (Scheme 1).

$$E + I \underset{k_{-1}}{\overset{k_1}{\rightleftharpoons}} EI^* \xrightarrow{k_2} EI$$

Scheme 1 Reaction pathway

Definitions:

$$K_i = \frac{k_{-1}}{k_1} \qquad (1)$$

$$k_{ass} = \frac{k_2}{K_i} \quad (2)$$

The velocities of the reactions are described by the rate constants k_1 (bimolecular, second-order) and k_{-1} and k_2 (unimolecular, first-order).

Scheme 6.1 describes an irreversible reaction. Irreversible inhibitors, e.g. serpins (18) and baculovirus p35 (19), form covalent bonds with the enzyme and often these bonds remain when the enzyme is denatured. Inhibitors that show significant reversibility (aprotinin, ovomucoids, cystatins and TIMPs) can be evaluated by truncating the latter part of the above scheme, that is, by omitting the rate governing the formation of the EI complex (k_2). Sometimes an equilibrium reversible inhibitor may bind so tightly that it is not possible to measure a binding constant. Under this condition the inhibitor is sometimes classified as 'pseudo-irreversible' and it is then appropriate to analyse it as an irreversible inhibitor.

4.4 Reversible inhibitors

An equilibrium between the complexed form of an enzyme and free enzyme characterizes reversible inhibition. Although it is sometimes possible to evaluate more complex schemes, for most purposes the following simple relationship describes the process.

$$E + I \underset{k_{-1}}{\overset{k_1}{\rightleftharpoons}} EI$$

Scheme 2

The equilibrium constant for the reaction, K_i, is given by

$$K_i = \frac{[E][I]}{[EI]} = \frac{k_{-1}}{k_1} \quad (3)$$

Thus, the quantities that define a reversible inhibitor are K_i and the two rate constants.

4.4.1 Importance of $[E]_o/K_i$ and tight-binding inhibitors

Many protein inhibitors of proteinases exhibit tight binding. In this case, the ratio of $[E]_o/K_i$, where $[E]_o$ is the initial enzyme concentration, influences the binding behaviour of an enzyme and an inhibitor. The importance of $[E]_o/K_i$ is discussed by Bieth (20,21).

Figure 3 demonstrates the influence on inhibition with various inhibitor concentrations on constant enzyme concentrations at various $[E]_o/K_i$ values, and illustrates the following points:

(1) When $[E]_o/K_i \geq 100$, the inhibition curves are almost linear and intercept the abscissa at $[I]_o/[E]_o = 1$. Under these conditions, the binding is so tight that the inhibitor titrates the active site of the enzyme. In this case, K_i cannot be determined by conventional derivations of the Michaelis–Menten relationship.

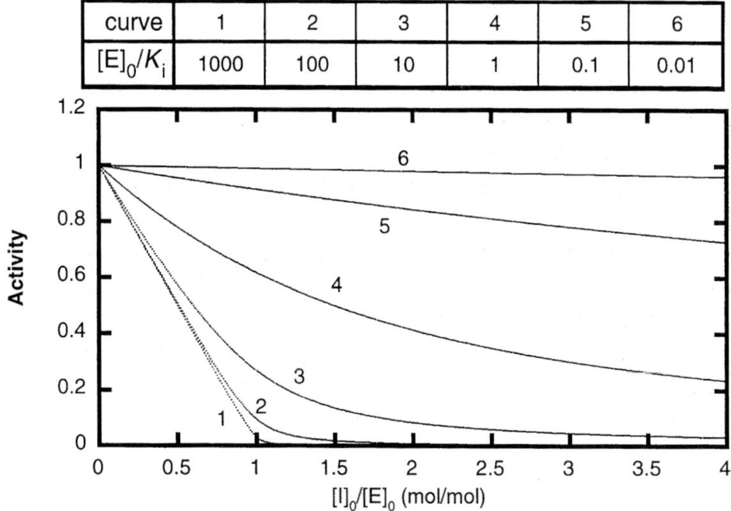

Figure 3 Theoretical plots of fractional enzymic activity as a function of the molar ration of $[I]_o$ to [E] at various $[E]_o/K_i$ ratios. Reproduced with permission from J. G. Bieth in *Bayer-Symposium V: Proteinase Inhibitors* (H. Fritz, H. Tscheschi, L.J. Greene and E. Truscheit, eds), Springer-Verlag, Berlin, 1994, pp. 463–469.

(2) When $[E]_o/K_i \leq 0.01$, there is virtually no EI complex formation at $[I]_o = [E]_o$. In this case, the inhibition is a 'classical' one, which assumes that $[I] = [I]_o$ (i.e. the enzyme does not deplete the inhibitor) and conventional derivations of the Michaelis–Menten relationship may be used to calculate K_i. Calculation of K_i can be made only when the inhibition is truly reversible. It is determined under the equilibrium conditions that give concave inhibition curve (curves 4–6 in *Figure 3*).

Tight-binding reversible inhibitors exhibit significant inhibition with equimolar concentrations of enzyme and inhibitor. In this case, [I] is no longer equal to $[I]_o$ for tight-binding inhibitor, and it must be computed from,

$$\begin{array}{c} [I]_o \;=\; [I] \;+\; [EI] \\ \text{Total} = \text{free} + \text{bound} \end{array} \quad (4)$$

4.4.2 Determination of K_i (tight-binding inhibitors)

Many effective inhibitors have very low K_i values ($10^{-8}-10^{-10}$ M) and it is often the case that adequate substrates are not available to measure activity at such a low enzyme concentration. This problem of tight binding can be circumvented by the method of 'progress curves' (22) in which inhibitor is added to a steady-state enzyme–substrate reaction (23). The method has the advantages that enzyme concentration need not be determined and that a reasonable estimate of K_i can be made from a single assay. The method is described in *Protocol 4*, and *Table 1* summarizes some numerical limits for kinetic constants.

Table 1 Some numerical limits for kinetic constants

1. The diffusion-rate-limited encounter of two molecules in aqueous solution at 37 °C is 10^8–10^9/M.s.

2. Some proteinase–inhibitor reactions are among the fastest known productive protein–protein interactions. Very sensitive substrates are required to measure such rates, even when using stopped-flow techniques. For example, to measure a k_{ass}, of 10^7/M.s by the techniques described here requires a substrate capable of revealing about 10^{-11} M enzyme in 2 min.

3. Some proteinase–inhibitor complexes are also among the tightest reversible interactions known. At the 10^{-11} M enzyme detection limit noted above, it is not possible to measure a K_i of much less than 10^{-12} M.

4. Since $K_i = k_{-1}/k_1$, and the maximum theoretical value of k_1 is 10^8/M.s, a K_i of 10^{-12} M can result from a k_{-1} of at most 10^{-4}/s ($t_{1/2}$ dissociation > 2 h). Inhibitors with K_i values below 10^{-12} M can therefore be thought of as essentially irreversible.

Protocol 4

Method for determining K_i

1 Follow the enzyme-catalysed hydrolysis of a substrate in the absence of an inhibitor to establish the uninhibited rate of substrate hydrolysis (v_o) (*Figure 4*). Ensure that this rate is linear and that the enzyme does not inactivate spontaneously under the given assay conditions.

2 Add inhibitor (at least 10-fold molar excess over enzyme) in no more than 5% of the total assay volume. Allow the reaction to proceed until the rate of hydrolysis has relaxed to a new steady state (*Figure 4*), thus establishing the inhibited rate (v_i).

3 Ensure that the substrate concentration is kept essentially constant by allowing no more than 5% hydrolysis. For example, 4-nitroanilide substrates should be used at concentrations above 0.5 mM when measuring A_{410} up to 0.2 absorbance units/min.

4 Under these conditions $K_{i(app)}$ (i.e. K_i in the presence of substrate) is given by:

$$v_o/v_i = 1 + [I]_o/K_{i(app)} \quad (5)$$

where $K_{i(app)}$ can be calculated from the known inhibitor concentration and the ratio of the two hydrolysis rates (expressed in arbitrary units).

5 To be more accurate, the experiment should be repeated at several different $[I]_o$ values. A plot of $(v_o/v_i) - 1$ against $[I]_o$ will have a slope of $1/K_{i(app)}$.

6 K_i, the true equilibrium constant, can be calculated from the relationship.

$$K_i = K_{i(app)}/(1 + [S]_o/K_m) \quad (6)$$

Therefore, when working at $[S]_o \ll K_m$, $K_i = K_{i(app)}$.

7 Some inhibitors bind so tightly that v_i is close to zero at all practical inhibitor and enzyme concentrations, and only an upper limit for K_i can be estimated. This depends on the detectability and stability of the enzyme at low concentration.

8 If attempts to determine K_i are unsuccessful because the value obtained is below

Protocol 4 continued

the detection limit of the enzyme–substrate system employed, then a more sensitive substrate should be used. If still unsuccessful, this supports an irreversible or very tight-binding reversible reaction. Either of these reactions can be analysed by pseudo first-order or second-order plots to determine k_{ass}.

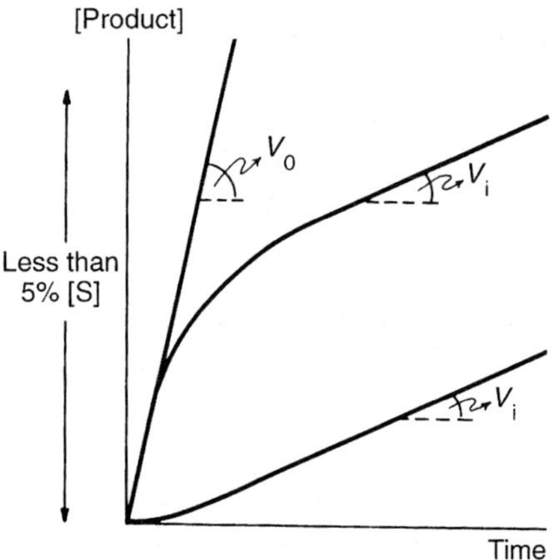

Figure 4 Progress curves for determination of $K_{i(app)}$. The rate of substrate hydrolysis is linear with time in the absence of inhibitor (v_o). When inhibitor is added the rate of hydrolysis will 'relax' to a new value (v_i). If enzyme is incubated with inhibitor before the addition of substrate the rate of hydrolysis will increase to the same value of v_i (lower curve). Reaction conditions should be chosen so that the two inhibited rates (v_i values) reach equivalence. When this condition is reached the true value of v_i is attained.

4.4.3 Association rate k_1

The association rate constant k_1 for a reversible inhibitor may be calculated exactly as for an irreversible one (see below) provided that the dissociation rate k_{-1} is not too fast ($k_{-1} < 10^{-3}$/s). If plots of ln[E] or 1/[E] against time show deviations from linearity at low [E], one should suspect that dissociation of the enzyme-inhibitor complex contributes significantly to [E]. When K_i and k_1 are known, k_{-1}, the dissociation rate constant, is given by

$$k_{-1} = K_i \cdot k_1 \quad (7)$$

4.5 Irreversible inhibitors

The parameters that govern the reaction of a proteinase with an irreversible inhibitor are the three rate constants (see Scheme 1). The most important quantity in this type of inhibition is the apparent rate of inhibition (or association), sometimes called k_{ass}. For inhibitors following the above mechanism, the meaning of k_{ass} is given by the relationship

$$k_{ass} = \frac{k_1 \cdot k_2}{k_{-1}} \quad (8)$$

Under appropriate conditions all three rate constants can be measured: however, for practical purposes, k_{ass} is sufficient to describe the reaction rate of irreversible inhibitors. This quantity can be readily measured by combining proteinase with inhibitor and assaying residual proteinase activity at time intervals under second-order conditions ($[I] \approx [E]_o$) or under pseudo first-order conditions ($[I]_o \gg [E]_o$). If a fast rate is expected ($k_{ass} > 10^5$/M.s), one should work under second-order conditions. Some practical advice for calculating k_{ass} is given below.

(1) When making proteinase–inhibitor titrations (prior to working under second-order conditions) ensure that mixtures are incubated for times sufficient for the reaction to be at least 95% complete. This may vary from 1 min to 1 h, depending upon the association rate and the concentration of the reactants. Employ high reactant concentrations if a slow reaction is suspected.

(2) The measurement of fast-acting inhibitors is limited by the ease of detection of the enzyme activity. Therefore, substrates capable of detecting low concentrations of enzyme must be employed. For example, the half-time for enzyme depletion at a reactant concentration of 10^{-9} M and k_{ass} of 10^7/M.s will be 1 min when a substrate capable of measuring 10^{-10} M [E] is used.

(3) Working under pseudo first-order conditions mitigates the need to know the enzyme concentration. We define this condition as $[I]_o > 50[E]_o$.

(4) A tenfold dilution of reaction mixture into assay buffer containing substrate is sufficient to halt the association of proteinase with inhibitor, provided that enough activity can be detected in 2–5 min.

4.5.1 Pseudo first-order conditions

Pseudo-first-order conditions occur when an enzyme is mixed with a large excess of inhibitor and the rate of consumption of enzyme is given by

$$\ln[E] = -k_{obs} \cdot t \qquad (9)$$

Samples are withdrawn from the mixture at timed intervals and the residual enzyme concentration [E] is measured. A plot of ln[E] (ln [residual enzyme activity]) against t (time of sampling) will have a slope of $-k_{obs}$, (pseudo first-order rate constant); k_{ass} is calculated from the relationship.

$$k_{ass} = \frac{k_{obs}}{[I]_o} \qquad (10)$$

4.5.2 Second-order conditions

For measurements of k_{ass} under second-order conditions one must first determine the active concentrations of both proteinase and inhibitor. These are then combined at equimolar concentrations. Samples are withdrawn and assayed at timed intervals, e.g. every 5 min, and k_{ass} calculated according to the following equation:

$$\frac{1}{[E]} = \frac{1}{[E]_o} + k_{ass}t \qquad (11)$$

A plot of 1/[E] (reciprocal of residual enzyme concentration) against t (time of sampling) will have a slope of k_{ass} and y-intercept of $1/[E]_o$ (initial enzyme concentration).

4.5.3 Determination by progress curve

As in the case of reversible inhibitors, the profile of product [P] vs. time after addition of $[I]_o$ describes the time-dependent approach to inactivation (see *Figure* 4). After obtaining several progress curves with at least six different inhibitor concentrations at fixed $[S]_o$ and $[E]_o$, these are analysed by non-linear regression according to the following equation:

$$[P] = \frac{v_z}{k_{obs}} (1 - e^{-k_{obs} \cdot t}) \qquad (12)$$

where (P) is the product and v_z is the velocity at time zero; k_{obs} is a pseudo first-order rate constant.

The feature of progress curves of irreversible inhibition is a simple exponential instead of the biphasic behaviour observed with reversible inhibition (see *Figure* 4). Progress curve analysis allows one to distinguish one-step (Mechanism 1) and two-step (Mechanism 2) irreversible mechanisms (see Box 1).

Box 1

Mechanism 6.1K_m

$$\begin{array}{l} \quad\; K_m \\ E + S \;\; ES \rightarrow E + P \\ + \\ I \\ k_1 \downarrow \\ EI \end{array}$$

Mechanism 6.2

$$\begin{array}{l} \quad\; K_m \\ E + S \;\; ES \rightarrow E + P \\ + \\ I \\ k_1 \downarrow\uparrow k\text{-}1 \\ EI^* \\ k_2 \downarrow \\ EI \end{array}$$

In the case of Mechanism 1, k_{obs} is a linear function of $[I]_o$

$$k_{obs} = \frac{k_1[I]_o}{1 + [S]_o/K_m} \qquad (13)$$

In the case of Mechanism 2, k_{obs} is a hyperbolic function of $[I]_o$

$$k_{obs} = \frac{k_2[I]_o}{[I]_o + K_i^*(1+[S]_o/K_m)} \qquad (14)$$

where,

$$K_i^* = \frac{k_{-1}}{k_1} \qquad (15)$$

and the reversible binding is assumed to be very rapid compared with the irreversible k_2 step.

Thus, the second-order association constant k_{ass} can be obtained graphically by a plot of k_{obs} vs. $[I]_o$. In Mechanism 1, k_{ass} will depend linearly on $[I]_o$ (Eq. 13) and can used to determine k_1 ($k_1 = k_{ass}$) by linear regression analysis. In Mechanism 6.2, non-linear regression analysis based on Eq. (14) will yield k_2 and K_i^*. The refined values of the parameters should be calculated by fitting the data to Eq. (12) and Eq. (14). This may be done easily using the ENZFITTER software (Biosoft, Cambridge, UK) (21). The second-order inactivation constant k_{ass} for Mechanism 6.2 is k_2/K_i^*. In the one-step mechanism, the formation of EI depends on only $k_{ass}[I]_o$. In the two-step mechanism, the lower the K_i^* the slower the inhibition process (formation of an irreversible EI complex). However, the lower K_i^*, the higher [EI*]; i.e., the more effective inhibition of proteolysis (formation of [P]) is observed. Many protein inhibitors of proteases probably form EI* complex with their target enzymes. The detection of the intermediates depends on the concentration of the inhibitor used in the progress curve experiments. To see the formation of EI*, $[I]_o$ may need to be an order of magnitude greater than K_i^*.

5 Practical applications of protein inhibitors

The discovery of endogenous protease inhibitors and characterization of their potencies towards target enzymes does not only allow us to predict possible biological and pathological roles of the inhibitor in regulating proteolytic events in cells and tissues, but also does provide useful reagents to inhibit particular proteases. Many protein inhibitors of proteases inhibit a group of related enzymes. However, because many mutagenesis methods to engineer proteins have become available, it is possible to generate selective inhibitors. Those mutated inhibitors are useful for probing the roles of specific proteases in biological and pathological systems. Another advantage of protein inhibitors is that they may be overexpressed locally in the tissue. A cell-based TIMP-1 gene transfer has allowed investigation of the role of matrix metalloproteinases in a rat xenograft model of aortic aneurysm (24). It will be also possible to use engineered inhibitors to develop tissue-targeted gene therapy under regulation of a tissue specific promoter. In certain diseases, these approaches may be advantageous over systemic administration of synthetic inhibitors of proteases, which may cause inadvertent side-effects.

Acknowledgements

The authors thank Dr Raja Khalifah for helpful discussion. This work was supported by NIH grants AR39189 (to H.N.), AR40994 (to H.N.) and HL51399 (to G.S.).

References

1. Bieth, J. (1984). *Biochem. Med.* **32**, 387.
2. Knight, C. G. (1986). In *Proteinase Inhibitors* (ed. Barrett A. J. and Salvesen G.), p. 23. Elsevier, Amstredam.

3. Huang, W., Suzuki, K., Nagase, H., Arumugam, S., van Doren, S. R., and Brew, K. (1996). *FEBS Lett.* **384**, 155.
4. Sen, L. C. and Whitaker, J. R. (1973). *Arch. Biochem. Biophys.* **158**, 623.
5. Herron, G. S., Banda, M. J., Clark, E. J., Gavrilovic, J., and Werb, Z. (1986). *J. Biol. Chem.* **261**, 2814.
6. Staskus, P. W., Masiarz, F. R., Pallanck, L. J., and Hawkes, S. P. (1991). *J. Biol. Chem.* **266**, 449.
7. Pavloff, N., Staskus, P. W., Kishnani, N. S., and Hawkes, S. P. (1992). *J. Biol. Chem.* **267**, 17321
8. Roberts, B. L., Markland, W., Ley, A. C., Kent, R. B., White, D. W., Guterman, S. K., and Ladner, R. C. (1992). *Proc. Natl. Acad. Sci.* **89**, 2429.
9. Tanaka, A. S., Sampaio, C. A., Fritz, H., and Auerswald, E. A. (1995). *Biochem. Biophys. Res. Commun.* **214**, 389.
10. Markland, W., Ley, A. C., Lee, S. W., and Ladner, R. C. (1996). *Biochemistry* **35**, 8045.
11. Scheidig, A. I., Hynes, T. R., Pelletier, L. A., Wells, J. A., and Kossiakoff, A. A. (1997). *Protein Sci.* **6**, 1806.
12. Markland, W., Roberts, B. L., and Ladner, R. C. (1996). *Methods Enzymol.* **267**, 28.
13. Ke, S. H., Combs, G. S., Tachias, K., Corey, D. R., and Madison, E. L. (1997). *J. Biol. Chem.* **272**, 20456.
14. Anastasi, A., Brown, M. A., Kembhavi, A. A., Nicklin, M. J.H., Sayers, C. A., Sunter, D. C., and Barrett, A. J. (1983). *Biochem. J.* **211**, 129.
15. Laskowski Jr, M., and Kato, I. (1980). *Annu. Rev. Biochem.* **49**, 593.
16. Lineweaver, H. and Burk, D. (1934). *J. Am. Chem. Soc.* **56**, 658.
17. Dixon, M. (1972). *Biochem. J.* **55**, 170.
18. Potempa, J., Korzus, E., and Travis, J. (1994). *J. Biol. Chem.* **269**, 15957.
19. Zhou, Q., Krebs, J., Snipas, S. J., Price, A., Alnemri, E. S., Tomaselli, K. J., and Salvesen, G. S. (1998). *Biochemistry* **37**, 27084.
20. Bieth, J. (1974). In *Bayer-symposium V 'proteinase inhibitors'* (ed. Fritz, H., Tschesche, H., Greene, L. J., and Truscheit, E.), p. 463. Springer-Verlag, Berlin.
21. Bieth, J. G. (1995). *Methods Enzymol.* **248**, 59.
22. Henderson, P. J.F. (1972). *Biochem. J.*, **127**, 321.
23. Nicklin, M. J.H. and Barrett, A. J. (1984). *Biochem. J.*, **223**, 245.
24. Allaire, E., Forough, R., Clowes, M., Starche, B., and Clowes, A. W. (1998). *J. Clin. Invest.* **102**, 1413.

Chapter 7

Mass spectrometry of proteolysis-derived peptides for protein identification

Bernhard Küster, Andrej Shevchenko and Matthias Mann
Center for Experimental BioInformatics (CEBI), Odense University, Staermosegaardsvej 16, DK–5230 Odense M, Denmark

1 Introduction

Proteolytic enzymes are used in biological mass spectrometry for countless purposes ranging from, e.g. the solubilization of membrane proteins to epitope mapping and protein conformational studies. In particular, proteases are indispensable tools for peptide mapping and protein primary structure elucidation. Until a few years ago, the latter application was usually performed either by *N*-terminal sequencing via the Edman chemistry or by HPLC separation of peptide mixtures again followed by Edman sequencing. This type of analysis was quite time consuming and protein amounts in the order of 100 pmol were frequently needed for 'internal sequencing'. However, only much lower amounts are frequently available from biochemical preparations involving multiple purification stages, thus more sensitive characterisation techniques are needed. MS has gradually become a viable alternative to Edman sequencing and now generally offers superior performance. Sub-picomole amounts of peptides can routinely be detected and partially sequenced by MS. In addition, the molecular weight, as obtained by MS, is a highly characteristic physicochemical property of a peptide and can often be used for the identification of known proteins in sequence databases without the need for generating any peptide sequence. The recent success of protein identification and peptide sequencing strategies based on MS has been made possible by substantial improvements in instrument hardware, data analysis software and sample preparation techniques over the past few years. This chapter focuses on practical aspects of mass spectrometry in the analysis of peptides derived from proteins separated by PAGE. A general introduction to MS in biochemistry has recently appeared in this series (1) and the numerous fruitful applications of MS in biological research have been highlighted in several recent reviews (2–6). The reader is encouraged to refer to these publications for complementary information.

2 Mass spectrometry

The manipulation of gas-phase ions by appropriate electrostatic and/or magnetic fields allows the measurement of the mass to charge ratio (m/z) of an analyte ion in a mass spectrometer. Information regarding the charge state together with the ionizing agent yields the molecular weight of the analyte. Hence, a mass spectrometry experiment essentially consists of the following steps:

- generation of (intact) gas-phase ions
- ion separation
- ion detection and data acquisition.

Each of the above steps has been realized in a variety of different ways but only essential features will be discussed here. Detailed descriptions of the theory and operation of mass spectrometers can be found in series of reviews (7–11).

2.1 Methods of ionization

Peptides and proteins are thermally relatively unstable molecules. Therefore, it is crucial to employ a 'soft' ionization method in order to generate intact molecular ions. Out of the many methods developed for this purpose only two are of practical relevance for the analysis of peptides and proteins. In MALDI the analyte compounds are co-crystallized with a large excess of a light-absorbing matrix (12). Irradiation of the crystals by a pulsed laser beam results in the rapid sublimation of matrix and embedded analyte molecules and the generation of intact gas-phase ions. For peptides, protonated, singly charged molecular ions (frequently denoted as $[M + H]^+$) are usually formed, but alkali metal adduct ions, as well as doubly charged species can also sometimes be observed. Some of the more commonly used matrices are listed in *Table 1*.

Table 1 Commonly used MALDI matrices for the analysis of proteins, peptides and glycopeptides[a]

Matrix	Proteins	Peptides	Glycopeptides	Reference
Sinnapinic acid[b]	X			13
HCCA		X		14
DHB		X	X	15
Ferulic acid[c]	X			13

[a] All matrices can be obtained from Sigma/Aldrich and can generally be used as supplied. If impurities are encountered, the material should be further purified by recrystallization.

[b] 3,5-Dimethoxy-4-hydroxycinnamic acid.

[c] 4-Hydroxy-3-methoxycinnamic acid.

In ESI, a continuous beam of ions is produced by electrostatically dispersing the effluent from a capillary needle (16). Microscopic, highly charged droplets which rapidly evaporate are produced in this process. The increasing charge density is then thought to cause desorption of charged analyte molecules from the droplet in order to 'carry off' excess charges. An alternative explanation for the generation of charged analyte molecules is that they are simply 'left behind' as solvent molecules (but not the excess charges) evaporate. One characteristic feature of ESI of peptides and proteins is the production of a population of multiply charged species $[M + nH]^{n+}$ in which n is the number of protons attached to the peptide or protein. Among other factors, the number of charges depends on the size of the molecule and is in the range of one charge per kilodalton. Small peptides generally bear one or a few charges, whereas large proteins may carry in excess of 100 charges. Since ionization takes place from liquid phase, ESI is readily coupled to liquid chromatography. ESI-MS is semiquantitative, i.e. the signal intensity depends on the analyte concentration but not on the flow rate. Hence, lowering the flow rate can increase the overall sensitivity of the process. This feature of ESI has been exploited in a technique called nanoelectrospray MS in which the sample is sprayed from a fine, pulled capillary (illustrated in *Figure 1*). Sample flow rates in the order of 25 nl/min are afforded by this device which greatly increases available analysis time and thereby diminishes the need for peptide separation prior to MS (3,17). This proves to be of prime importance for the sequencing of peptides at low sample levels (see below).

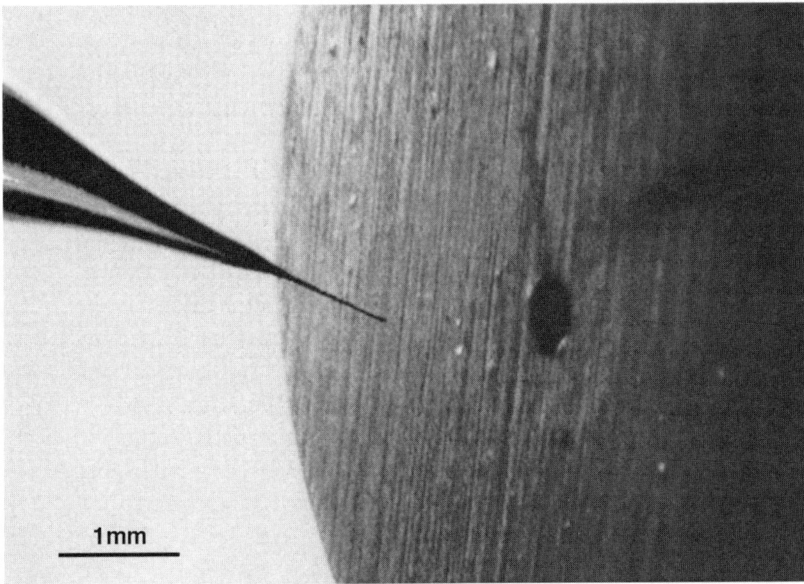

Figure 1 Nanoelectrospray capillary in front of the orifice of a mass spectrometer. Capillaries are pulled from borosilicate glass and coated with a thin layer of gold (14). The shape of the capillary should closely resemble the one shown here and is crucial for generating a low yet stable flow.

2.2 Mass analysis

Since pulsed lasers generate bursts of ions at the firing rate of the laser, MALDI is most commonly coupled to TOF mass analysers. Ions are typically accelerated to kinetic energies of 20–25 keV and subsequently allowed to drift through a field-free flight tube before they impact on the detector and produce a current, which is recorded as the signal. The flight time of an ion in the field free region is proportional to the square root of its mass to charge ratio. Hence, having identical kinetic energy, light ions travel faster through the instrument than heavy ones. The mass to charge ratio of any ion can then be determined by comparing its flight time with those of compounds of known mass either in the same spectrum (internal calibration) or using a calibration spectrum (external calibration). In theory, the mass range of TOF analysers is unlimited. However, owing to the high abundance of matrix-related ions at low mass and the relatively inefficient detection of heavy (i. e. slow) ions, the practical mass range is between 500 and 150 000 Da, with a drop-off in mass spectrometric performance starting from 15 kDa. A distinct advantage of TOF analysers is the capability to detect the entire ion population produced by every laser shot, thus rendering the technique extremely sensitive. For masses below 5000 Da advances in TOF technology, namely the introduction of reflectron-TOF and pulsed ion extraction, have led to much increased resolution and mass measurement accuracy (18,19). The latter feature especially is crucial in protein identification strategies employing MALDI peptide mass mapping as discussed below.

As outlined above, a continuous beam of multiply charged ions is produced in ESI. Thus, the ions generated are usually mass analysed by scanning instruments such as quadrupoles. In this type of mass analyser, ions are separated based on their trajectories in an oscillating electric field created between two pairs of parallel metal rods. At a given oscillation frequency and amplitude of the electric potentials applied to the rods, only ions of a certain m/z value are allowed to pass through the quadrupole field to reach the detector. By continuously varying the frequency and amplitude parameters, ions of different m/z values can be recorded sequentially at the detector to produce a mass spectrum. In ES-MS the limited m/z range of quadrupole analysers is compensated for by the production of multiply charged ions. Scanning instruments have the inherent disadvantage that only a small fraction of all ions is recorded at any one time. Therefore, non-scanning mass analysers such as ion traps have, in principle, a sensitivity advantage over quadrupoles. The ion trap accumulates ions for a given time and then ejects them in a mass-dependent manner. It can also perform multiple stages of MS-MS (see below) by repeatedly fragmenting, trapping and ejecting ions. More recent developments have focused on interfacing ES with TOF analysers, previously predominantly used with the MALDI technique. In addition to recording all ions quasi-simultaneously, this geometry takes advantage of the fact that TOF analysers are capable of producing both high mass resolution and very high mass accuracy. A particularly interesting new geometry consists of a combination of a quadrupole and TOF analyser to enable MS-MS. This coupling holds considerable potential for the analysis of peptides and proteins (20,21).

2.3 Tandem mass spectrometry

In order to generate peptide sequence information by mass spectrometry, two mass analysers have to be coupled in tandem. The concept of MS-MS is illustrated in *Figure 2* for ESI-MS on a triple quadrupole mass spectrometer. For sequencing by MS-MS, the first mass analyser (Q_1) is used to isolate one ion of interest from the peptide mixture. This so-called precursor ion is guided into a collision cell (Q_2) where it undergoes multiple collisions with inert gas molecules (typically argon or nitrogen). The excess energy imparted by this process results in cleavage of peptide bonds and thereby generates a series of *C*-terminal (Y″-type) and *N*-terminal (A- and B-type) fragment ions (the nomenclature for fragment ions is described in ref. 22). The *m*/*z* values of these fragment ions are determined using the third quadrupole (Q_3). At least a partial amino acid sequence of the peptides can be derived by interpretation of the fragment ion spectrum.

An additional MS-MS experiment can be employed to detect peptide signals in the presence of contamination or high chemical background. In a technique called precursor ion scanning, all analyte compounds are fragmented inside the

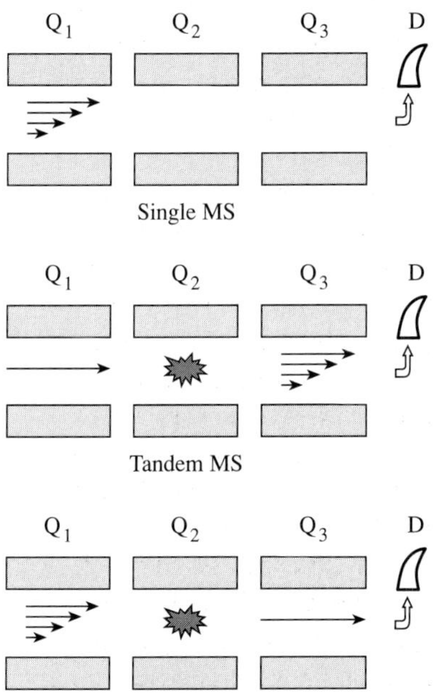

Figure 2 Modes of operation of a triple quadrupole mass spectrometer for peptide analysis. In single MS mode, the first quadrupole (Q_1) is scanned over a set mass range. All ions are transmitted by quadrupoles 2 and 3 (Q_2, Q_3) and are subsequently detected at the plane of the detector (D) to produce a mass spectrum. In MS-MS mode, Q_1 is used to transmit a selected mass into the collision cell (Q_2), where it is fragmented by collision induced dissociation using an inert collision gas such as argon or nitrogen. The charged fragment ion masses are determined by scanning the third (Q_3) quadrupole. In precursor ion scanning mode, Q_1 is scanned and all ions are fragmented inside the collision cell in turn. Only fragment ions of a specific mass are transmitted by Q_3 and allowed to reach the detector.

Table 2 Database searching tools accessible via the World Wide Web

URL	Name of programme	Database(s)	Searching features
http://www.mann.embl-heidelberg.de/Services/PeptideSearch/PeptideSearchIntro.html	PeptideSearch	NRDB	List of peptide masses Peptide sequence tags
			Edman-type data
http://prowl.rockefeller.edu/cgi-bin/ProFound	ProFound	OWL	List of peptide masses
http://prowl.rockefeller.edu/PROWL/pepfragch.html	PepFrag	SwissProt, PIR, NRL3D, Genpept, NR, dbEST, *S. cerevisiae*, *E. coli*, *B. subtilis*	List of fragment ion masses
http://prospector.ucsf.edu/	MS-Fit MS-Tag MS-Edman	NCB, Genpept, SwissProt, OWL, dbEST List of fragment ion masses Edman-type sequence data	List of peptide masses
http://gserv1.dl.ac.uk/SEQNET/mowse.html	MOWSE	OWL	Peptide sequence and composition data
http://www.mdc-berlin.de/~emu/peptide_mass.html	Peptide Mass Search	SwissProt, Protein Identification Resource (PIR) and SwNew	List of peptide masses
http://www.lsbc.com:70/Lutefisk97.html	Lutefisk97	OWL	List of fragment ion masses
http://cbrg.inf.ethz.ch/subsection3_1_3.html	MassSearch	SwissProt, EMBL	List of peptide masses

collision cell in turn. In this process, amino acid-specific fragment ions are formed (notably immonium ions of amino acids). The second mass analysing quadrupole (Q_3) is set to transmit a specific 'marker' ion only. The resulting precursor ion spectrum displays only those species which have produced the marker ion monitored in Q_3. For example, the immonium ion of Ile/Leu (*m/z* 86) is usually monitored in order to distinguish peptide species from chemical background. Once peptide ions have been identified in this way, they can be subjected to sequencing by MS-MS as outlined above. In the nanoelectrospray technique, precursor ion scanning is an essential tool for achieving ultimate sensitivity.

3 Interrogation of sequence databases using mass spectrometry data

The rapid progress of genome sequencing projects of model organisms and the concurrent growth of sequence databases provide the biological sciences with an invaluable source of information. In particular, ORFs and ESTs can reveal potential proteins and scrutiny of sequences for homologies to known proteins or protein families of related species might predict functions for the putative gene products. As a result, the characterization of proteins is often reduced to identification in a sequence database. The molecular weight and partial sequence of peptides, as obtained by MS, provide highly specific data for the identification of proteins in databases. Several search programs are accessible via the World Wide Web and a compilation of addresses and services available is shown in *Table 2*.

3.1 Choice of protease

Identification of gel-separated proteins begins with their proteolytic degradation in order to extract peptides for subsequent mass spectrometry analysis. The protease employed has to be compatible with both MS analysis of the resulting peptides and database searching in order to allow straightforward and unambiguous identification. In choosing the right enzyme one sometimes has to trade off one factor for the other. Some features of commonly used proteases are compiled in *Table 3*.

For database searching, the most important requirements are the absence of unspecific cuts and the specific cleavage of as few amino acids as possible. Hence, commonly used enzymes such as chymotrypsin or pepsin are not very useful for database searching. The choice of enzyme will also affect the average size of the generated peptides. Trypsin, for example, will give rise to a relatively large number of small peptides whereas Lys-C digestion generally yields fewer but larger peptides. For peptide mass mapping it is therefore advantageous to use trypsin for small and highly modified proteins to take full advantage of the many peptides generated by this enzyme such that at least some of them can be used for mass spectrometric protein identification. Lys-C, on the other hand, can be advantageous for large and unmodified proteins because the larger peptides are more specific probes for database searching.

Table 3 Choice of protease for in-gel digestions followed by MS and database searching

Protease[a]	Cleavage specificity	Suitability for MS analysis	Suitability for database searching
Trypsin	Selective cleavage at *C*-terminus of Arg and Lys; very high specificity	Excellent; peptides generally between 500 and 2500 Da; very informative MS-MS spectra; limited autolysis	Good; only two amino acids have to be considered for the *C*-terminus.
Lys-C	Exclusive cleavage at *C*-terminus of Lys; very high specificity.	Good; peptides are fairly large; well suited for the recovery of phosphopeptides; no autolysis; MS-MS performance compromised due to the presence of internal Arg residues	Good; only one amino acid has to be considered for the *C*-terminus; but fewer peptides usually available for sequencing and searching.
Glu-C	pH dependent cleavage at *C*-terminus of Glu (pH 4-5) and Glu + Asp (pH >7); moderate specificity	Moderate; reproducible pattern of autolysis products; internal Lys and Arg residues compromise performance in MS-MS	Good; only two amino acids have to be considered for the *C*-terminus
Chymotrypsin and pepsin	Cleavage at *C*-terminus of hydrophobic amino acids; low specificity	Poor; many autolysis products; internal Lys and Arg residues compromise performance in MS-MS	Poor; cleavage specificity too low

[a]All proteases are supplied in good quality by Boehringer Mannheim

Different considerations have to be taken into account when peptides are to be sequenced by MS-MS. In this case, the resulting peptides should not contain basic internal residues such as Lys or Arg because these amino acids localize charges on their side-chains and thereby render tandem MS experiments less predictable and much harder to interpret. Hence, Glu-C would not generally be a desirable enzyme, although its cleavage specificity is relatively high. Complete sequences are easier to obtain from relatively small peptides than from large ones. Conversely, complete sequences of longer peptides are desirable for the design of low-degeneracy primers in cloning strategies. Taking all requirements for MS analysis and database searching into consideration, trypsin is a good first choice for almost any application. Tryptic peptides are generally in the range of 500–2500 Da, allowing their masses to be measured with high accuracy by MALDI-MS. Furthermore, all basic amino acids, except His, are localised at the *C*-terminus of the peptides, which greatly simplifies interpretation of MS-MS spectra. A second enzyme can be used to generate a data set 'orthogonal' to the first analysis or to verify additional parts of the primary structure.

3.2 Peptide mass mapping

Protein identification by peptide mass mapping is based on the idea that while any one of a list of experimentally determined peptide masses might be found in a number of proteins, it is very unlikely that many peptide masses would all be found within the same protein by chance (23–27). In this approach, a list of peptide masses is generated by proteolytic degradation of a protein followed by MS (peptide mass mapping). These measured peptide masses are compared with a theoretical protease digest—according to the enzyme specificity—of all the proteins present in the database. The protein is then identified as the one with the best agreement between the measured and calculated peptide mass maps. In order for this statistical approach to work, mainly two parameters need to be considered. These are the number of peptide masses measured which match the calculated masses of peptides from proteins in the database and the mass accuracy used in the search. Every peptide mass represents an independent data point. Therefore, the larger the number of peptides used in the search the higher the search specificity (i.e. the power to discriminate between correct and random matches). For the same reason, searches exhibit a considerable error tolerance since generally only a few peptides are needed to call a significant match. Hence, the presence of modified peptides and the presence of spurious peaks in the experimentally determined mass list does not compromise the search results to any great extent.

The search-input data essentially only consists of a list of measured peptide masses. Therefore, mass measurement accuracy is of prime importance. In this respect, instrument performance can be a limiting factor. State-of-the art MALDI instruments equipped with pulsed ion extraction and reflectron-type TOF mass analysers are ideal for this kind of database search since mass accuracy in the 30–50 p.p.m. range can be obtained routinely. If high mass accuracy is not available, considerably more peptide masses are needed to call a significant match as statistically, the search specificity increases approximately by a factor of (improvement in mass accuracy) to the power of (number of peptides measured). An example of protein identification by peptide mass mapping is given in section 4.4.

When searching the database with a given set of peptide masses, large proteins in the database are much more likely to produce a false positive than smaller ones. The extensive collection of peptides resulting from a theoretical proteolytic cleavage of large proteins is more likely to contain a random match for any peptide mass than the small number of peptides derived from small proteins. This fact is illustrated in *Table 4*. Within the chosen search parameters, 10 peptides matched NIP 29 from yeast, a 29 kDa protein. The next best matches all correspond to proteins of molecular masses well in excess of 200 kDa. Some searching programs account for the tendency of large proteins to match many peptide masses randomly by scoring the results based on empirical rules. Generally, it is advisable to search with no restrictions on the mass range of the protein to avoid typical 'false negatives' such as proteins that migrate on the gel as a covalent dimer, or proteins that are fragments of larger precursors. After the search results are obtained, an estimate of the protein mass—such as deduced from its migration on a gel—can verify a match.

Another search parameter that can be chosen in most search programs is the number of 'missed cleavage sites', i.e. the number of uncleaved sites a peptide can contain. Such peptides are almost always present in protein digests and would go unnoticed if the search would not accommodate this feature. However, the more missed cleavage sites that are allowed in the search, the more ambiguous the results may become because the number of theoretical peptide masses is greatly increased. Therefore, it is highly desirable to optimize experimental digestion conditions in order to minimize incomplete digestion as far as possible. Peptides with uncleaved sites and modified peptides can also be accommodated in a 'second-pass search': after a potential match has been retrieved from the database, any unmatched peptides are compared against the sequence, allowing for less stringent enzyme specificity, lower mass accuracy and common peptide modifications.

The practical advantages of peptide mass mapping using MALDI-MS are speed, sensitivity and the simplicity of the approach. The technique is also amenable to automation, thus allowing the analysis of large numbers of proteins (28). An obvious limitation of this type of search is that essentially the full coding sequence of the protein in question has to be represented in the sequence database. This is not yet the case for most proteins. Sometimes a highly homologous protein can be identified from other species especially from those whose complete genome has been sequenced. However, even with good data, homology in the order of 90% is needed for this approach. Conversely, protein variants displaying only slight differences in their amino acid sequences might not be distinguishable in a database search.

A general observation is that fewer peptides are recovered when small amounts of proteins are digested. Hence, below a certain limit too few peptides will be available for protein identification using peptide mass mapping, effectively determining the sensitivity of the technique.

3.3 Database identification via tandem mass spectrometry

The fragmentation pattern of a peptide in an MS-MS experiment contains information about the sequence of that peptide. Thus, a tandem mass spectrum is much more discriminating than a mass alone. A single tandem mass spectrum generally does not enable deduction of the complete sequence of a peptide and even when this is possible it is very time consuming. Therefore, approaches have been developed that allow screening of the mass spectrometry data against sequence databases without complete sequence assignment. Once the peptide sequence has been found, it can be verified with a high degree of confidence by comparison with the tandem mass spectrum.

Although a complete sequence can rarely be deduced, a set of fragment ions corresponding to a stretch of consecutive amino acids can almost always be obtained. In the peptide sequence tag approach (29) (*Figure 3*) the partial sequence and mass information derived from MS-MS spectra is grouped into three modules corresponding to three regions of the peptide sequence about which there is information. The derived sequence itself (region m_2) is flanked by the masses of both of the two unsequenced parts of the peptides (m_1 and m_3). Together with the protease cleavage specificity and the intact peptide mass five search criteria are available for searching. These are:

- *N*-terminal cleavage specificity
- mass of region 1
- short stretch or 'tag' of sequence
- mass of region 3
- *C*-terminal cleavage specificity.

For convenience, the syntax of the peptide sequence tag can be read from the computer screen in an MS-MS experiment and entered directly into a search program: (lowest mass of fragment ion series)sequence(highest mass of fragment ion series). Using trypsin as the protease and a triple quadrupole, the ion series can be assumed to be Y″ ions (charge retention on the *C*-terminus). If this search does not yield a match then it can be repeated under the assumption of a B ion series (charge retention on the *N*-terminus of the peptide). The specificity of this kind of database search is very high since the determined sequence is 'locked' between two mass values. Very few or unique matches can generally be retrieved for peptide sequence tags containing a sequence stretch of two or three amino acids. However, the specificity of the approach stems mainly from the subsequent verification step: the sequence retrieved from the database is compared with the tandem mass spectrum. Since only a small part of the information content of the spectrum is used to retrieve the sequence, the calculated fragmentation spectrum verifies the peptide sequence with very high confidence. Mass measurement accuracy is less critical compared with peptide mass mapping and errors in the range of 1–2 Da are quite acceptable. The modular design of the peptide sequence tag also renders the approach very error tolerant. For example, not all regions have to be used for searching. Alternatively, errors in any one of the three regions can be allowed which enables retrieval of database sequences that either contain sequence errors or amino acids that are post-translationally modified. This extensive search flexibility proves particularly valuable when EST databases are searched. These databases contain short cDNA sequences generated by one pass sequencing of various cDNA libraries from model species. The full coding region of a protein is rarely represented in this database and sequences frequently exhibit errors due to frame shifts or incorrectly assigned nucleotides. A peptide sequence tag can still identify the peptide even in the presence of these errors.

In principle, a sequence tag derived from a single peptide can uniquely identify a protein. Further confirmation should, however, be obtained by independent identification of the same protein from a different peptide. The latter is of prime importance when sample amounts are low because contaminating proteins such as keratins become quite prominent and peptides of the target protein may represent a small minority among these 'contaminating' peptide peaks.

In addition to the peptide sequence tag approach, several other techniques have been developed to correlate MS-MS data with sequence databases. For example, a predicted tandem mass spectrum can be calculated for all peptides in the database and correlated and scored against the measured spectrum (30). This approach has the advantage of being completely automated. Other algorithms search the database by the peptide mass and a few fragment ions, without specifying whether these are of the B or Y″ type (31).

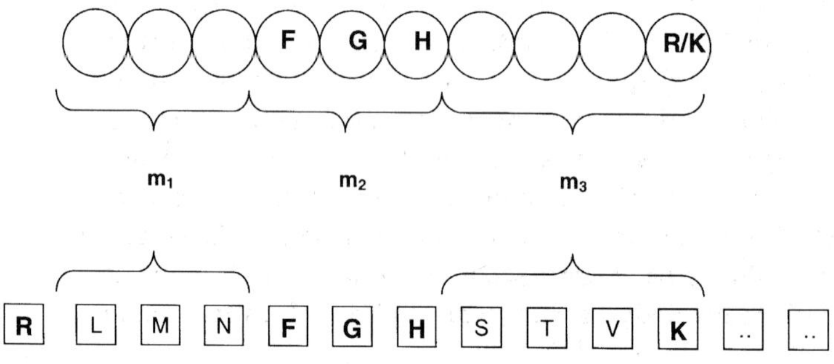

Figure 3 Principal of the peptide sequence tag approach. The information present in tandem MS spectra of peptides is organized into three regions: the mass up to a short stretch of sequence (m_1), the determined partial amino acid sequence of the peptide (FGH in this case) and the mass to the *C*-terminus of the peptide (m_3).

4 Analytical strategy for the identification or cloning of proteins using mass spectrometry

A general three stage analytical strategy for the identification or cloning of proteins using mass spectrometry is outlined in *Figure 4* (32). This strategy has been developed over a number of years in our laboratory and has been found to be generally applicable to the identification of proteins. Protein mixtures are separated by gel electrophoresis and digested *in situ* to yield a collection of peptides. The first analytical stage takes advantage of the speed and simplicity of MALDI peptide mass mapping in conjunction with searching of a comprehensive NRDB. Only a small part of the peptide solution is needed for this analysis. If no conclusive result is obtained, peptides are subjected to partial sequencing by nanoelectrospray MS-MS as the more specific second stage analytical technique. Peptide sequence tags are generated and used to search the NRDB and if necessary EST database. If an EST hit is obtained, the physical EST clone can be used to clone the corresponding gene. If no significant EST hit is identified, *de novo* peptide sequencing has to be undertaken. This third analytical stage involves peptide derivatization and exhaustive interpretation of the two sets of MS-MS in order to design appropriate degenerate oligonucleotide primers for subsequent cloning.

In the following text, practical aspects of the different steps in the strategy are discussed. Special attention is paid to sample preparation as this is of prime importance for the analysis of low-abundance proteins.

4.1 Low-level protein preparation for characterization by mass spectrometry

When attempting low-level (i.e. sub-picomole) protein identification or sequencing, a number of precautions have to be taken in order to avoid sample contamination and losses. All buffers and reagents should be made fresh using chemicals of the highest available purity. Stocks of chemicals should be kept in a clean environment. Dust is a common source of contamination and every effort should be made to clean thoroughly all equipment used in sample handling (e.g. microcentrifuge tubes, gloves, spatulas, glassware, etc.). Gloves should be worn consistently to minimize contamination by human and sheep keratins (originating from skin and clothing). The overall level of contamination can be greatly reduced if all sample handling steps are carried out in a laminar flow hood. Most detergents and polymeric compounds are detrimental to MS analysis and should therefore be avoided.

The most commonly used low-level protein separation and purification technique in biological research is PAGE, and considerable effort has been expended to enable the characterization of gel-separated proteins. Until recently, this generally took the form of electroblotting the proteins onto membranes followed by Edman degradation. However, the efficiency of blotting small amounts of proteins onto membranes is variable and adsorption to the membrane can lead to

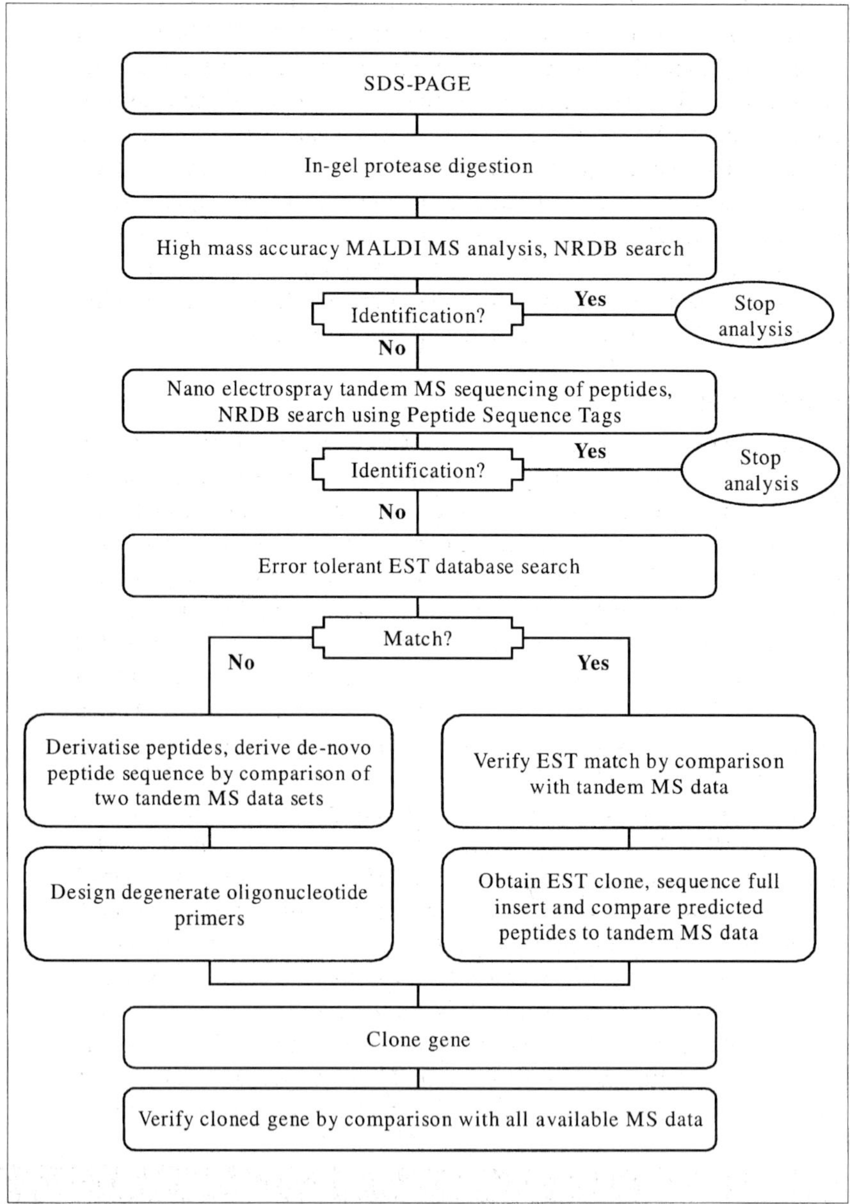

Figure 4 General strategy for the identification and cloning of proteins using MS. Gel-separated proteins are digested in gel with a sequence-specific protease. The set of peptides obtained is mapped by MALDI-MS and the resulting list of masses is used to search an NRDB. If a protein is not identified at this stage, the peptides are subjected to partial sequencing using nanoelectrospray MS-MS. Peptide sequence tags are generated and used to search the NRDB and EST databases. Novel genes can be cloned from EST clones or following *de novo* peptide sequencing and subsequent design of degenerate oligonucleotide primers. The identity of the cloned gene can be verified by comparing the complete MS data to the oligonucleotide sequence.

additional sample loss. An attractive alternative for low to sub-picomole amounts of proteins is to perform all necessary sample manipulations while the proteins are still inside the gel matrix. This not only circumvents the blotting step but the gel also provides a 'safe container' for the embedded proteins, which results in greatly reduced sample losses. The procedures given below were optimised to enable the sensitive characterization of proteins directly from one- or two-dimensional polyacrylamide gels of varying thickness (0.5–1.5 mm) and acrylamide concentration (7–18%). For best results, gels should be used no earlier than 1 h after casting to reduce acrylamidation of cysteine residues during electrophoresis. Gels can be run in standard buffers containing 0.1% SDS with or without urea but polymeric detergents or stabilizers must be omitted.

4.2 Protein visualization

A number of staining procedures is available for the visualization of protein bands or spots on polyacrylamide gels. Of these, Coomassie Brilliant Blue is probably the most commonly used when protein amounts in excess of 100 ng (0.5–5 pmol) are available. The protocols for in-gel digestion described below are fully compatible with this staining method and good MS results are generally obtained if gels contain enough protein for Coomassie staining. If in doubt concerning the amount of protein present in a particular gel, it is recommended that the gel is stained with Coomassie first. If staining results are not satisfactory, the gel can still be stained with silver. Silver staining affords ≈ 10–20-fold higher detection sensitivity and has no adverse effect on subsequent MS analysis if carried out as described below (33). It is essential to avoid the use of glutaraldehyde in the staining protocol. This agent chemically crosslinks proteins and thereby usually precludes further characterization. Since silver staining is very sensitive, special attention has to be paid to sample loading onto gels. Prestained molecular weight markers should not be used because they will obscure small amounts of analyte protein after staining. Instead, the amount of marker should be adjusted (i.e. lowered) to approximately the level expected for the proteins of interest. Spilling of small amounts of sample across loading wells can also become a problem. Therefore, it is recommended that at least one well is left blank between the marker lane and the first sample lane. If samples contain very variable amounts of protein, further gaps should be left between sample lanes.

Protocol 1

Silver staining of polyacrylamide gels

Reagents

- Fixation solution: methanol–acetic acid–water 50:5:45 v/v/v
- Sensitizing solution: 0.02% sodium thiosulphate
- 0.1% $AgNO_3$
- Developing solution: 0.04% formaldehyde in 2% sodium carbonate
- 1% acetic acid

Protocol 1 continued

Method

1 After the gel has been run, fix the protein by incubating the gel slab in fixation solution for 20–30 min.

2 Rinse the gel slab with water (two changes, 2 min per change) and then leave it further in water for 1 h on a shaking platform.[a]

3 Sensitise the gel with sensitizing solution for 1–2 min. Discard solution and quickly rinse the gel slab with two changes of water (10 s each).[b]

4 Incubate the gel in chilled 0.1% $AgNO_3$ for 30 min at 4 °C (in a refrigerator).

5 Discard silver nitrate solution and quickly rinse the gel with two changes of water (30 s each change).

6 Develop the gel with developing solution. Discard the developing solution as soon as it turns yellow and replace it with a fresh portion.

7 When a sufficient degree of staining has been obtained, quench staining by discarding the developing solution and replacing it with 1% acetic acid. Wash the gel with 1% acetic acid several times and store in the same solution.[c]

[a] Extended washing time helps to eliminate yellowish background usually observed after long developing of the gel (see step 6 of this protocol).

[b] Agitate gently to make sure that the gel slab is covered evenly.

[c] Silver stained gels can be stored at 4 °C in 1% acetic acid for months. In some cases, the colour of the stained protein bands might slightly change in time. However these changes do not affect the results of mass spectrometric sequencing.

An intermediate visualization sensitivity is achieved by reverse staining with zinc/imidazole (34) and *Protocol 2* ensures full compatibility with subsequent MS analysis. However, caution has to be exercised during destaining since proteins are mobilized in this step and might diffuse out of the gel matrix.

Protocol 2

Negative staining of polyacrylamide gels[a]

Reagents

- 0.2 M Imidazole in 0.1% SDS
- 50 mM Zinc sulphate
- methanol–water–acetic acid 50:48:2 v/v/v

Method

1 After electrophoresis, rinse the gel slab with two changes of water (1 min per change).

2 Discard water and incubate the gel in 0.2 M imidazole in 0.1% SDS for 5 min.

3 Discard imidazole solution and quickly rinse the gel with water.

Protocol 2 continued

4 Discard water and develop the gel with a 50 mM zinc sulphate solution.

5 In the course of developing, proteins emerge as transparent bands (spots) on a white opaque background of the gel. Therefore, it is recommended that gels are developed in transparent glass or plastic laboratory dishes and viewed against a black background.

6 When sufficient staining has been obtained, quench the developing by discarding the zinc solution and rinsing the gel with water.[d]

7 Excise protein band (spot) with a clean scalpel and place in a 0.5 ml microcentrifuge tube.

8 Incubate for 5 min in methanol–water–acetic acid.

9 Continue at step 2 in *Protocol 3*.

[a] Negative staining is performed essentially as described in (34) with minor modifications.

[b] Note that zinc sulphate is commercially available as a crystallohydrate containing seven water molecules.

[c] Do not incubate the gel for a long time. At this stage, proteins have not yet been fixed and may migrate out of the gel.

[d] The stain is stable in water and gels can be stored at 4 °C. If desired, zinc/imidazole stained gels can be rapidly destained with chelating agents as EDTA (0.1 mg/ml), citric acid (0.1%) or Tris-glycine buffers (gel running buffer without SDS). Note that in this case proteins get 'mobilized' and small proteins might readily diffuse out of the gel matrix.

4.3 Proteolytic cleavage of gel separated proteins

In order to ensure complete digestion of proteins inside the gel matrix, aggressive proteases at high concentrations are normally used. The total amount of enzyme present in a gel piece is typically in the range of 5–10 pmol, which, in most cases, represents a large excess over the protein substrate. Therefore, it is crucial that the protease employed is largely resistant to autoproteolysis or at least yields a simple, predictable pattern of autolysis products. As can be seen from *Table 3*, this is the case for some but not all proteases.

If protein bands of interest appear to be contaminated, a 'blank' spot can be excised from below the band of interest in order to obtain a pattern representative of the contamination present.

Protocol 3

In-gel protein reduction and alkylation

Reagents

- Acetonitrile
- Reduction buffer: 10 mM DTT in 100 mM NH_4HCO_3
- 100 mM NH_4HCO_3
- Alkylation buffer: 55 mM iodoacetamide in 100 mM NH_4HCO_3

Protocol 3 continued

Method

1 Rinse the gel with water. Excise bands of interest with a clean scalpel. Cut as close to the edge of the band as possible in order to reduce the amount of 'background' gel.

2 Chop the excised bands into pieces of $\approx 1 \times 1$ mm and transfer the gel particles into a 0.5-ml microcentrifuge tube.

3 Wash the gel particles with 100–150 μl of water (5 min). Vortex briefly, spin down gel particles and discard the liquid.

4 Add acetonitrile (corresponding to ≈ 3–4 times the total volume of gel pieces) and incubate for 10–15 min until the gel pieces shrink, become opaque and stick together. Spin down the gel particles and discard all liquid.

5 Dry down gel particles in a vacuum centrifuge.

6 Swell the gel pieces in reduction buffer and incubate for 30 min at 56 °C to reduce the protein.[a]

7 Spin down gel particles and discard all liquid. Shrink the gel pieces with acetonitrile.

8 Replace acetonitrile with alkylation buffer. Incubate for 20 min at room temperature in the dark.

9 Discard iodoacetamide solution. Wash the gel particles with 150–200 μl of 100 mM NH_4HCO_3 for 15 min. Spin down gel particles and discard all liquid. Shrink the gel pieces with acetonitrile.[b]

10 Dry down gel particles in a vacuum centrifuge.

[a] Note that we recommend *in-spot* reduction even if proteins were reduced (but not alkylated) prior to electrophoresis.

[b] Extended destaining might be necessary for Coomassie Brilliant Blue-stained gels. If gel pieces remain blue after the reduction and alkylation step has been performed, rehydrate gel particles in 100–150 μl of 100 mM NH_4HCO_3. Incubate for 10 min and add an equal volume of acetonitrile. Vortex briefly, incubate for 10 min, spin down gel particles and discard all liquid. Shrink gel pieces with acetonitrile. Repeat washing until all stain is removed.

Protocol 4

In-gel trypsin digestion[a]

Reagents

- Digestion buffer: 50 mM NH_4HCO_3 containing 5 mM $CaCl_2$[b]
- Trypsin, sequencing grade (Boehringer Mannheim)[c]
- 25 mM NH_4HCO_3
- Acetonitrile
- 5% Formic acid

Protocol 4 continued

Method

1 Rehydrate gel particles in the digestion buffer containing 12.5 ng/μl trypsin at 4 °C (ice bucket).

2 Incubate for 30–45 min at 4 °C. After 15–20 min check the samples and add more digestion buffer (including trypsin) if all liquid has been absorbed by the gel pieces.

3 Remove and discard remaining supernatant. Add 5–25 μl of the same buffer but without trypsin to cover gel pieces and to keep them wet during enzymatic cleavage (37 °C, overnight).

4 After overnight incubation, spin down the water droplets condensed inside the lid of the microcentrifuge tube. Take up a small (0.3–0.5 μl) aliquot of the supernatant for MALDI analysis (see *Protocol 5*).

5 Extract peptides from the gel matrix by adding 10–15 μl of 25 mM NH_4HCO_3. Vortex briefly, incubate for 10 min, spin down the gel particles and recover the supernatant. Add acetonitrile (1–2 times the volume of the gel particles) and incubate at 37 °C for 15 min with agitation. Spin down the gel particles and collect the supernatant. Add 40–50 μl of 5% formic acid and extract for 15 min at 37 °C with agitation. Spin down the gel particles, collect supernatant and add acetonitrile (1–2 times the volume of the gel particles). Incubate at 37 °C for 15 min with agitation. Spin down the gel particles and recover the supernatant. Pool all extracts and dry down in a vacuum centrifuge.[d]

[a] The same protocol also applies to digestions using Lys-C and Glu-C (Boehringer Mannheim).

[b] For C-terminal ^{18}O-labelling of tryptic peptides, the digestion buffer contains 50% v/v $H_2{}^{18}O$. Since trypsin stock solutions are usually prepared in 1 mM HCl in normal ($H_2{}^{16}O$) water, the volume of the aliquot of trypsin stock solution should be subtracted from the volume of $H_2{}^{16}O$ used to prepare the digestion buffer. Note that commercially available $H_2{}^{18}O$ is chemically not pure enough and should be distilled prior to use.

[c] Although sequencing-grade trypsin produces more autolysis products than modified trypsin, using the unmodified enzyme is preferred because it is very well characterized.

[d] Dry digests can be stored in a freezer at −20 °C for months.

4.4 Protein identification by MALDI peptide mass mapping

Several features of MALDI-MS make this technique an excellent tool for the initial analysis of peptides derived from in-gel digestions. First, sample preparation is very simple but nonetheless highly efficient. Homogeneous matrix surfaces, necessary for obtaining high mass accuracy and low femtomole sensitivity (HCCA, sinnapinic acid and ferulic acid) can easily be prepared by the fast evaporation method described in *Protocol 5* (35,36). The addition of nitrocellulose to the matrix (33,37) further improves the sensitivity and generality of the fast evaporation method. Excess salts encountered in the digestion process are removed by matrix rinsing. As a result, a high-quality MALDI spectrum can be obtained directly from a small (i.e. 1–2%) aliquot of digest supernatant. In our experience, direct analysis of digest supernatant affords superior results over the analysis of extracted peptides. At the time of writing, the best sensitivities reached with MALDI peptide mass mapping correspond to the low silver-stained range of gel separated protein.

A further advantage of MALDI is the speed of data acquisition and the simplicity of data interpretation. Both data acquisition and interpretation can be automated (22) to enable high sample throughput if required. The mass accuracy needed for high-confidence protein identification can be obtained by internal mass calibration. For HCCA the mass range between 800 and 1100 contains a number of reproducibly occurring matrix ions (notably m/z 855.10 and m/z 1066.10) which can be distinguished from peptide ions by their different fractional mass value. Peptide mixtures derived from in-gel digestion using unmodified (i.e. non-alkylated) trypsin display a small number of well-defined autolysis products (notably m/z 2163.057) which, together with matrix ions, can be used for an accurate two-point calibration.

Protocol 5

Sample preparation for MALDI peptide mass mapping

Reagents

- HCCA
- Acetone
- Isopropanol
- Nitrocellulose (Transblot, BioRad)
- 10% formic acid
- 5% formic acid

Method

1 Prepare a saturated solution of HCCA in acetone. Vortex briefly and centrifuge to remove insoluble material.

2 Prepare a solution of nitrocellulose (10 g/l) in 1:1 (v/v) acetone–isopropanol.

3 The matrix solution is prepared by mixing four volumes of HCCA solution with one volume of nitrocellulose solution.

4 Deposit 0.3 μl of matrix solution onto the MS probe. The solution will spread out (no confining edges) and evaporate within a few seconds to produce a thin film.[a]

5 Deposit 1 μl of 10% formic acid onto the matrix film. Inject 0.3–0.5 μl of the supernatant of an in-gel protein digest (see Protocol 4) into the acidic droplet and allow to dry at room temperature.[b]

6 Rinse the sample by depositing 5–10 μl of 5% formic acid onto the film. Shake off the liquid immediately and allow the sample to dry. This procedure lowers the salt content of the sample and can be repeated several times as long as the matrix film is not visibly damaged.[c]

Protocol 5 continued

7 Insert probe into the mass spectrometer and acquire spectrum.

[a] The matrix film should be homogeneous and no crystals should be visible to the naked eye. If somewhat thicker matrix layers are desired 1–3% (v/v) water should be added to the matrix solution.

[b] Sample injection into an acidic droplet is necessary to prevent dissolution of the thin film by alkaline buffers such as NH_4HCO_3.

[c] The nitrocellulose retains peptides during the washing procedure but does not hamper subsequent analysis by MALDI MS.

As an example of protein identification by peptide mass mapping, *Figure 5* shows a MALDI spectrum of tryptic peptides derived from a yeast protein complex following separation by one-dimensional SDS-PAGE. The inset shows the resolved $^{12}C/^{13}C$ isotope pattern for one of the peptide peaks. The masses obtained were used to search an NRDB consisting of (at the time of writing) ≈ 300 000 entries. The search results are compiled in *Tables 4* and *Table 5*. Yeast NIP29 was unambiguously identified in the search. Several criteria can be applied to corroborate the match:

(1) The target protein should be among the top candidates on the list based on the number of peptides matched.

(2) At least five peptides must be detected to claim a significant match. This might be difficult to achieve for some small, extensively modified or very hydrophobic proteins, as well as for low protein amounts, which generally results in the observation of fewer peptides.

(3) The majority of peptide masses should match within 50 p.p.m. mass accuracy. If the instrument does not allow this mass accuracy more peptide masses have to match in 2, above.

(4) There should be a gap in peptide mass matches between the top and second best candidate. In our example (*Table 4*), 10 peptides matched for the top candidate whereas only six matched for the second best match. The second best protein from yeast only matched with five peptides.

(5) A second pass search should be performed (*Table 5*) in which as many of the remaining peptide masses as possible are matched to the sequence, considering common modifications (oxidation of methionine, acrylamidation of cysteines), common adducts (sodium (+ 22 Da) or potassium (+ 38 Da)), missed cleavages and lower mass accuracy. In *Figure 5*, ions corresponding to peptides from NIP29 are marked with bullets. All major ions in the spectrum are assigned to this protein and cover ≈ 38% of the protein sequence, a typical value obtained in MALDI peptide mass mapping of gel-separated proteins.

(6) While it is not recommend that the mass range searched is restricted, a rough estimate of the molecular weight of the target protein (as judged by SDS-

PAGE) can be used to distinguish significant hits from false positives after the search. As is evident from *Table* 4, most of the second-best matches correspond to proteins with molecular sizes in excess of 200 kDa. Proteins of this mass range are more likely to yield four to six random matching peptides for any set of peptide masses. In such cases, searching with slight changes in the allowed peptide mass deviation can result in a completely different ranking of randomly matched proteins, whereas the target protein will be largely unaffected.

(7) Secondary information, such as apparent isoelectric point and species of origin, can further substantiate a match. However, these criteria should not serve as primary search information because of the variability of two-dimensional gel electrophoresis and the occurrence of common contamination such as human keratins and virus proteins in yeast preparations.

(8) Many protein sequences contain consecutive tryptic cleavage sites (Lys-Lys, Lys-Arg, Arg-Lys and Arg-Arg). These sequence patterns are often reflected in MALDI peptide maps as peaks differing by 128.095 Da (Lys) or 156.101 Da (Arg). These pairs of signals can be used to construct a 'terminal peptide sequence tag'. Although there is only a single amino acid to search by, the high mass accuracy of the mass measurement can reduce the number of random matches significantly.

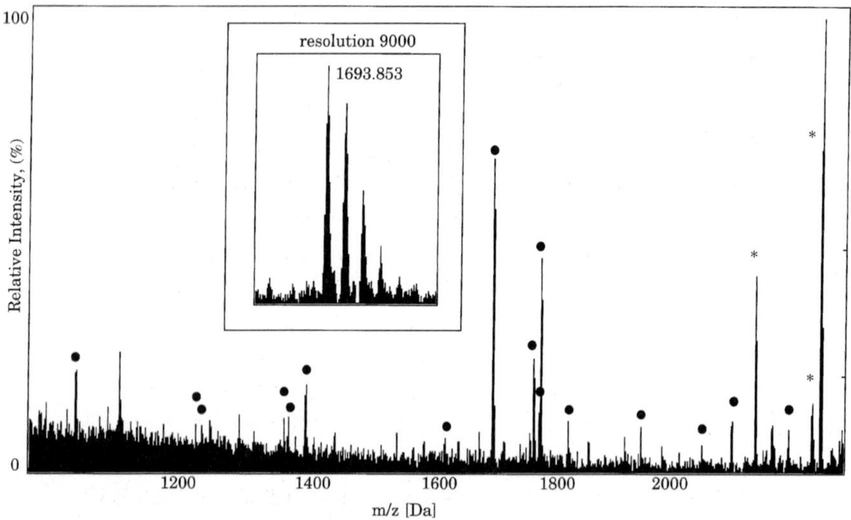

Figure 5 Protein identification by MALDI peptide mass mapping. The unseparated peptide mixture derived from an in-gel proteolytic digestion was analysed by MALDI using the sample preparation procedure described in *Protocol 5*. The spectrum was calibrated using trypsin autolysis products (marked with an asterisk) and low mass matrix-related ions. The masses obtained were used to search an NRDB. Search results are compiled in *Tables 4* and *5*. All major signals in the spectrum correspond to sequence specific peptide ions of yeast NIP29 (marked with bullets) and thereby unambiguously identify the protein. The inset shows the resolved $^{12}C/^{13}C$ isotope pattern for one of the peptide signals (monoisotopic mass 1693.853 ± 0.02Da and illustrates the high mass accuracy and resolution that can be obtained by MALDI-TOF MS.

Table 4 Protein identification by MALDI peptide mass mapping: search results[a]

Peptides matched	Mass (kDa)	Database accession	Protein name
10	29.28	swissprot: P33419	NI29_YEAST NIP29 PROTEIN
6	222.38	spt:Q02456	Q02456 MYOSIN HEAVY CHAIN
6	262.17	spt:Q23081	Q23081 CODED FOR BY *C. elegans*
6	232.59	spt:Q02015	Q02015 MYOSIN HEAVY CHAIN
6	231.47	trembl:M93676	GGNMHCB_3 product
6	230.17	trembl:M93676	GGNMHCB_2 product
6	223.86	swissnew:P12847	MYSE_RAT MYOSIN HEAVY CHAIN
6	229.04	trembl:M93676	GGNMHCB_1 product
6	110.14	swissnew:P37202	DIS3_SCHPO MITOTIC CONTROL PROTEIN
5	126.87	swissprot:P38990	PAK1_YEAST SERINE/THREONINE-PROTEIN

[a] Search parameters: protein mass range 0–300 kDa; cleavage agent, trypsin; peptide mass accuracy 0.005%; methionine is native; cysteine is Carbamidomethyl-Cys; peptide charge state, protonated; number of peptides required for match, 5; number of uncleaved sites, one; number of peptides used in search, 27.

Table 5 Protein identification by MALDI peptide mass mapping: second pass search[a]

Measured mass (Da)[b]	Calculated mass (Da)	Difference (Da) (p.p.m.)	Residues	Sequence
1060.498	1060.518	0.019 (18)	180–188	(K)NFASLEHSR(S)
1238.686	1238.686	0.0 (0)	22–31	(R)VRKEHEEALK(K)
1292.643	1292.624	0.019 (15)	128–138	(K)FASQNVIDDQR(L)
1360.62	1360.661	–0.04 (29)	177–188	(R)GDKNFASLEHSR(S)
1366.814	1366.781	0.033 (24)	22–32	(R)VRKEHEEALKK(L)
1393.682	1393.733	–0.05 (36)	165–176	(R)LESFILNSISDR(G)
1636.781	1636.793	–0.012 (7)	125–138	(R)TDKFASQNVIDDQR(L)
1693.854	1693.876	–0.021 (12)	165–179	(R)LESFILNSISDRGDK(N)
1763.931	1763.966	–0.034 (19)	162–176	(R)ITRLESFILNSISDR(G)
1773.830	1773.863	–0.032 (18)	94–109	(R)LPPPPFSSYGMPPTNR(S)
1775.884	1775.929	–0.045 (25)	128–142	(K)FASQNVIDDQRLEIK(Y)
1824.837	1824.873	–0.035 (19)	33–48	(K)LREENFSSNTSELGNK(K)
1952.972	1952.968	0.004 (2)	32–48	(K)KLREENFSSNTSELGNK(K)
2064.033	2064.109	–0.075 (36)	162–179	(R)ITRLESFILNSISDRGDK(N)
2120.065	2120.099	–0.033 (16)	125–142	(R)TDKFASQNVIDDQRLEIK(Y)
2228.129	2228.142	–0.013 (6)	14–31	(K)FQDDTLNRVRKEHEEALK(K)

[a]These 16 peptides cover 96 out of 253 amino acids (37.94 %). 11 masses were not identified.

Sequence coverage: amino acids represented in the mass map are printed in bold:
MDYSNFGNSASKK**FQDDTLNRVRKEHEEALKKLREENFSSNTSELGNK**KHYRAQERMSSPLHRLSPTGKSDDRKVKSPLDDKLRRQLREGNTR**LPPPPFSSYGMPPTNR**SNLDRIRRRTSSPVR**TDKFASQNVIDDQRLEIK**YLERIVYDQGTVIDNLTSR**ITRLESFILNSISDRGDKNFASLEHSR**SFSGFPTNKTYGLQMGGLYENDMPYRRSSDNINKEGAREDRSSQIHIENESTEDILKILSSSFHN

[b]All masses correspond to the monoisotopic masses of the protonated molecules.

4.5 Peptide sequencing by nanoelectrospray tandem mass spectrometry

If a protein cannot be identified by MALDI mass mapping, a partial peptide sequence has to be generated by MS-MS in order to construct a peptide sequence tag for subsequent database searching. Nanoelectrospray MS has proven to be a powerful technique for this task. As mentioned above, the main characteristic of nanoelectrospray MS is the very low sample flow rate (25 nl/min) which translates directly into prolonged available analysis time on small volumes of peptide solution (40 min/μl). This allows the direct analysis of unseparated peptide mixtures by sequentially fragmenting the peptide ions in turn. At low sample levels in particular, this increased experimental time frame allows careful inspection of spectra for the selection of appropriate peptide signals for tandem MS and the optimization of experimental parameters during MS-MS experiments. In contrast to MALDI, peptides derived from in-gel protease digestions have to be concentrated and purified prior to analysis by nanoelectrospray MS. This can be accomplished by a simple one-step procedure as illustrated in *Figure* 6 and described in *Protocol* 6.

Protocol 6

Sample desalting and concentration for nanoelectrospray MS analysis

Equipment and reagents

- Nanoelectrospray capillaries (custom made,[a] or commercially available from Protana A/S, Danmark)
- Custom-made capillary holder (*Figure* 6)
- Table micocentrifuge (e.g. PicoFuge, Statagene)
- POROS R2 reversed-phase resin (PerSeptive Biosystems)
- 5% formic acid
- 50% methanol in 5% formic acid

Method

1. Mount a microcolumn (i.e. a nanoelectrospray needle) in a custom-made holder as outlined in *Figure* 6.
2. Prepare a slurry of POROS R2 reversed-phase resin in methanol and remove fines by repeated agitation/settling of the slurry with subsequent removal of the supernatant.[b]
3. Pipette 5–7 μl of the slurry into the microcolumn and spin down the particles by gentle centrifugation using a table microcentrifuge. Repeat loading if necessary until the cone of the capillary is filled with sorbent.
4. Break the tip of the capillary by gently touching the tabletop and spin out the remaining liquid. The liquid should pass freely through the sorbent layer, but the sorbent particles themselves must remain firmly within the microcolumn.

Protocol 6 continued

5 Redissolve the dried peptide mixture obtained by in-gel digestion (see *Protocol 4*) in 5–15 μl of 5% formic acid.[c] Equilibrate the sorbent with 5% formic acid (1–2 μl) and load the sample onto the microcolumn using gentle centrifugation. After loading, wash the microcolumn with 2–3 μl of 5% formic acid and spin down the liquid until the sorbent layer is completely dry.

6 Mount the nanoelectrospray capillary and the microcolumn in the holder (see *Figure 6*) and check their alignment. Elute the sample with two portions of 0.5 μl of 50% methanol in 5% formic acid.[d] Remove the spraying capillary from the holder, mount it in a nanoelectrospray ion source and acquire spectra.

[a] The fabrication of nanoelectrospray capillaries is described in ref. (14).

[b] POROS R2 resin is generally used for peptides but R1 or R3 material may also be used for very hydrophobic or hydrophilic peptides, respectively.

[c] Note that higher initial formic acid concentrations may be needed in order to recover large, hydrophobic peptides.

[d] Most peptides will be eluted under these conditions but higher or lower percentages of organic phase might also be used.

Owing to the small dimensions of the nanoelectrospray ion source, operation parameters differ somewhat from conventional electrospray ion sources. A stable spray can normally be generated at a needle potential of 400–800 V instead of 3–5 kV and by applying a gentle back-pressure to the capillary by means of a syringe filled with air. If the pressure does not start the flow, it may be necessary to open the spraying capillary by gently touching the interface plate. Very small droplets are produced in the nanoelectrospray process which rapidly evaporate. Hence, the distance between the spraying needle and the sampling orifice can be decreased to a few millimetres.

An example for protein identification by nanoelectrospray MS-MS is given in *Figure 7*. The experiment essentially consists of three steps. First, peptides have to be mass measured. Selected peaks then need to be fragmented in order to generate the required sequence information. Finally, the tandem mass spectra have to be interpreted for the construction of the peptide sequence tag and subsequent database searching.

When analysing sub-picomole amounts of in-gel digested proteins, it is a common observation that peptide signals can be obscured by high levels of chemical noise in single MS spectra (Q_1 spectrum, *Figure 7A*). In such cases, precursor ion scanning for characteristic amino acid ions (notably the immonium ion of Leu/Ile at m/z 86) is a highly discriminating experiment which enables facile detection of peptide ions, despite the presence of non-peptide contaminants (*Figure 7B*). Once peptide masses have been obtained in this way, individual peptides can be selected for MS-MS (*Figure 7C*). Again, at low sample levels it may be difficult to make a straightforward choice since spectra are frequently dominated by trypsin autolysis products or peaks corresponding to contaminating

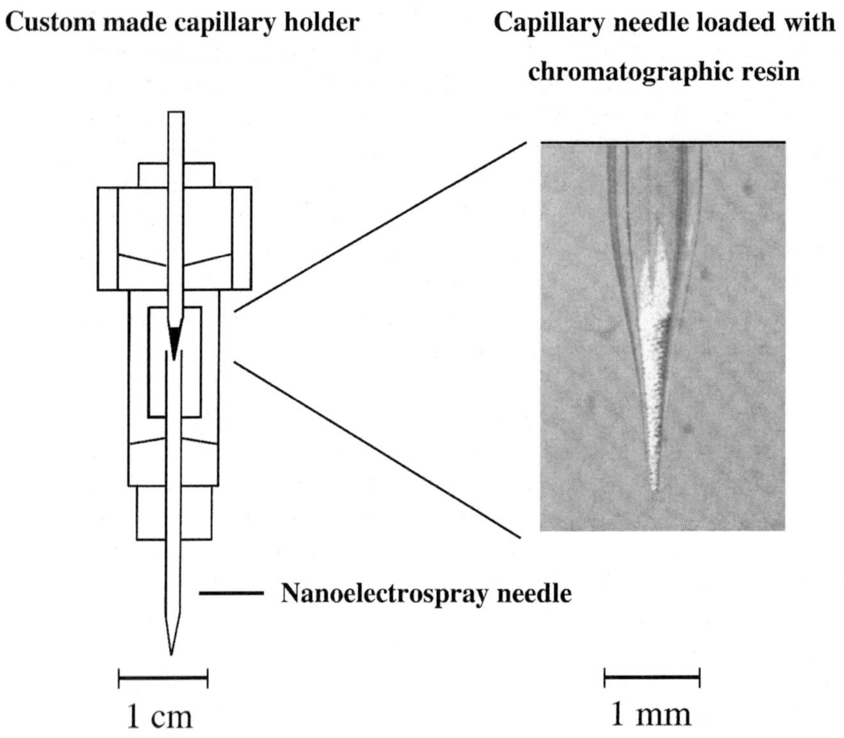

Figure 6 Small-scale peptide purification. Peptides obtained by in-gel proteolytic cleavage are concentrated and purified using a small amount of POROS R2 resin. First, the sample is applied to a microcolumn as shown on the right and washed with a few microlitres of 5% formic acid. The column is then transferred into the capillary holder and aligned with the nanoelectrospray needle. Finally, peptides are eluted into the nanoelectrospray needle with two 0.5-μl aliquots of 50% methanol in 5% formic acid using a table centrifuge.

proteins, such as human and sheep keratins. It is therefore advisable to analyse 'blank' gel pieces and to tabulate the masses and sequences of common peptides in order to avoid repeated sequencing of trypsin and keratin peptides. Tryptic peptides yield mainly doubly charged peptides because of the basicity of the *C*-terminal amino acid and the free *N*-terminus. However, higher charge states might also be observed and hence peptide precursor ions should be 'micro-scanned' at high resolution to determine their charge states for simplified data interpretation.

The behaviour of peptide ions under CID conditions depends on the collision energy used, the amino acid composition and mass and, to a lesser extent, the peptide charge state. Generally, fragmentation occurs by cleavage of amide bonds between adjacent amino acids. After cleavage, the charge can reside on the *N*-terminal (A- and B-type) or *C*-terminal (Y″-type) fragment ion or on both. Doubly charged precursor ions generally form singly charged fragments, and triply charged precursors yield both singly and doubly charged fragments. Tryptic peptides are similar in their chemical nature since the most basic amino acids (Lys, Arg) are located at the *C*-terminus. Hence, their fragmentation behaviour is also similar and can, to some extent, be predicted for a given type of mass spectrometer. Empirically, CID of tryptic peptides on triple quadrupole mass spectrometers mainly (but not exclusively) yields Y″-type ions. For such peptides, the most important experimental parameter is the peptide mass. As a general rule, the collision energy needed for fragment ion generation increases with increasing mass. At the same time, lower energy is needed to generate large fragment ions whereas more energy is required for the formation of small fragment ions. Thus, in order to generate long stretches of peptide sequence, the collision energy may have to be adjusted during an MS-MS experiment. Alternatively, the tandem mass spectrum can be acquired in two or three segments.

As can be seen by comparing *Figures 7A* and *7B*, selection of a particular peptide mass for MS-MS will also always include chemical noise. The non-peptide species are mainly singly charged and give rise to singly charged fragment ions which will populate the low m/z region of the tandem mass spectrum. For low abundance peptides, the presence of these background fragment ions can substantially hamper interpretation of the low m/z region of the spectrum. Therefore, precursor ions should be selected with 'high' (e.g. 1.5 Da) resolution when recording the low m/z region. The high m/z range is normally very clean, i.e. contains few if any contaminating ions and hence data interpretation is relatively simple. The following rules can be applied to find a consecutive ion series for the construction of a peptide sequence tag.

(1) Chose an abundant ion in the high m/z region of the tandem mass spectrum as a reference point. For MS/MS of tryptic peptides on a triple quadrupole mass spectrometer, this is likely to be a Y" ion. Note that B-ion series can dominate this part of the spectrum if internal basic residues are present or when the *C*-terminal peptide of the protein is sequenced. Ion trap MS-MS also generally yields abundant B-type fragment ion series.

(2) The next sequence ion can be found by searching for a peak (up or down in mass) with a mass difference of precisely one amino acid residue. Frequently, satellite peaks corresponding to the loss of water (−18 Da) or ammonia (−17 Da) accompany respective sequence ions. Note that more than one possibility might sometimes have to be considered.

(3) With the next sequence ion identified, (2) is repeated until no further unambiguous sequence assignments can be made. Generally, two to four amino acids in a peptide sequence tag are sufficient for database searching and such tags can be generated even from very weak MS-MS spectra.

From the spectrum shown in *Figure 7C* the peptide sequence tag (600.2)ED(I/L)(957.4) was assembled. Searching of the database using this tag, together with the mass of the intact (neutral) peptide identifies VMA2 from yeast. As in MALDI peptide mass mapping, the database match should be confirmed by independent information. This can be achieved by

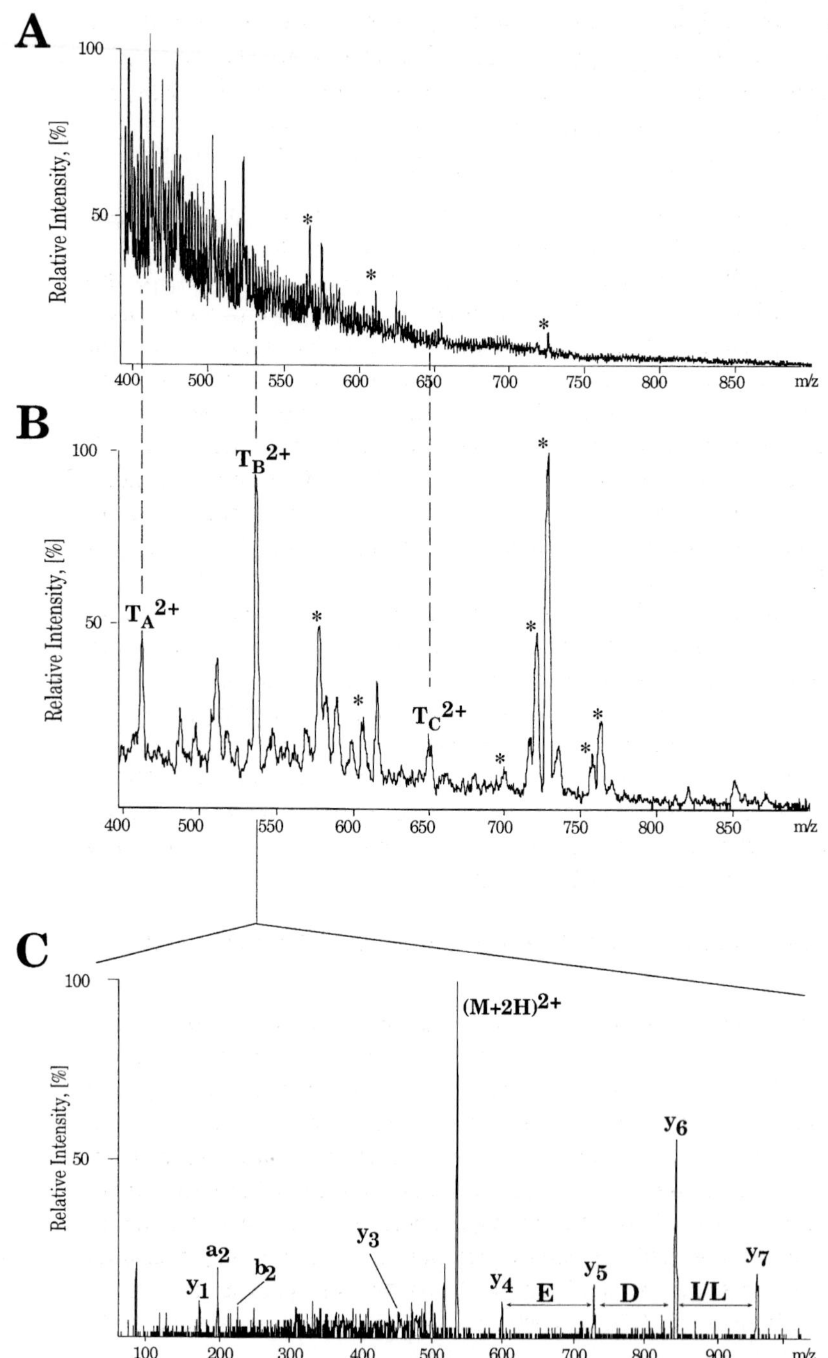

Figure 7 Protein identification by nanoelectrospray MS-MS. (A) Mass spectrum (Q_1 scan) of an in-gel tryptic digest of a yeast protein. Signals corresponding to trypsin autolysis products are marked with asterisks. (B) Precursor ion mass spectrum for the immonium ions of Leu/Ile (*m/z* 86) acquired from the same sample. Tryptic peptides of the target protein are

(1) Identification of the same protein using peptide sequence tags derived from MS-MS spectra of other peptides (e. g. peptides T_A and T_C in *Figure 7B*).

(2) Comparing the experimentally obtained tandem mass spectrum with the predicted tandem mass spectrum of the sequence retrieved.

(3) Peptide masses as obtained by Q_1 scans or MALDI MS.

Data interpretation may not always be as straightforward as in the example discussed here. Internal proline residues, for example, usually yield an intense Y''_n ion corresponding to cleavage of the *N*-terminal proline peptide bond. In this case, the following sequence ion (Y''_{n-1}) is either of very low abundance or totally absent, which can complicate further data interpretation. However, proline normally gives rise to an 'internal' fragment ion series—consisting of fragments starting with the proline residue—which can often confirm the sequence.

Two pairs of amino acids have the same nominal mass. While Leu and Ile (113 Da) have identical elemental compositions and therefore cannot be distinguished by the low-energy CID conditions afforded by most mass spectrometers, Lys can usually be distinguished from Gln (128 Da) by the fact that tryptic peptides rarely contain internal Lys residues. If, however, internal Lys residues are present, the precursor ion is normally triply charged and gives rise to an abundant doubly charged fragment ion series. Methionine residues are sometimes oxidized during sample preparation and thereby become isobaric with Phe (147 Da). However, MS-MS spectra of peptides containing an oxidized Met exhibit satellite peaks (−64 Da) for every Met containing Y″ ion due to the loss of the sulphoxide group (38). Methionine oxidation is also rarely quantitative. Hence, the mass spectra (Q_1) frequently contain the respective unmodified peptide (−16 Da). Other chemical modifications such as acrylamidation should also be considered when interpreting MS-MS spectra.

4.6 Towards cloning of proteins using mass spectrometry data

Before concluding that a protein under investigation is unknown, every possible effort should be made to find it in a database. In particular, error-tolerant searches should be performed, i.e. allowing for peptide modifications or considering mass and sequence errors in the experimentally obtained data. If still no protein is identified in comprehensive amino acid sequence databases, searching should be extended to EST databases. If a match has been found, it should be corroborated in the same way as described in the previous section. The physical EST clone can then be obtained and used to clone the cognate gene. The full oligonucleotide insert should be sequenced and the predicted peptides compared with the experimentally obtained MS-MS spectra in order to verify the identity of the clone.

---

designated T_A to T_C. (C) MS-MS of the doubly charged ion T_B^{2+}. From this spectrum, the sequence tag (600.2)ED(L/I)(957.4) was derived and searched against an NRDB. Yeast protein VMA2 was unambiguously identified by comparison of the predicted fragmentation pattern with the experimentally obtained tandem mass spectrum.

If extensive, error-tolerant database searching does not produce any significant match, then *de novo* peptide sequencing has to be attempted. This takes the form of careful and complete interpretation of MS-MS spectra in order to assemble long amino acid sequence stretches. Although tryptic peptides predominantly produce Y″-type fragment ions in the high mass part of MS-MS spectra obtained on triple quadrupole instruments, their presence can be difficult to establish in the molecular weight region below the parent ion. In order to recognize the full Y″ ions series the peptides are methyl esterified on all carboxylic acid groups (*C*-terminus, side-chains of Asp and Glu), as outlined in *Protocol 7*. As a result, a mass increase of 14 Da is observed for each derivatized site. Since Y″ ions (but not B ions) contain the *C*-terminus, a respective mass shift is observed in tandem mass spectra and an interpreted sequence can be confirmed by comparing the two data sets. Furthermore, the esterification simplifies the differentiation between Glu/Gln and Asp/Asn residues which, in their native state, differ only by one mass unit and are sometimes difficult to distinguish in weak MS-MS spectra. In the example given in *Figure 8*, a total of four methyl groups were incorporated into the peptide and the localization of the three acidic amino acids is apparent by comparing the tandem mass spectra of the native and the esterified peptide.

An elegant alternative way to recognize *C*-terminal fragment ions (i. e. Y″ ions) selectively, is by isotopic labelling of the *C*-terminus. This can easily be accomplished by performing the in-gel trypsin digest in the presence of 50% $H_2{}^{18}O$ (see footnote of *Protocol 4*). This labelling results in the detection of peptide ions in a mass spectrum as doublets with a 2 Da spacing. The same isotope pattern will be observed for *C*-terminal fragment ions in MS-MS spectra as can be seen in *Figure 9* (lower panel). *N*-terminal ions (B type) do not exhibit this isotope effect and are therefore readily distinguished from the corresponding *C*-terminal fragment ions. It should be noted, however, that the mass spectrometer employed has to provide sufficient resolution to enable the unambiguous detection of the 2 Da spacing. Hybrid instruments consisting of a quadrupole-TOF geometry are ideally suited for this kind of labelling since they afford high resolution for both MS and MS-MS experiments, whereas sensitivity may be compromised when triple quadrupole mass spectrometers are used.

It should be stressed here, that *de novo* peptide sequencing by MS is not trivial. Complete interpretation of mass spectra is not as straightforward as assembling a sequence tag and requires substantial expertise and experience. Generally, more material is also needed because additional experiments have to be performed in order to assign complete peptide sequences (methyl esterification). When peptides are labelled with ^{18}O, signals are reduced by 50% because the signal is split between the $^{16}O/^{18}O$ isotopes.

For cloning, oligonucleotide primers have to be designed on the basis of the obtained peptide sequence. It is advantageous to select peptide sequences which translate into low degeneracy oligonucleotides. Therefore, it is not always possible to use peptides which were originally detected by high-sensitivity precursor ion scanning of Leu/Ile immonium ions since these amino acids have the same mass and are encoded by a total of nine (6 + 3) nucleotide triplets. Hence, sufficient material ought to be available such that peptides can be recognized in single MS experiments. Furthermore, it should be attempted to sequence as many peptides as possible in order to be able to choose the lowest degeneracy peptides for primer design. In this scenario, the long analysis time afforded by nanoelectrospray MS unfolds its full potential.

Finally, when the gene of interest has been cloned either by the use of EST clones or degenerate oligonucleotide primers, the cloned sequence should be compared with all available MALDI and nanoelectrospray MS-MS data. Since only a small part of the sequence information was used in designing the primers this comparison verifies the identity of the cloned protein.

Protocol 7
Methyl esterification of peptides

Reagents

- Distilled anhydrous methanol
- Distilled acetylchloride
- 5% acetic acid

Method

1 To prepare the esterification reagent, cool 1 ml of freshly distilled methanol to −40°C to −80°C in a 1.5-ml microcentrifuge. Quickly add 100 μl of freshly distilled acetylchloride,[a] close the lid of the tube and allow the reaction mixture to warm up to room temperature.

2 An unseparated peptide mixture, as recovered after in-gel digestion (see *Protocol 4*) is dried down in a vacuum centrifuge. Add 3–10 μl of the reagent to ensure that salts or residual buffer components are completely covered. If a large solid precipitate is observed, a 3-min sonication step is recommended.[b]

3 Incubate the reaction mixture at room temperature for 30 min.

4 Dry down the reaction mixture in a vacuum centrifuge. Redissolve esterified peptides in 5% formic acid[c] and desalt/concentrate prior to nanoelectrospray MS as described in *Protocol 6*.[d]

[a] Take appropriate precautions as the mixture may react violently.

[b] The presence of salts normally does not affect the completeness of the derivatization reaction, provided that a sufficiently large volume of the reagent has been added.

[c] Higher formic acid concentrations might be necessary for good recovery of some peptides.

[d] Esterified peptides are stable under acidic conditions but may undergo rearrangement at mild alkaline pH.

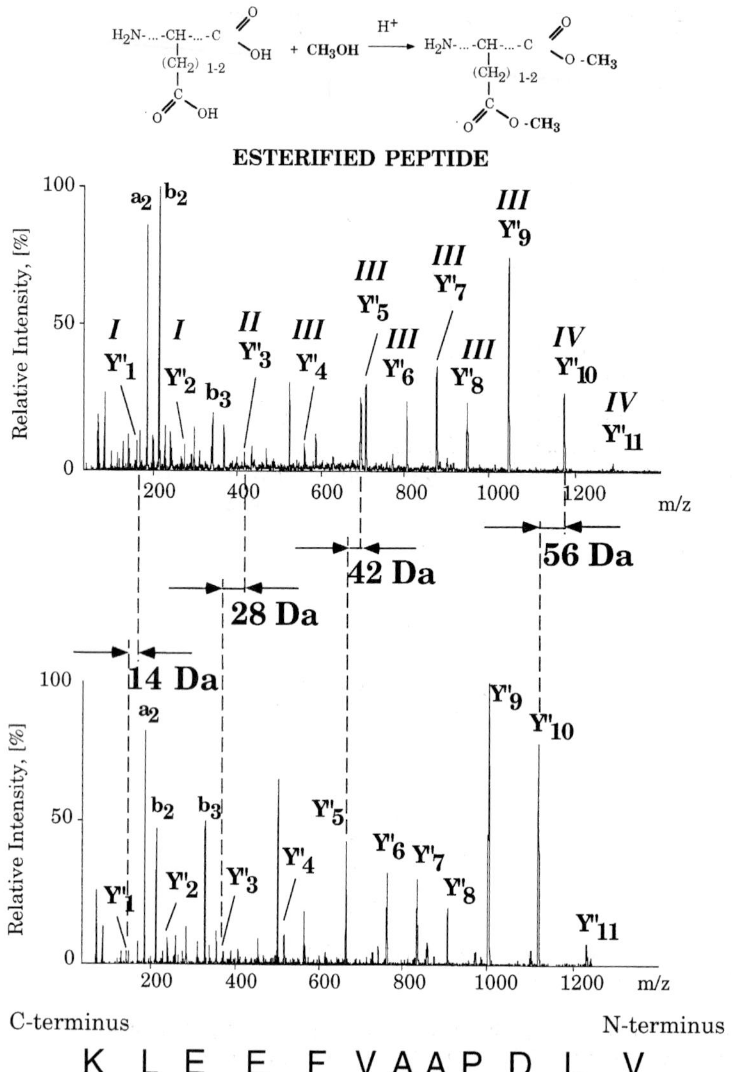

Figure 8 *De novo* peptide sequencing using peptide esterification. The unseparated mixture of tryptic peptides recovered after in-gel digestion of thrombomucin was split in two portions. The first portion was analysed by nanoelectrospray MS-MS. The second portion was subjected to methyl esterification as outlined in the top panel and described in Protocol 7. The derivatized sample was subsequently also analysed by nanoelectrospray MS-MS. The lower panel shows the spectrum obtained for the doubly charged peptide ion at *m/z* 665.6. The corresponding spectrum of the derivatized peptide (*m/z* 721.6) is shown in the middle panel. Esterification results in a 14 Da mass increase for all *C*-terminal fragment ions plus an additional increment of 14 Da for each Asp or Glu residue. The actual number of methyl groups attached to the *C*-terminal fragment ions (Y" ions) is shown in roman numerals above the respective sequence ion. Consideration of precise mass differences between adjacent Y" ions and observation of characteristic mass shifts between corresponding Y" ions in the spectra of the native and esterified peptide allowed the unambiguous determination of the full peptide sequence.

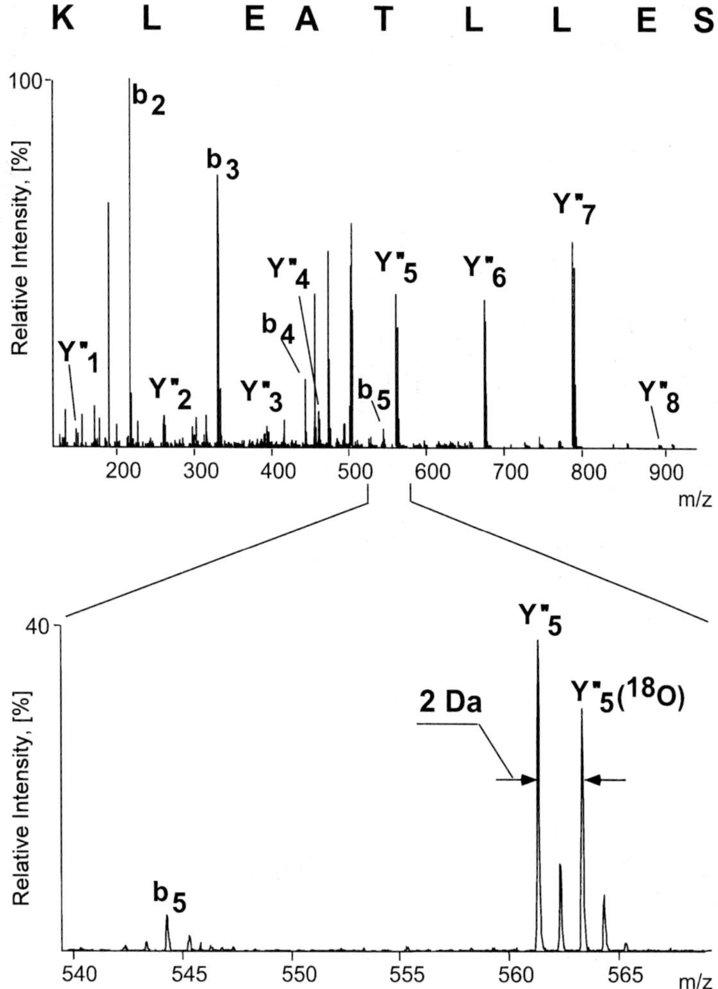

Figure 9 *De novo* peptide sequencing using *C*-terminal ^{18}O labelling. If the in-gel trypsin digestion is carried out in the presence of 50% $H_2{}^{18}O$, the heavy oxygen isotope is partly incorporated into the carboxylic acid group at the *C*-terminus of the peptide. Hence, the *C*-terminal (Y″) fragment ions series in a tandem mass spectrum (upper panel) exhibits a characteristic isotopic pattern (lower panel) corresponding to the incorporation of the label which readily distinguishes it from the unlabelled *N*-terminal (B type) fragment ions. Precise mass differences between isotopically labelled peaks in the tandem mass spectrum allowed the determination of the full peptide sequence.

5 Post-translationally modified proteins

To some extent, the great success of the strategies for protein identification discussed above, lies in the use of partial yet specific information. Any collection of peptide signals in a MALDI peptide map or a peptide sequence tag derived from even a single peptide may suffice for a successful identification of the protein. However, identification of a modification requires the specific detection of the particular modified peptide, a considerably more difficult task. Furthermore, protein phosphorylation and particularly glycosylation is associated with an often high degree of microheterogeneity. As a result, identification of such modifications is less universal and sensitive. Nonetheless, MS is a key tool in the identification and characterisation of modified peptides since virtually all modifications are associated with an increase or decrease in molecular weight.

The most successful approach for the detection of glycopeptides or phosphopeptides in complex peptide mixtures employs on-line LC electrospray MS. In a technique known as collision excitation scanning, glycopeptide- and phosphopeptide-specific fragment ions are produced by CID while scanning the low mass range, whereas the mass of the intact molecules is determined under non-CID conditions while scanning the high mass range (39,40). Comparison of single ion chromatograms of these marker ions with the UV trace of the LC separation reveals the location and molecular weight of the modified peptides. Further characterization of the separated glycopeptides or phosphopeptides can then be carried out by specific enzyme treatments or MS-MS. The same concept was adapted for several other modifications including sulphatation and myristoylation and further developed to allow the simultaneous detection of multiple modifications in one LC-MS experiment (41).

Alternatively, phosphorylation and glycosylation can be detected at the low to sub-picomole level by precursor ion scanning of unseparated peptide mixtures or protein samples using nanoelectrospray MS (42,43). For phosphopeptides in particular, this approach has since proven to be of great value because of its high sensitivity. Only minimal alterations have to be made to the procedures described in *Protocol* 6. In addition, MS-MS sequencing of the modified peptide in the same experiment can reveal the site of phosphate attachment. Unfortunately, it was noted that for unseparated peptide mixtures, the glycosylation marker ions are not specific enough to assign glycopeptides with certainty.

Information regarding the glycosylation of proteins separated by gel electrophoresis has recently been obtained using in-gel proteolytic digestion followed by MALDI-TOF MS of the unseparated peptide mixture. Although some heterogeneity of the glycan moiety was apparent from the spectra, HPLC separation of the peptide mixture was necessary for a more detailed account of the site-specific glycosylation pattern (44). Alternatively, analysis of the global glycosylation pattern of gel separated glycoproteins was described employing in-gel deglycosylation using peptide-*N*-glycosidase F followed by profiling the released glycans by MALDI-TOF MS (45). Since the protein remains intact during deglycosylation, further protein characterization can be carried out using peptide mass mapping or sequencing by MS-MS. Despite these developments which promise improved sensitivity, protein glycosylation analysis is still most comprehensively addressed by LC-MS.

6 Concluding remarks

High-sensitivity protein and peptide characterization requires advanced analytical tools as well as efficient sample manipulation procedures. The protocols described in this chapter have proven to be very robust yet simple enough that they should work reproducibly in every laboratory. Gel-separated proteins can be analysed at the sub-picomole level when the precautions described are taken, but the sensitivity that can be realized also depends on the performance of the mass spectrometer available.

Both MALDI peptide mass mapping and peptide sequencing by nanoelectrospray MS-MS are powerful but, unfortunately, still rather expensive tools for the identification of proteins. Ideally, one would have access to both methods in order to realize the full benefits of two complementary techniques. However, if forced to decide on one or the other, it currently seems that nanoelectrospray MS-MS is the more versatile of the two methods. Known proteins can be identified with ease using peptide sequence tags and the ability to generate *de novo* peptide sequences aids in the process of cloning novel genes. However, the technique is also currently more laborious and requires more technical expertise than MALDI-MS. As a result, the number of samples that can be processed is limited. If a higher sample throughput is required or if a limited number of proteins have to be analysed repeatedly, the assets of MALDI-MS, notably speed, throughput and ease of operation, come into play. Since most of the more abundant proteins are represented in sequence databases already, and more and more genomic sequence information becomes available, MALDI-MS can be the primary workhorse in many applications. The majority of proteins from frequently studied organisms such as human, mouse, rat or *Drosophila* and almost all yeast proteins can already by identified by MALDI-MS alone.

Acknowledgements

We thank A. Podtelejnikov, M. Wilm and J. Andersen for the provision of data and for valuable discussion during the preparation of the manuscript. B.K. is supported by a long-term fellowship granted by the European Molecular Biology Organisation (EMBO). M.M.'s laboratory at EMBL is supported by a generous grant from the German technology ministry (BMFT).

References

1. Jensen, O., Shevchenko, A., and Mann, M. (1997). In *Protein structure – a practical approach* (ed. T. E. Creighton), p. 29. Oxford University Press, New York.
2. Winston, R. L. and Fitzgerald, M. C. (1997). *Mass Spectrom. Rev.* **16**, 165.
3. Wilm, M., Shevchenko, A., Houthaeve, T., Breit, S., Schweigerer, L., Fotsis, T., and Mann, M. (1996). *Nature* **379**, 466.
4. Roepstorff, P. (1997). *Curr. Opin. Biotechnol.* **8**, 6.
5. Nguyen, D. N., Becker, G. W., and Riggin, R. M. (1995). *J. Chromatogr. A*, **705**, 21.

6. Burlingame, A. L., Boyd, R. K., and Gaskell, S. J. (1996). *Anal. Chem.* **68**, 599R.
7. Gaskell, S. J. (1997). *J. Mass Spectrom.* **32**, 677.
8. Guilhaus, M., Mlynski, V., Selby, D. (1997). *Rapid Commun. Mass Spectrom.* **11**, 951.
9. Jennings, K. R. and Dolnikowski, G. G. (1990). In *Methods in enzymology* (ed. J. A. McCloskey), Vol. 193, p. 37. Academic Press, San Diego.
10. Jonscher, K. R. and Yates 3rd, J. R., (1997). *Anal Biochem,* **244**, 1.
11. Mann, M. and Talbo, G. (1996). *Curr. Opin. Biotechnol.* **7**, 11.
12. Karas, M. and Hillenkamp, F. (1988). *Anal Chem,* **60**, 2299.
13. Fenn, J. B., Mann, M., Meng, C. K., Wong, S. F., and Whitehouse, C. M. (1989). *Science,* **246**, 64.
14. Wilm, M. and Mann, M. (1996). *Anal Chem,* **68**, 1.
15. Mamyrin, B. A. (1994). *Int. J. Mass Spectrom. Ion Proces.* **131**, 1.
16. Brown, R. S. and Lennon, J. J. (1995). *Anal. Chem.* **67**, 1998.
17. Morris, H., Paxton, T., Dell, A., Langhorne, J., Berg, M., Bordoli, R. S. *et al.* (1996). *Rapid Commun. Mass Spectrom.* **10**, 889.
18. Shevchenko, A., Chernushevich, I., Ens, W., Standing, K. G., Thomson, B., Wilm, M., and Mann, M. (1997). *Rapid Commun. Mass Spectrom.* **11**, 1015.
19. Roepstorff, P. and Fohlman, J. (1984). *Biomed Mass Spectrom,* **11**, 601.
20. Henzel, W. J., Billeci, T. M., Stults, J. T., Wong, S. C., Grimley, C., and Watanabe, C. (1993). *Proc. Natl. Acad. Sci. USA* **90**, 5011.
21. James, P., Quadroni, M., Carafoli, E., and Gonnet, G. (1993). *Biochem. Biophys. Res. Commun.* **195**, 58.
22. Mann, M., Hojrup, P., and Roepstorff, P. (1993). *Biol. Mass. Spectrom.* **22**, 338.
23. Pappin, D., Højrup, P, and Bleasby, A. J. (1993). *Curr. Biol.* **3**, 327.
24. Yates 3rd, J. R., Speicher, S., Griffin, P. R., and Hunkapiller, T. (1993). *Anal. Biochem.* **214**, 397.
25. Jensen, O. N., Mortensen, P., Vorm, O., and Mann, M. (1997). *Anal. Chem.* **69**, 1706.
26. Mann, M. and Wilm, M. (1994). *Anal Chem,* **66**, 4390.
27. Eng, J., McCorrmack, A. L., and Yates 3rd, J. R. (1994). *J. Am. Soc. Mass. Spectrom.* **5**, 976.
28. Qin, J., Fenyo, D., Zhao, Y. M., Hall, W. W., Chao, D. M., Wilson, C. J. *et al.* (1997). *Anal. Chem.* **69**, 3995.
29. Shevchenko, A., Jensen, O. N., Podtelejnikov, A. V., Sagliocco, F., Wilm, M., Vorm, O. *et al.* (1996). *Proc. Natl. Acad. Sci. USA.* **93**, 14440.
30. Shevchenko, A., Wilm, M., Vorm, O., and Mann, M. (1996). *Anal. Chem.* **68**, 850.
31. Fernandez-Patron, C., Calero, M., Collazo, P. R., Garcia, J. R., Madrazo, J., Musacchio, A. *et al.* (1995). *Anal. Biochem.* **224**, 203.
32. Vorm, O. and Mann, M. (1994). *J. Am. Soc. Mass Spectrom.* **5**, 955.
33. Vorm, O., Roepstorff, P., and Mann, M. (1994). *Anal. Chem.* **66**, 3281.
34. Jensen, O. N., Podtelejnikov, A., and Mann, M. (1996). *Rapid Commun. Mass Spectrom.* **10**, 1371.
35. Jiang, X., Smith, J., and Abraham, E. (1996). *J. Mass Spectrom.* **31**, 1309.
36. Carr, S. A., Huddleston, M. J., and Bean, M. F. (1993). *Protein Sci.* **2**, 183.
37. Huddleston M. J, Annan, R., Bean M. T., and Carr S. A. (1993). *J Am. Soc. Mass Spectrom.* **4**, 710.
38. Jedrzejewski, P. T. and Lehmann, W. D. (1997). *Anal Chem,* **69**, 294.
39. Wilm, M., Neubauer, G., and Mann, M. (1996). *Anal. Chem.* **68**, 527.
40. Neubauer, G. and Mann M (1997). *J. Mass Spectrom.* **32**, 94.
41. Mortz, E., Sareneva, T., Haebel, S., Julkunen, I., and Roepstorff, P. (1996). *Electrophoresis* **17**, 925.

42. Kuster, B., Wheeler, S. F., Hunter, A. P., Dwek, R. A., and Harvey, D. J. (1997). *Anal. Biochem.* **250**, 82.
43. Beavis, R. C. and Chait, B. T. (1989). *Rapid Commun. Mass Spectrom.* **3**, 432.
44. Beavis, R., Chaudhary, T., and Chait, B. (1992). *Org. Mass. Spectrom.* **27**, 156.
45. Strupat, K., Karas, M., and Hillenkamp, F. (1991). *Int. J. Mass Spectrom. Ion Proces.* **111**, 89.

Chapter 8

Using proteinases for Edman sequence analysis and peptide mapping

John Shannon

Department of Microbiology, University of Virginia, Charlottesville, Virginia 22908, U.S.A.

1 Introduction

1.1 The need for digesting proteins to peptides

When identifying or comparing proteins, it is usually necessary to analyse the component peptides, despite the availability of analytical techniques for intact proteins. One and two-dimensional electrophoresis, mass spectrometry, chromatography, *N*-and *C*-terminal sequencing and amino acid analysis all characterize proteins but if differences between proteins are small, the differences may be lost when analysing a whole protein. However a difference of one amino acid between two peptides from a protein can be detected,

Sequencing a protein from the *N*-terminus may provide sufficient information to identify a protein, or enable construction of oligonucleotide probes. However, many proteins loaded into an Edman sequencer give no data because of modifications to the terminal amino group; one estimate is estimate that 80% of cytosolic proteins are blocked at the amino terminus (1). The amount of data that can be obtained in one sequencing series is limited by inefficiencies of the Edman chemistry. In some cases, 50 or more amino acids of sequence data can be obtained, but usually the amount of data is less. *C*-terminal sequencing gives less data and is less sensitive, and is mainly used to determine where a polypeptide ends. Thus, to obtain comprehensive sequence data or to make a detailed comparison of different proteins, it is necessary to digest the protein into peptides. Proteolytic digestion is reliable, efficient and offers several choices of digestion specificity, and thus is used more often than the harsher chemical cleavages.

This chapter emphasizes analysis by Edman sequencing. MS sequencing techniques are described in Chapter 7. Both require digestion of proteins to peptides, and are complementary rather than alternative techniques. Edman sequencing requires pure peptides before analysis, whereas MS can select one peptide from a mixture. MS sequencing is unaffected by modifications to the *N*-terminus, is highly sensitive, can better identify post-translational modifications, and analyse sequences which are difficult for Edman sequencing. Comparing peptides from digestion of proteins by MS may be sufficient to identify differences between proteins. Edman sequencing can analyse long peptides, can distinguish leucine and isoleucine (unlike most mass spectrometers), and can obtain sequence data from some peptides that are difficult to analyse by MS. Obtaining data by Edman sequencing is relatively slow, but with a ideal sample, interpretation is relatively simple, whereas MS data is acquired rapidly, but interpretation is more complex. Both techniques are required to obtain maximal protein sequence data.

2 Substrate preparation

2.1 Purification techniques

Methods of purifying proteins include subcellular fractionation, ion exchange, gel filtration, hydrophobic interaction, affinity chromatography, immunoprecipitation, reverse-phase chromatography, and SDS-PAGE. A description of these techniques is beyond the scope of this chapter and can be found elsewhere (2–4). SDS-PAGE polyacrylamide gel electrophoresis can prepare proteins for sequence analysis; the basic techniques are described by Hames and Rickwood (5).

If a protein is dilute, there are several techniques for concentrating it. Concentrated solutions are easier to work with and suffer less loss of material. There are membrane concentrators, in which losses of protein can be decreased by passivation with Tween-20 or a polyethylene glycol-bisphenol A diglycidyl ether (6), precipitation techniques (7,8) (*Protocol 1*) and evaporation, in which sample losses may be decreased by inclusion of 0.02% Tween-20 and avoiding complete dryness. Note that detergents can interfere with MS analysis. Proteins in gels can be concentrated by electrophoresing from several gel pieces into another gel (9,10) using modified loading techniques. To reduce losses of protein, keep solutions concentrated, minimize contact with new surfaces and keep solutions cold or frozen. During chromatography, small columns (see Section 4.4.1) keep concentrations high.

Prevention of proteolysis during purification gives higher purity and higher yields; see Chapter 9. A starting point if proteolysis is a problem is to buy a commercial mixture of proteinase inhibitors.

Protocol 1

Chloroform–methanol precipitation of Wessel and Flügge (7)

Reagents

- Methanol
- Chloroform

Protocol 1 continued

Method

1. Add 0.4 ml methanol to 0.1 ml of sample, mix and centrifuge at 9000 g for 10 s.
2. Add 0.1 ml chloroform, mix and centrifuge at 9000 g for 10 s.
3. Add 0.3 ml water, mix vigorously and centrifuge at 9000 g for 1 min. Carefully remove the upper layer and discard it.
4. Add 0.3 ml methanol to the remaining liquid and precipitated protein. Mix and then centrifuge the tube at 9000 g for 2 min.
5. Remove the supernatant from the precipitated protein.

2.2 Sample requirements

Before digesting a protein, it is necessary to know the following: sample amount and concentration, other materials present, purity.

2.2.1 Quantification

Quantification can be done by Coomassie binding assays, other colorimetric assays, or amino acid analysis. If none of these techniques are being used in the lab, the simplest procedure may be to use a commercial kit. Amino acid analysis can quantify 1 μg of protein when the common PTC chemistry is used and is less affected by a protein's composition than some colorimetric methods. The amino acid composition can be helpful in selecting a proteinase for digestion (Section 3.1). However, salts, free amino acids and other materials, can interfere. If using amino acid analysis, consult with the laboratory which will perform the analysis about sample preparation.

In addition to direct protein assays, the amount of protein can be roughly estimated by its absorbance during chromatography or staining on a gel by comparison with a known amount of protein. However, if the protein is concentrated after the final purification step, the estimates made at the final step may be optimistic. It is common in laboratories performing sequence analysis to find that the amount of protein that reaches sequence analysis is much less than what was estimated earlier.

2.2.2 Interfering substances in the protein preparation

Some purification steps add materials that interfere with subsequent separations, digestion and sequence analysis. Some common reagents are listed in *Table 1*. Techniques for removing interfering materials are listed in *Table 2*. In some cases, dilution may reduce the concentration of an interfering material to an acceptable level e.g., diluting urea to a concentration that does not denature a proteinase. Choosing appropriate procedures may make it unnecessary to remove materials in a protein preparation, e.g. performing a purification step in a buffer compatible with the proteinase used for digestion. For this reason, lysyl pep-

Table 1 Materials which may be present in protein solutions and cause interference with further analysis

Material	Problem
SDS and other detergents.	SDS fatal to MS, Edman sequencing, and reverse-phase chromatography; non-ionic detergents can interfere but tolerable in reverse-phase chromatography and at times in MALDI-TOF MS; high concentrations will inhibit proteolysis
Organic solvents	Can interfere with proteolysis and reverse-phase chromatography; can denature protein and destroy biological activity
Buffers, guanidine, urea	May change pH needed for proteolysis; may inhibit proteinase; interfere with Edman sequencing and electrophoresis
Coomassie Brilliant Blue	May interfere with proteolysis, causes artefact peaks on reverse-phase separation

Table 2 Methods for removal of undesired materials in protein solutions

Removal method	Advantages of removal method	Disadvantages of removal method
Detergent removal cartridges[a]	Fairly simple and fast	Possible precipitation if protein needs detergent for solubility
Precipitation	Simple	requires careful manipulation
Membrane concentrators	Simple, and concentrates sample	relies on extensive dilution; losses common
dialysis	Simple	May dilute sample, introduce contaminants
Evaporation in vacuum centrifuge or by lyophilization	Simple, concentrates protein	only suitable for removing organic solvents and other volatile materials
Reverse-phase chromatography	Removes salts, non-ionic, detergents, may give purification and concentration	Not suitable for SDS removal, adds TFA and organic solvent, requires major equipment

[a] Vendors include Michrom BioResources, Pierce, PolyLC. Use of the columns, following the manufacturers instructions, is straightforward.

tidase is suggested for digesting a protein because it functions in 6 M urea, which does not need to be removed after alkylation (see *Protocol 2*). All techniques for removing interfering materials can cause major loss of sample so that designing an experiment to avoid a cleanup step simplifies sample preparation.

2.2.3 Purity

When comparing proteins and obtaining sequence data, it is obviously essential to have pure proteins. Some analytical techniques are listed below. Every different analysis increases confidence that the protein is pure, but uses more of the sample.

2.2.3.1 Analytical methods

Sodium dodecyl Sulphate-polyacrylamide gel electrophoresis is a standard analytical technique that gives a size estimate and can be used as a preparative technique. However co-migration can occur. Two-dimensional gels give higher resolution, but prepare limited quantities of protein.

MS gives accurate molecular weight measurements, but proteins give different responses, so the technique is not quantitative for unknown proteins. The protein must be in a suitable solvent for analysis (See Chapter 7).

Chromatography by a method not used previously during the purification can be both an analytical and preparative technique. Reverse-phase chromatography as a final purification step eliminates some interfering materials but often destroys biological activity and adds acid and organic solvent; recoveries of some proteins are poor.

2.2.3.2 Assessment of purification data

All analytical techniques have different selectivities for separation, and may preferentially detect different proteins. Proteins in gels do stain to different extents, but the more common reason for weak staining is a small quantity of protein. When protein appears pure, it is possible that an irrelevant protein is being seen by staining or absorbance, while a protein of interest with high biological activity is not seen by these physical detection methods. The source of the protein may give some clues; for example, fetuin was found in a sample purified from cell culture medium, albumin and transferrin in a protein purified from serum, immunoglobulin in a protein isolated on an immunoaffinity column and tropomyosin in a protein isolated from muscle. In all cases, the biological or enzymatic activity was present, but not sufficiently closely linked to the major protein seen. During isolation of proteins by gels, keratin is a common contaminant. Estimation of specific activity may indicate if an apparently pure protein is mainly the one of interest or a small fraction of a preparation of an irrelevant protein.

2.3 Reduction and alkylation of proteins

Reduction of disulphide bonds helps denature the protein to enhance digestion, and to eliminate possible disulphide-linked peptides. Alkylation of cysteine helps identify it during sequencing by converting it to a stable derivative; unprotected cysteine converts in low yield to dehydroalanine, which also comes from serine. Alkylation also prevents disulphide bonds between peptides reforming. In *Figure 1C*, bovine serum albumin was digested with lysyl peptidase. Panels B and C show that the reduction and alkylation in C increases the number of peptides produced. During electrophoresis, free acrylamide can react with some cysteine residues (11).

For Edman sequence analysis, the usual alkylating reagents are *N*-isopropyliodoacetamide and 4-vinylpyridine; iodoacetate and iodoacetamide give derivatives which elute close to other amino acids and thus may not be correctly identified.

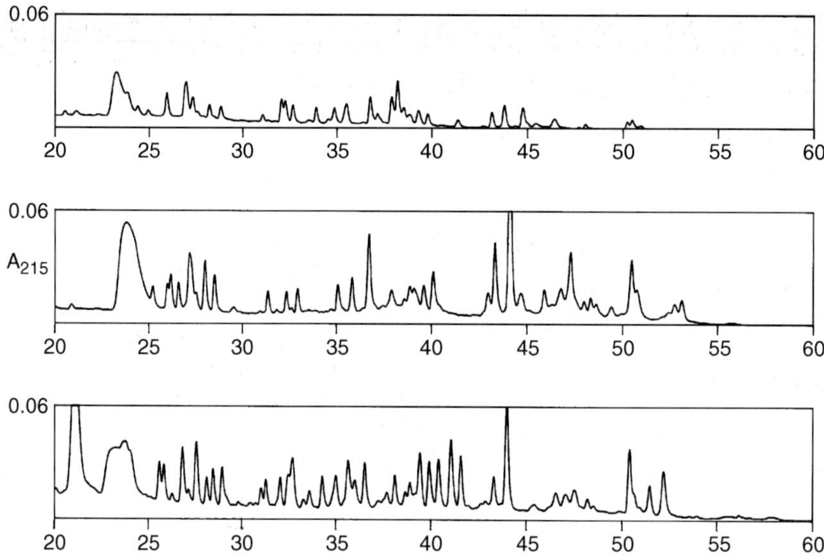

Figure 1 Digestion of BSA—effect of denaturation and alkylation. Three micrograms of BSA was digested with 0.3 μg of lysyl peptidase at 37 °C for 16 h, in 0.1 M Tris-HCl, pH 8, with or without 2 M guanidine-HCl, and the peptides were separated on a Vydac C-18 column, 1 × 150 mm. Solvent A, 0.1% TFA, solvent B, 0.09% TFA in 70% acetonitrile; 7% B to 10 min, 93% B at 70 min, flow rate 50 μl/min. (A) Tris-HCl buffer without denaturing agents or alkylation; (B)the protein was heated in 6 M guanidine-HCl and then digested in 2 M guanidine-HCl as described in section 3.2.6 (list item 1); (C) as in (B) but the protein was alkylated prior to digestion after denaturation with guanidine-HCl.

4-Vinylpyridine is prone to polymerization on storage, and is troublesome to work with, so *N*-isopropyliodoacetamide is the reagent of choice. *N*-Isopropyliodoacetamide was designed for compatibility with PTH amino acid separation systems used on Edman sequencers (12) and like the cysteine derivative of 4-vinylpyridine, elutes ahead of proline in common PTH-amino acid separations. *N*-Isopropyliodoacetamide is easily removed during reverse-phase chromatography, and does not necessarily need to be removed before adding a proteinase for digestion. Alkylation of residues besides cysteine can complicate peptide maps (13) but if long incubations are avoided, side-reactions do not cause major problems; in *Protocol 2*, extra mercaptoethanol is added after the alkylation step as a precaution against side-reactions.

A Protocol for alkylating proteins in solution is provided in *Protocol 3* below; protocols for alkylating proteins for gels are given in Chapter 7. Urea in solution forms ammonium cyanate which reacts with amino groups; methylamine is included to protect the amino groups.

Protocol 2

Alkylation of proteins in solution with *N*-isopropyliodoacetamide (12)

Reagents

- 1 M methylamine acetate or hydrochloride
- *N*-Isopropyliodoacetamide (Molecular Probes); 2.3 mg in 20 μl of methanol, 80 μl of water
- 7 μl of 2-mercaptoethanol in 1 ml of water, and a solution with 22 μl of 2-mercaptoethanol in 1 ml of water, to be made during step 3
- 1 M Tris-HCl pH 8 of 1 M ammonium bicarbonate (no pH adjustment needed)
- 1 ml of an 8 M urea solution made from 480 mg urea, 100 μl water, 320 μl of 1 M ammonium bicarbonate or Tris-Cl pH 8 (recommended for lysyl peptidase), 80 μl of 1 M methylamine acetate or hydrochloride

Method

1. Dissolve the protein in 50 μl of the urea solution and add 5 μl of the fresh 7 μl/ml 2-mercaptoethanol solution.
2. Incubate for 1 h at 60 °C under inert gas.
3. Add 7 μl of the *N*-isopropyliodoacetamide solution to the protein solution. Incubate for 30 min at 25 °C under inert gas.
4. Add 5 μl of the second (22 μl/ml) mercaptoethanol solution.
5. Before digestion, dilute the urea to prevent denaturing the proteinase, although some proteinases are active in 8 M urea. It is not usually necessary to remove the alkylating reagents before digestion.

3 Digestion

3.1 Choice of proteinase

The most commonly used proteinases are trypsin, lysine specific proteinases (endoproteinase Lys-C and lysyl peptidase), endoproteinase Glu-C and endoproteinase Asp-N. These proteinases have a restricted primary specificity and have proven useful for producing peptides for sequence and other analyses. Less-specific proteinases may give many peptides that are too small to give much data. Amino acid analysis can help choose a suitable proteinase by showing how often a particular amino acid, e.g. lysine, occurs in a protein. Keil (14) has compiled extensive information about the specificity of proteinases. Occasionally, proteinases will not cleave at an expected site and/or they may cleave at unexpected sites. These anomalies may produce desirable peptides for sequence analysis.

The first choice for a proteinase is trypsin or a lysine specific proteinase, unless there is information suggesting otherwise. Lysyl peptidase is very convenient

Table 3 Activity of proteinases in denaturing agents[a]

Solvent	Concentration	Trypsin	Chymotrypsin	Endoproteinase Lys-C	Endoproteinase Glu-C	Lysyl peptidase (*Achromobacter*)	Endoproteinase Asp-N
SDS	0.1%	109[b]	+[c]	109[b]	77[b], R[c]	100 at 30 °C[b]	10[b]
	1%					+[c]	
Urea	1 M	90[b]		90[b]	82[b]		105[b]
	2 M		+[c]		R[c]		
	4 M	+[c]					
	8 M					active[c]	
Gu-HCl	1 M	12[b]			94 [a]		80[b]
	2 M		+[c]			+[c]	
Acetonitrile	10%	161[b]			118[b]		125[b]
	20%	+[d]	R[d]		R[cd]	R[d]	
	30%		+[c]				
	40%	+[c]				+[c]	

[a] + Indicates activity, amount not specified; R indicates restricted activity, amount not specified; Gu-HCl, guanidine hydrochloride. The sources quoted used different measures of activity, hence there are some apparent discrepancies.

[b] Manufacturer's literature

[c] Reference 26

[d] Reference 27

for producing peptides for Edman sequencing or comparative peptide maps because its stability in urea enables it to be added to a denatured, alkylated substrate protein from *Protocol 2* without further manipulation. Because lysyl peptidase produces fewer and larger peptides than, say, trypsin, purification of peptides is less complex and obtaining sequence information by Edman chemistry is more efficient when sequencing a smaller number of longer peptides.

Proteinases vary in their tolerance to denaturing agents, as shown in *Table 3*. For example, lysyl peptidase is useful because it can work in strongly denaturing conditions which are needed to digest some proteins. According to anecdotal reports, proteinases from different suppliers and batches vary in their activity. Several proteinases are now available in a high-purity sequencing grade, which is useful for demanding applications. However, regular grade proteinases can perform adequately (15).

3.1.1 Trypsin

Trypsin is still the most commonly used enzyme because it is specific, readily available, usually gives suitable sized peptides for analysis (15–20 amino acids) and is now available in forms resistant to autolysis (Boehringer Mannheim, Promega). Because of the specific cleavage, lysine or arginine is found at the *C*-terminus of the peptide, unless the peptide is the *C*-terminal peptide of the protein.

The specificity of trypsin is relied on for identification of radiolabelled phosphorylation sites (16). Tryptic peptides from the digestion of a phosphorylated protein are sequenced with a modified Edman sequencing cycle (17) to determine the position of the phosphorylation site. Examination of the sequence of the protein shows which peptide(s) may be phosphorylated at the positions found by Edman sequencing; however, incomplete and non-specific cleavages can complicate the analysis.

3.1.2 Lysine-specific proteinases

There are two enzymes which digest at lysine residues only, namely endoproteinase Lys C from *Lysobacter enzymogenes* (18) (suppliers include Boehringer Mannheim, Calbiochem and Sigma) and lysyl peptidase from *Achromabacter lyticus* (Wako). When the specificity is compared in detail, the two enzymes show differences (19) but for a first digestion, lysyl peptidase has the advantage of working in the denaturing conditions used for reduction and alkylation (*Protocol 2*). There are few autolysis products from lysyl peptidase, which is stable in a freezer for at least 2 years.

3.1.3 Arginine-specific proteinases

Arginine-specific proteinases are available (20), but are used less often. Enzymes include mouse submaxillary gland endoproteinase Arg-C (Calbiochem, Pierce, Sigma), or clostripain, from *Clostridium histolyticum* (Boehringer Mannheim, Promega). An alternative approach to digesting specifically at arginine residues is to citraconylate lysine residues, digest with trypsin at arginine, then remove the citraconyl groups (17). Endoproteinase Arg-C may exhibit non-specific proteolysis (21).

3.1.4 Aspartate- and glutamate-specific proteinases

Endoproteinase Glu-C from *Staphylococcus aureus* V8 cleaves only at glutamoyl bonds in phosphate buffer, pH 7.8, or glutamate and aspartate in ammonium acetate or bicarbonate (22). This proteinase is commonly used, although it does not cleave all the expected bonds, and cleaves some non-specific sites (14). With careful optimization of the digestion conditions, non-specific cleavages and incomplete digestion can be eliminated (13). Several vendors sell this enzyme (e.g. Boehringer Mannheim, Calbiochem, Pierce, Sigma).

Endoproteinase Asp-N from *Pseudomonas fragi* cleaves on the amino side of aspartyl residues and at higher concentrations of enzyme, on the amino side of glutamyl residues. This enzyme is relatively expensive, but is popular with workers seeking specific digestions. The cost of the enzyme should be balanced against the cost of performing a digestion that does not generate the required data. This is a metalloproteinase so EDTA must be avoided. Suppliers include Boehringer Mannheim, Calbiochem and Sigma.

3.1.5 Chymotrypsin

Chymotrypsin has preferential specificity for aromatic and large hydrophobic amino acids, but contrary to that which is suggested by some programs used to predict digestion patterns, the specificity is not tight (14). This proteinase is used less often than those described above.

3.1.6 Other proteinases

When the above proteinases fail to produce a satisfactory digestion, or any digestion, other proteinases, such as pepsin and subtilisin may be used.

3.2 Conditions for proteolytic digestions

When performing a digestion to generate peptides for sequencing or mapping, the user wants complete, specific digestion of the substrate protein without unexpected digestions of the substrate or contributions of peptides from the digesting proteinase. In some situations, optimizing digestion conditions must and can be done (13), but most users intend to perform only one digestion because of limited amounts of sample and hence choose a standard protocol for digestion.

Proteinases are relatively hardy enzymes, in part because they must resist their own proteolytic activity and hence they tolerate a variety of reaction conditions. Often the supplier will recommend conditions for use.

Protocol 3
Standard digestion using lysyl peptidase

Reagents

- 100–200 pmole of protein freshly alkylated using *Protocol 2*
- Lysyl peptidase
- 2 μg of cytochrome *c* or other standard protein alkylated using *Protocol 2*
- TFA

Method

1. Add 0.2 μg (or as appropriate for amount of protein) of lysyl peptidase to the alkylated protein, control digestion (no substrate) and a tube containing the alkylated cytochrome *c* or other standard protein
2. Incubate for 16 h at 37 °C
3. Add 2 μl of trifluoroacetic acid to stop the reaction and freeze tubes until analysis

3.2.1 Amount of proteinase

A typical ratio is proteinase:substrate is 1:100 to 1:30 by weight. For proteinase-resistant proteins, more proteinase may be needed, at the risk of non-specific proteolysis and significant autolysis. Increasing the amount of endoproteinase Asp-N may increase the number of cleavages at gluyamyl residues. When a protein is difficult to digest, autolysis resistant trypsin can be used at ratios of 1:5 proteinase–substrate.

3.2.2 Reaction conditions

The temperature is commonly 37 °C although lower temperatures may be recommended to decrease enzyme inactivation (13). Overnight incubations are common to ensure complete digestion although for limited digestion, the incubation times may be as short as 10 min.

Ammonium bicarbonate or Tris-HCl, pH 8, are common buffers; the supplier usually includes recommendations. Note that ammonium salts inhibit lysyl peptidase so Tris-HCl should be used with this enzyme. The buffer also alters the specificity of endoproteinase Glu-C. Ammonium bicarbonate is volatile and thus can be removed under vacuum, although drying may cause peptides to dry irreversibly onto tubes.

3.2.3 Controls

A control reaction containing everything except the protein being digested should be included. This is to show which peptides or other materials come from the proteinase or other reagents. A positive control of a protein that is known to be digested, such as cytochrome c, should also be included to verify that the proteinase was active under the reaction conditions. The amount of the test protein should be similar to the protein of interest to determine what amounts of peptides to expect.

After a digestion, there are sometimes no peptides or only peptides derived from the proteinase. These results emphasize the importance of knowing how much substrate protein was present and of running controls. If there is sufficient material, a gel will show if the protein was digested; on reverse-phase chromatography, undigested protein usually elutes late in the gradient (at over 40% acetonitrile).

3.2.4 Autolysis

Proteinases can also be substrates for proteolytic digestion, including self digestion. Autolysis cannot be entirely prevented, but can be kept to acceptable levels by keeping amounts of proteinase relatively low. The control digestion mentioned above and knowledge of the sequence of the proteinase help in recognizing autolysis products. With substrate–proteinase ratios of 30:1 (w/w), if digestion is complete, autolysis products are not a significant problem.

The best-studied autolytic reaction is that of trypsin. Autolysis of trypsin occurs on prolonged digestion and is especially noticeable if the ratio of trypsin to substrate is high (e.g. > 1:30). In addition to producing peptides from trypsin, autolysis produces pseudotrypsin (23), which has chymotryptic-like specificity (24) and may explain some non-tryptic cleavages that are occasionally observed. Reductive methylation of lysine residues of trypsin (25) greatly decreases autolysis, allowing high ratios of trypsin to substrate to be used. Autolysis-resistant trypsin is available commercially from Promega and Boehringer Mannheim.

3.2.5 Denaturing agents

Proteinase-resistant proteins may require use of denaturants for digestion. *Figures 1A* and *B* show the effect of incubating bovine serum albumin in 8 M guanidine-HCl and then diluting to 2 M guanidine-HCl, using the procedure of Riviere *et al.* (26). See *Table 3* or the manufacturer's literature for what conditions a particular proteinase will tolerate.

For proteinase resistant proteins, one or more of the following procedures may work (26):

(1) Heat the protein in 6 M guanidine-HCl at 50 °C for 30 min, then dilute to 2 M with digestion buffer and lysyl peptidase, chymotrypsin or subtilisin, which are active in 2 M guanidine-HCl.

(2) Incubate the protein in 8 M urea at 37 °C for 30 min, then add lysyl peptidase or subtilisin. See *Protocol 2* for alkylation and how to prepare the 8 M urea. Note that urea can block *N*-terminal amino groups because of the formation of ammonium cyanate; ammonium bicarbonate buffer or methylamine will block this reaction.

(3) Heat the protein in 1% SDS at 95 °C and digest with lysyl peptidase or subtilisin; removal of SDS is necessary before performing reverse phase chromatography (see *Protocol 1* and *Table 2*).

(4) Incubate in 40% acetonitrile and digest with lysyl peptidase or trypsin; 40% acetonitrile alters the digestion a little relative to a digestion in 20% acetonitrile. A combined sample preparation method is to alkylate the protein (*Protocol 2*) and then remove the reagents by reverse-phase chromatography. Proteins usually elute in 40–60% acetonitrile, so after neutralizing the TFA, and adding buffer to dilute the acetonitrile to 40%, add proteinase to digest the protein. Before injecting the digestion mixture on a reverse-phase column, reduce the acetonitrile concentration by evaporation and/or adding 0.1% TFA. In organic solvents often a temperature of 37 °C causes loss of activity whereas 22 °C does not (27).

These different digestion conditions do not give identical peptide maps but should give enough peptides to obtain sequence data,

3.3 Digestion of proteins isolated on polyacrylamide gels

With modern techniques, protein bands on polyacrylamide gels yield sufficient purified protein for sequence analysis. This technique is described in Chapter 7, so details are not given here. Again, the preferred reagent for alkylation for Edman sequencing is *N*-isopropyliodoacetamide; 4-vinylpyridine is also suitable but is less desirable for MS analysis. For Edman sequencing, 100 pmol is a desirable amount of protein, with the lower limit being about 20 pmol; a fixed amount of protein appears to be lost during handling, so that halving the amount of protein on a gel will more than halve the amount of peptide sequenced. The protein concentration should be as high as possible; Williams and Stone (28) state that the yield during sequencing is better correlated with sample density on the gel than the total amount of sample. With adequate amounts of sample, an in gel protein digest can be analysed by MS, which uses only a small portion of the sample, and Edman sequencing. Techniques for comparison of proteins by peptide mapping have been described by Cleveland *et al.* (29)

Proteins can be successfully electroeluted from gels and digested to give peptides for sequence analysis. However, this technique involves procedures for elution of the protein, concentration from the elution buffer and removal of SDS used during elution, offering opportunities for loss and contamination. Less involved techniques for removing proteins from gels are centrifugal elution, using Amicon Micropure separators from Millipore, and passive elution (30) (one system is commercially available from Promega). However, the most popular techniques are digestion in the gel, or after transfer to PVDF or less often, nitrocellulose. Proteins fixed in gels or transferred to PVDF are stable for long periods, even at room temperature.

Digestion on PVDF is also quite successful, but transferring to PVDF involves another handling step, which can result in loss of protein. In addition, the detergent used can interfere with any MS that is needed. A procedure for digestion of proteins on PVDF is given by Fernandez *et al.* (31).

3.4 Monitoring a reaction

Commonly, an investigator only has enough protein to perform one digestion and long incubation times are commonly used. If the digestion needs to be run the minimal length of time, samples can be withdrawn for analysis, either by starting or stopping reactions at different times, or withdrawing samples during the incubation. For analysis during the digestion, gels may be too slow; alternatives are MALDI-TOF MS and HPLC. MALDI-TOF is fast and tolerates some salt. Reverse-phase HPLC can be performed in minutes, using higher than normal flow rates on either conventional columns (32) or perfusive columns (Perseptive Biosystems), sold for their high flow rate.

4 Analysis of the proteolytic digestion

After a proteolytic digestion, the peptides produced need to be separated for analysis. The analysis may also show if there is undigested substrate protein. For isolation of peptides by Edman sequencing, reverse-phase chromatography is the normal technique. To determine whether the expected peptides were produced, MALDI-TOF MS can give rapid data using small quantities of material.

4.1 Mass spectrometry

MALDI-TOF can be useful in giving data about purity and peptide size. Estimation of molecular weights may be sufficient to identify an individual protein or compare it to another. This subject is addressed in Chapter 7. For Edman sequencing, molecular weight information is used to decide how long a sequence is expected, and may aid in interpreting the sequence data. Because of the variable ionization of peptides in MALDI-TOF, the appearance of only one species does not ensure purity.

4.2 Electrophoresis

Polyacrylamide gels with Tricine buffers can separate peptides with sizes over 2 kDa, but the technique is not commonly used for high sensitivity isolation of peptides for sequencing. A popular procedure for such gels is that of Schägger and von Jagow (33) or precast gels are available from several vendors.

Two-dimensional TLC is popular for separating radiolabeled phosphopeptides produced by trypsin digestion of regulatory proteins (16).

Capillary electrophoresis is another option for separation of peptides and has different selectivity from reverse-phase chromatography; thus it can be used to assess the purity of peptides separated by chromatography (34). Because of the limited quantities of material it can handle, it is not widely used.

4.3 Ion exchange

Cation exchange chromatography can separate peptides by charge, and other interactions with the column can separate peptides with the same charge, but has lower resolution than reverse-phase chromatography. In addition, the salts and buffers used must be removed before sequencing. Despite its appeal, this technique is less common than reverse-phase chromatography.

4.4 Reverse-phase chromatography

This is the most common method of separating peptides from a proteolytic digest. It offers high resolution, good recovery, removal of salts, usually prepares the peptides in volatile solvents compatible with Edman sequencing, can prepare the peptide in volumes of 100 μl or less and can be modified to separate co-eluting peptides. Standard conditions are a C-18 column eluted with a gradient of acetonitrile in water with 0.1% trifluoroacetic acid.

4.4.1 Equipment

Currently, 2.1 mm diameter columns are commonly used. Compared with standard 4.6 mm columns, they are four times more sensitive for a given sample load, elute the peptide in a smaller volume and probably give better recovery by reducing the surface area to which the peptide is exposed. Columns of 2.1 mm or smaller diameter are standard for separating the amounts of peptides produced by digesting a protein isolated from a gel or from a 100-pmol digest. The sample volume from a 2.1 mm column is about 100 μl, which is acceptable for loading on to a sequencer in aliquots of 30 or 15 μl. In contrast, a 4.6 mm column elutes peptides in a volume of about 500 μl.

A suitable HPLC pumps a binary gradient at the flow rate appropriate tot he size of column used, namely 1 ml/min for 4.6 mm diameter columns, 200 μl/minute for 2.1 mm diameter columns, and 50 μl/min for 1 mm diameter columns. The type of pumps and the way the gradient is made is not critical. Pumps designed for 1 ml/min flow rates may not be suitable for the lower flow rates of 2.1 mm and smaller columns, because (i) the pumps may not be accurate at flow rates that are a small fraction of what they were designed for, e.g., a pump system running at 200 μl/min instead of 1 ml/min; (ii) a large volume mixer causes large delays between the gradient produced by the pump(s) and that being delivered to the column; and (iii) large volumes in the flow cell and tubing after the column reduce separation of peptides because of mixing in the tubing and flow cell. In some cases, a pump can be adapted to lower flow rates by changing the mixer, flow cell and tubing. Generally, the manufacturers of the equipment can advise what flow rates a system is suitable for, and if it can be modified for lower flow rates. Successful chromatography on 1-mm columns can be demanding (35) and smaller columns need more specialized equipment. PEEK tubing and polymeric ferrules and nuts are often more convenient than stainless steel for plumbing although there are anecdotal reports that they may contribute spurious peaks to a chromatogram. A desirable addition to the chromatography equipment is an uninterruptible power supply, to reduce the risk of losing a whole digestion experiment because of the stopping of a chromatography run by a momentary power outage.

Normally, a high pressure limit of 20 700 kPa (3000 p.s.i.) is adequate; a higher pressure probably indicates a blockage in the system. The volume of sample loaded at one time should not exceed half the injector loop volume because of laminar flow—if larger, some sample may be lost; thus an injector loop of 500 μl is convenient to accommodate samples of up to 200 μl. Ahead of the column, place an in-line filter to remove any particulates or precipitates from the sample, and fragments of pump or injector seals. If the pressure in an HPLC system is too high, the filter is the first place to look for an obstruction.

For peptide separations, a UV detector set at or close to 215 nm is standard. It is common to collect fractions by hand, based on visual monitoring of the absorbance signal on the detector. Ensure that there is not a long delay in displaying the current absorbance signal. A trial digest, perhaps using a control protein, gives a feel for what peaks are significant and the mechanics of collecting many fractions. Automated fraction collectors are convenient but require calibration to ensure that they collect the peaks of interest, and must be selected so that they do not add dead volume after the flow cell with extra tubing, because dead volume decreases the quality of the separation achieved on the column. The detector outlet tubing should be no longer than necessary to minimize dead volume after the detector. However, if it is too short, outgassing in the detector cell will produce air bubbles and spurious signals. Air bubbles will cause rapid, random fluctuations in the signal. To remove them, flush the system with 90% or 100% organic solvent with a higher flow rate than normal. With some detectors, it is possible to change to a visible wavelength, expose the detector cell and inspect for bubbles directly.

To recording the absorbance signal, the options are strip chart recorders, integrators, and computer based data systems. Integrators mark peak elution times, which is one way of identifying fractions; some can replot data, which is useful if the wrong absorbance range was chosen before the run. Data systems give more flexibility in replotting data, and allow changing the display scale during the run. Watching the display during the run helps determine if an absorbance change is an interesting peak or not. Both integrators and data systems can lag behind the display on the detector, so cannot always be relied on to decide when to collect a fraction. One system of identifying fractions is to label them with the elution time shown on an integrator; an assistant to label the tubes makes this task easier.

The standard columns for separating peptides are end-capped, silica-based C-18 columns, with 5 μm particles and 300 Å pores. Most vendors sell C-18 columns recommended for peptide separations. There are many C-18 columns, but only some are recommended for peptides. Although columns from different vendors give different separations, no column is the best for all separations. The standard column length is 150 or 250 mm, but 30 mm long cartridge columns are attractive because of the relatively low cost, so that one column can be dedicated to one sample or group of samples. There are not big differences in selectivity or recovery between C-4, C-8 and C-18 columns, although shorter carbon chains give better recovery of hydrophobic proteins, and longer chains

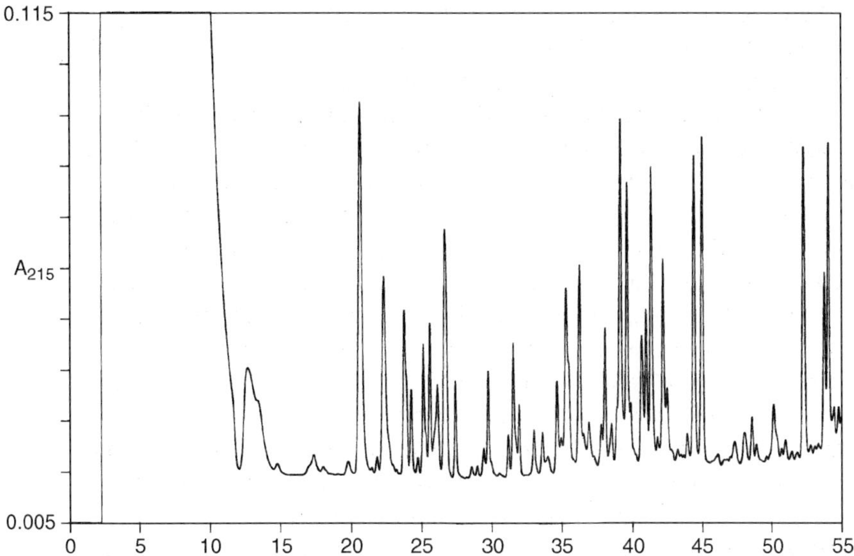

Figure 2 Standard digestion of bovine serum albumin.
6 μg of bovine serum albumin was alkylated according to *Protocol 2*, digested according to Protocol 3 and peptides separated according to *Protocol 5*. The column was a Phenomenex Jupiter C18, 150 mm × 2.1 mm. Solvents were prepared according to *Protocol 4*; gradient, 5% B to 5 min (end of loading), 40% B at 50 min, 70% B at 55 min, 25 °C.

retain peptides more strongly. A longer column can separate more peptide peaks than shorter columns (36).

The high sensitivity of modern analytical techniques makes a clean working area necessary. Preparing electrophoresis buffers in the same room may cause contamination of fractions while they are being collected. Weighing out a protein in the same room may also contaminate samples, especially if a fume hood directs the air flow towards the samples. Some workers rinse tubes before use and store them capped to minimize contaminants.

The standard tubes for collecting fractions are small polypropylene centrifuge tubes. Tubes with etched surfaces stay labelled better after freezing and thawing—with potentially dozens of fractions from a protein digestion, smeared or missing labels can ruin an experiment. Some workers find that for demanding work, some brands of tube give better recovery of peptides than others. At low levels of peptide, storing fractions cold can reduce losses, so an ice bucket or even colder material should be on hand for storing collected peptides.

Protocols 4 and *5* are a set of standard conditions; modifications may improve the separation of particular peptides. In *Figure 2* digest of bovine serum albumin is shown. The substrate was alkylated using *Protocol 2*, digested according to *Protocol 3* and the peptides separated using *Protocol 5*.

Achieving a satisfactory baseline for high sensitivity chromatography can be frustrating. Sometimes the baseline slope is excessive even using concentrations of TFA that have been satisfactory in the past, and after changing the TFA trifluoroacetic acid concentration, the baseline changes again. This may be due to the column absorbing TFA.

Protocol 4

Preparation of solvent for reverse-phase chromatography of peptides.

Equipment and reagents

- High-purity water acetonitrile and TFA; store TFA under inert gas
- Sonicator

Method

1 ***Degas***[a] the water and acetonitrile by applying vacuum while the solvents are being sonicated for 2 min; usually, about 1.5× more water than acetonitrile is needed.

2 Do not filter the solvents because the filters may add contaminants to solvent.[b]

3 Add TFA[c] to give 0.1% (v/v) in water, 0.09% (v/v)[d] in acetonitrile.

[a] Degassing is not always necessary but decreases the chances of outgassing when acetonitrile and water mix, leading to bubbles in the detector flow cell. Some workers use a mixture of 70% acetonitrile for solvent B to reduce this problem. Some HPLC systems, especially those with low pressure gradient mixing, come with a helium sparging system which makes prior degassing unnecessary.

[b] High-purity solvents should be free of particulates and thus do not require filtering, but the vessels used for the solvents must be clean. The inlet frits used in solvent reservoirs on the pumps usually filter out particles bigger than 10 μm, but frits in the column are usually 2 μm, so the inlet filters should not be used to remove particulates.

[c] Add the TFA after degassing to avoid removing it during degassing.

[d] The concentration of TFA is lower in acetonitrile to reduce the baseline rise seen at 215 nm; 0.09% is a starting point and the concentration needed to achieve the flattest baseline obtainable may vary—a few gradients may be needed to get the flattest baseline.

Protocol 5

Separation of peptides by reverse-phase HPLC.

Equipment and reagents

- HPLC system for running binary gradients, and equipped with a detector for detection at 215 nm
- Integrator and/or data system to record absorbance
- Solvents prepared as in *Protocol 4*
- C18 or other column suitable for peptide fractionation
- Mixture of peptides for testing purposes e.g. angiotensins I and II, des-Asp angiotensin I

Protocol 5 continued

Method

1. Purge or prime the pumps to eliminate air bubbles and old solvent, sending the solvent to waste.
2. Pump the starting solvent mixture, usually 5% B, through the column, at 1 ml/min for a 4.6 mm diameter column, 200 μl/min for a 2.1 mm diameter column with 5% solvent B (0.09% TFA in acetonitrile) for 10 min.
3. Run a blank gradient of 5% acetonitrile to 70% B with a change of 1% acetonitrile/min to 50% acetonitrile (see *Table* 4). Start the gradient after injecting the sample, and switch the injector to the 'load' position after injection. To avoid loss of sample, the sample volume should not exceed about half of the volume of the injection loop.
4. Record absorbance at 215 nm, with a scale of 0.1 absorbance units.
5. If baseline rise is greater than 0.02 absorbance units, run another blank, or adjust TFA concentrations.
6. If the column has not been used recently, inject the peptide standards (2 μg for a 4.6 mm column, 0.2 μg for a 2.1 mm column).
7. If the separation of the test peptides looks satisfactory, load the sample from the digest. Running the control digestion of proteinase alone will help identify artefact peaks. Collect peaks by hand and place on ice. One way to identify peaks is to watch an integrator and label the peaks by the time recorded on the integrator.
8. Store the collected fractions at −70 °C.
9. If available, analyse the fractions by laser desorption MS to see if there is more than one peptide and determine the mass of the peptide.
10. Just before loading on to the sequencer, TFA can be added to give 20% acid, to help dissolve peptide from the tube (37).
11. After the last run, flush the column with 50% acetonitrile or other mixture of organic solvent and water to remove TFA. The absorbance will drop to −0.2 to −0.26 when the acid is washed out of the column.

[a] High pH dissolves silica, and low pH hydrolyses alkyl chains from the silica and for this reason, columns should not be stored in the acid solvents used for elution.

There may be artefact peaks in the blank chromatogram. Common sources of artefact peaks are proteins left on the column from previous samples. If a peak slowly decreases in successive blank runs, it is probably material left on the column. Standard clean-up procedures are inefficient at removing proteins stuck on columns; one successful, but harsh procedure is elution with a gradient from water to 37% acetonitrile, 37% 2-propanol, 26% formic acid (R. Burley, personal communication); 100 μl of trifluroethanol (for a 4.6 mm diameter column) has been suggested (38). Peaks that are roughly constant between runs, or even increase, are probably from water or TFA, or even the HPLC. To determine if a

Table 4 Sample gradient for separating peptides on a 2.1 mm diameter column using syringe pumps with 10 ml capacity

Time (min)	Acetonitrile (%)	Events
0 (after 9-min equilibration)	5	Injector to inject position
5	5	Injector to load position
50	50	Most peptides eluted
55	70	End of run

contaminant is from water, run blanks with a short equilibration time before injection, and a long equilibration time. If the contaminant is being adsorbed from water, it will be concentrated on the column and give a larger peak in the run with a long equilibration time.

Most peptides elute between 10% and 40% acetonitrile, so the gradient can be compressed at the end and the starting concentration of acetonitrile can be increased from 5% if the gradient capacity is limited, as occurs with syringe pumps. Most proteins elute below 70% acetonitrile so a gradient to 100% is unnecessary. Gradients changing at 1%/min are standard. A 100% aqueous solution can cause the C-18 chains to collapse onto each other, decreasing the performance of the column, so the solution passing through the column should have at least 5% acetonitrile at all times.

For 2.1 and 1 mm diameter columns, syringe pumps are common. These pumps have a 10-ml reservoir of each solvent per run, which limits the length of a gradient. To load the sample onto the column from the injector loop, pump at starting conditions (e.g. 5% acetonitrile) for at least 7 min before injecting the sample into the sample loop, and pump at least twice the sample volume through the sample loop before starting the gradient and moving the injector to the load position. There will be a large absorbance peak early in the run caused by salts, denaturants and other materials in the sample that do not stick to the column, and some absorbance changes associated with the mechanical act of switching the sample loop in and out of the fluid path due in part to refractive index changes.

4.4.2 Co-elution of peptides

A typical proteolytic digestion produces many peptides, some of which co-elute. Decreasing the number of peptides produced is one way to decrease this problem, which is a reason to choose one proteinase, e.g. lysyl peptidase instead of trypsin, for a first digestion of a protein. The problems to be addressed are how to determine if a fraction from a chromatography run contains more than one peptide, and then how to separate them.

The shape of a peak can be a clue that a fraction contains more than one peptide. A shoulder on the peak, a peak which is much higher than others, or an unusually wide peak suggest that more than one peptide is present, but basic peptides can tail, and glycopeptides can give broad peaks amidst other reasons for imperfect peak shapes. Analysis by MALDI-TOF MS may show if there is more

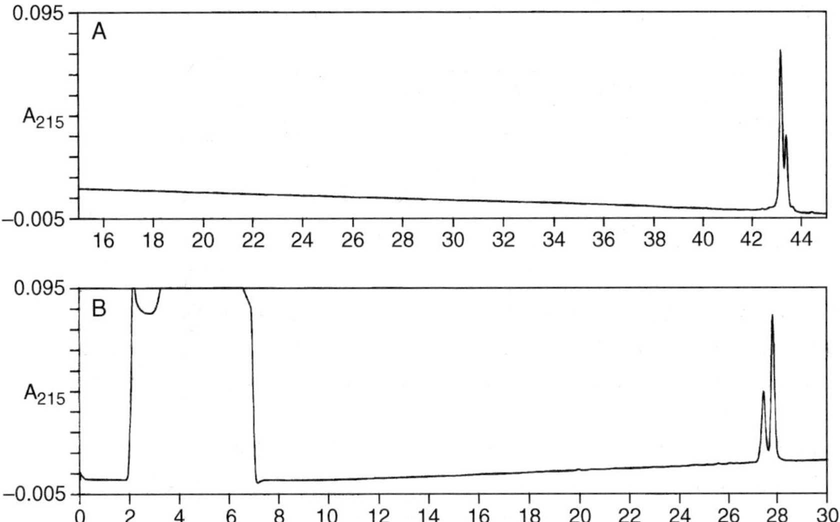

Figure 3 Separation of co-eluting peptides. Two peptides from a digestion co-eluted under the conditions described in *Figure 2* and were rechromatographed. Solvents were prepared according to *Protocol 4*. Both columns were 150 × 2.1 mm and run at 25 °C. (A) Zorbax® C-18 300SB column; gradient, 10% B to 5 min, 45% B at 50 min. (B) Zorbax® 300SB-CN column; gradient, 20% B to 5 min, 55% B at 50 min

than one peptide present, in addition to providing valuable molecular weight information. Unfortunately, sometimes impurity is only discovered during sequence analysis which is one reason for not using all of a sample for one analysis if there is sufficient material. If a sample gives more than one sequence, the data may still be useful (Section 4.5). Rechromatography of a fraction can show the presence of multiple peptides and separate them, although usually workers seek other evidence that rechromatography is warranted before performing it.

Faced with the prospect of co-eluting peptides, one strategy is selection of peaks which are likely to be pure. If only limited information from a digest is needed, late eluting, well-separated peptides with narrow symmetric peak shapes can be chosen for further analysis on the grounds that they are more likely to be pure and give a long sequence. However, often there are co-eluting peptides of interest which must be separated. Several conditions can be changed to achieve separation. *Figure 3* shows the separation of two peptides which co-eluted under the conditions used in *Figure 2*. Changing C-18 columns gave some separation (*Figure 3A*), but changing the gradient did not further improve the separation. Using a different column gave a satisfactory solution (*Figure 3B*), and increasing the temperature increased the separation further (data not shown).

4.4.2.1 Changing columns

Co-eluting peptides can be separated by reverse-phase columns with different ligands (e.g. cyanopropyl instead of C-18. or C-18 columns from different manu-

Table 5 Solvent systems to separate co-eluting peptides in reverse-phase chromatography

Solvent	Comments
Triethylamine acetate pH 5.5, 10 mM	Volatile, compatible with Edman chemistry
Formic acid, 0.1%	Gives broader peaks than TFA, volatile
Heptafluorobutyric acid, 0.05–0.1%	High absorbance, volatile
Propanol instead of acetonitrile	Causes baseline rise, higher pressures, not commonly used

facturers) or different C-18 columns from the same manufacturer. Using a cyanopropyl column instead of a C-18 seems an efficient way of separating co-eluting peptides. The cost of buying another column should be balanced against the potential costs of trying other approaches that may fail to separate the peptides or ultimately lose the peptides. One line of cyanopropyl columns suitable for separating peptides is the Zorbax Stable Bond 300 Å line from Hewlett-Packard.

4.4.2.2 Changing solvents

Changing the modifiers in the solvents is a classic method of separating co-eluting peptides. When successful, it is cheap and simple. However, larger peptides seem prone to poor recovery in some solvent systems, and some of the modifiers can interfere with subsequent analysis. Some systems are given in *Table 5*.

4.4.2.3 Changing gradient conditions

After using the column and gradient conditions in *Protocol 5*, the gradient and column temperature can be changed. Although theory and a number of chromatography experiments show that altering an acetonitrile gradient slope, or elution temperature can separate co-eluting peptides, these approaches seem less successful than those above. Using a shallow gradient may simply cause peptides to elute in more dilute form rather than separately. Increasing the operating temperature may separate peptides, and can improve the chromatography of hydrophobic peptides which give poor peak shapes at normal temperatures. The temperature of 60 °C, which may be needed for this type of experiment, is above the recommended range for some columns; one line that is advertised as suitable for high-temperature applications is the Zorbax Stable Bond series from Hewlett-Packard.

When comparing proteins by peptide mapping, one or more peptides may appear different between the proteins being compared. If the peptides are separated by chromatography and the differences are small, there should be some analysis to show that the peptides really are different, and do not appear different because of slight differences in chromatography. In addition to analysis by MS or Edman sequencing, the two peptides can be mixed and rerun to see if the two peptides separate when chromatographed under identical conditions. By mixing the peptides information as to which peptide comes from which pro-

Table 6 Sites for sequence similarity searches

Address	Notes
http://www.ncbi.nlm.nih.gov/BLAST/	BLAST searches at NCBI
http://fasta.bioch.virginia.edu/fasta/cgi/searchx.cgi	fasta. This site is run by one of the program, authors and allows downloads of the fasta programs, and has fasta programs for mixed sequences (Section 4.4.2)
http://www.ch.embnet.org/software/BottomBLAST.html?	BLAST searches from the European Molecular Biology network

tein will be lost, unless there is sufficient peptide to use for analysis in addition to the rerunning experiment.

4.5 Data interpretation

After obtaining a sequence, it should be compared with existing sequences to determine if the protein is a known protein, or related to a known protein. These comparisons will show if a peptide came from the proteinase used for digestion, or if the protein is an irrelevant contaminant. Some web sites for performing similarity searches are shown in *Table 6*.

The fastf program, which can use data from a mixed peptide sequence to identify a protein, is available at the University of Virginia site (http://fasta.bioch.virginia/edu/fasta/cgi/searchx.cgi). In the commercial GCG package, the find-patterns command can also be used with mixed sequence data (39) but using this command requires mastering the required file format.

Acknowledgments

I thank the following people for helpful discussions over the years: members of the laboratories of Richard J. Simpson, Kathryn Stone, and Ulf Hellman and Michael Kinter and Nicholas Sherman of the W. M. Keck Foundation Center for Biomedical Mass Spectrometry (University of Virginia). The work used as the basis of this chapter is supported in part by the Pratt Committee of the University of Virginia Medical School, under the guidance of Jay W. Fox.

References

1. Brown, J. L. and Roberts, W. K. (1976). *J. Biol. Chem.* **251**, 1009.
2. Doonan, S. (1996). *Protein purification protocols*, p. 1. Humana Press.
3. Karger, B. L. and Hancock, W. S. (1996). *Methods in enzymology* **270**, Academic Press, London.
4. Scopes, R. K. (1994). *Protein purification: principles and practice*, 3rd edn Springer Verlag, New York.
5. Hames, B. D. and Rickwood, D. (1981). *Gel electrophoresis of proteins; a practical approach* IRL Press, Oxford.
6. Amicon Application Note, document 2301
7. Wessel, D. and Flügge, U. I. (1984). *Anal. Biochem.* **138**, 141.

8. Sauvé, D. M., Ho, D. T., and Roberge, M. (1995). *Anal. Biochem.* **226**, 382.
9. Rider, M. H., Puype, M., van Damme, J., Gevaert, K., De Boeck, S., D'Alayer, J. *et al.* (1995). *Eur. J. Biochem*, **230**, 258.
10. Lombard-Platet, G. and Jalinot, P. (1993). *BioTechniques* **15**, 668.
11. Chiari, M., Righetti, P. G., Negri, A., Ceciliani, F., and Ronchi, S. (1992). *Electrophoresis* **13**, 882.
12. Krutzsch, H. C. and Inman, J. K. (1993). *Anal. Biochem.* **209**, 109.
13. Jones, M. D., Merewether, L. A., Clogston, C. L. and Lu, H. S. (1994). *Anal. Biochem*, **216**, 135.
14. Keil, B. (1992). In *Specificity of proteolysis*, p. 1. Springer-Verlag, Berlin.
15. Shannon, J. D., Baramova, E. N., Bjarnason, J. B. and Fox, J. W. (1989). *J. Biol. Chem.* **264**, 11575.
16. Boyle, W. J., van der Geer, P., and Hunter, T. (1991). In *Methods in enzymology* (ed. Hunter, T. and Sefton, B. M. Vol. 201, p. 110. Academic Press, San Diego.
17. Shannon, J. D. and Fox, J. W. (1995). In *Techniques in Protein Chemistry VI* (ed. Crabb, J. W.), p. 117. Academic Press, San Diego.
18. Jekel, P. A., Weijer, W. J. and Beintema, J. J. (1983). *Anal Biochem.* **134**, 347.
19. Brenner, M. J., Bahktyari, A., Martin, S. A, and Scoble, H. A. (1990). *J. Protein Chem.* **9**, 296.
20. Schenkein, I., Levy, M., Franklin, E. C. and Frangione, B. (1977). *Arch. Biochem. Biophys.* **182**, 64.
21. Proudfoot, A. E. I., Magnenat, E., Haley, T. M., Maione, T. E. and Wells, T. N. C. (1995). *Eur. J. Biochem.* **228**, 658.
22. Houmard, J. and Drapeau, G. R. (1972). *Proc. Natl. Acad. Sci.* **69**, 3506.
23. Smith, R. L. and Shaw, E. (1969). *J. Biol. Chem.* **244**, 4704.
24. Keil-Dlouhá, V., Zylber, N., Imhoff, J.-M., Tong, N.-T., and Keil, B. (1971). *FEBS Lett.* **16**, 291.
25. Rice, R. H., Means, G. E. and Brown, W. D. (1977). *Biochem. Biophys. Acta* **492**, 316.
26. Riviere, L. R., Fleming, M., Elicone, C. and Tempst, P. (1991). In *Techniques in Protein Chemistry II* (ed. Villafranca, J. J.), p. 171. Academic Press, New York.
27. Welinder, K. G. (1988). *Anal. Biochem.* **174**, 54.
28. Williams, K. R. and Stone, K. L. (1995). In *Techniques in protein Chemistry VI* (ed. Crabb, J. W.) p. 143. Academic Press, New York.
29. Cleveland, D. W., Fischer, S. G., Kirschner, M. W., and Laemmli, U. K. (1977). *J. Biol. Chem.* **252**, 1102.
30. Larson, G. A. and Schultz, J. W. (1993). *BioTechniques* **15**, 316.
31. Fernandez, J., Andrews, L., and Mische, S. M. (1994). *Anal. Biochem.* **218**, 112.
32. Moritz, R. L., Eddes, J., Ji, H., Reid, G. E. and Simpson, R. J. (1995). In *Tecniques in protein chemistry VI* (ed. Crabb, J. W.) p. 311. Academic Press, New York.
33. Schägger, H. and von Jagow, G. (1987). *Anal. Biochem.* **166**, 368.
34. Stromqvist, M. (1994). *J. Chromatog. A* **667**, 304.
35. Elicone, C., Lui, M., Geromanos, S., Erdjument-Bromage, H. and Tempst, P. (1994). *J. Chromatogr. A* **676**, 121.
36. Stone, K. L. and Williams, K. R. (1988). In *Macromolecular sequencing and synthesis: selected methods and applications* (ed. Schlesinger, D. H.), p. 7. Alan R. Liss, New York.
37. Erdjument-Bromage, H., Geromonas, S., Chodera, A. and Tempst, P. (1993). In *Techniques in Protein Chemistry IV* (ed. Angeletti, R. H.), p. 419. Academic Press, San Diego.
38. Bhardway, S. and Day, R. A. (1999). *LC GC* **17**, 354.
39. Vodkin, M. H., Novak, R. J. and McLaughlin, G. L. (1996). *BioTechniques* **21**, 1116.

Chapter 9
Prevention of unwanted proteolysis

Michael J. North
Department of Biological Sciences, The University of Stirling, Stirling FK9 4LA, Scotland

Robert J. Beynon
Department of Veterinary Preclinical Sciences, University of Liverpool, PO Box 147, Liverpool L69 3BX, UK

1 Introduction

To many biochemists and cell biologists, proteinases are not so much interesting enzymes or useful tools but persistent nuisances. While many proteins seem to be immune to the 'proteinase problem', the frequency with which artefacts arise indicates that a cautious approach is justified. Moreover, the availability of accurate mass from new mass spectrometric methods may well bring to light more examples of unwanted proteolytic degradation, especially terminal trimming that might have previously remained undetected through the low resolution of SDS-PAGE.

Unwanted proteolysis is a universal problem encountered in the analysis of proteins from every type of organism, and it is wise to be defensive and assume that proteolytic problems could arise (see *Table 1* for representative examples). Furthermore, if information derived from the study of isolated proteins is to be of relevance to the situation in whole cells, there must be no doubt that the protein molecule being examined represents the physiological form and does not arise as a consequence of sample preparation and analysis. Thus, it is just as important to be aware of when accidental proteolysis *in vitro* might occur and how it can be diagnosed as it is to know the means of prevention.

Intracellularly, proteolysis is highly controlled, by compartmentalization, the presence of endogenous proteinase inhibitors and the influence that a range of cellular components can have on the conformation of proteins. When tissues are disrupted during the course of protein isolation these controls are lost. Subcellular compartments are ruptured, enzyme–inhibitor complexes form or dissociate and interactions between proteins and other molecules are altered.

Table 1 Examples of proteolytic artefacts

System	Observation	Reference
C-terminal truncated forms of STAT5 protein in mammary gland	Protease inhibitors ineffective. Hot SDS lysis effective at prevention of artefacts	1
Arg tRNA synthetase artefacts	Whether artefact or not was unresolved	2
Porcine growth hormone during solubilization of inclusion bodies—*C*-terminal truncation	Inhibitor cocktail ineffective. Prevention of degradation by shifts in pH and temperature or by dissolution in denaturing agents	3
Artefactual degradation of periodontal pathogen proteins	High temperatures or low pH effective. Different proteinase inhibitors effective for different organisms	4
Interleukin β4 subunit fragmentation	Prevented by cysteine proteinase inhibitors and chelating agents.	5
Human follitropin fragmentation—*C*-terminal and N-terminal heterogeneity	Likely to be artefactual, cause unidentified	6
Propionomelanocortin degradation	Proteinase inhibitors give no added benefit over strong acid or heat inactivation	7
Glutamine synthetase degradation in cell free extracts of *Escherichia coli*	Proteolytic contamination in lysozyme preparation, or intrinsic peptide hydrolase activity of lysozyme	8
Tyrosinase in mouse melanoma cells	Cathepsins B and D major source of activity that generates artefactual soluble forms of enzyme	9
Fibrinolysis by allergen preparations	Allergens contain contaminating proteinases, and do not activate plaminogen	10
DNA topoisomerase degradation artefacts	Smaller variants are artefacts	11
Hydra protein preparations	Use of a protease cocktail to eliminate enzymes seen on gelatin zymograms, also increases yield of protein kinase C	12

Components of the extraction buffer may in turn affect protein conformation and so increase vulnerability to proteolysis. Cellular homogenates have protein concentrations that are usually between two and three orders of magnitudes lower than the intact cell. Thus, there are many changes that might lead to circumstances where adventitious proteolysis might occur.

Proteolytic problems do not always receive detailed coverage in the literature. Two articles written some years ago by Pringle (13,14) are notable exceptions, and more recently the topic has been discussed by Beynon (15) and Beynon and Oliver (16). Measures that purport to counter proteolysis are widely used, but the rationale for adoption of a particular strategy is rarely indicated. Details of experiments conducted to establish optimal conditions to avoid unwanted proteolysis are rarely presented (see ref. 17 for an exception).

In his 1975 article, Pringle (13) stated that 'it is not possible to give a magic antiprotease formula guaranteed to eliminate all artefacts in every case in which it is applied'. Detailed but specific protocols would be of little help in countering individual problems. The aim of this chapter is to make the reader aware of likely proteolytic problems and their origins and to demonstrate strategies to deal with them.

2 Proteolytic susceptibility of native proteins

2.1 Intrinsic factors determining the susceptibility of proteins to proteolysis

Proteins differ markedly in their stability both *in vivo* and *in vitro*. The proteins that are most rapidly turned over in the cell are often the most vulnerable to proteolysis in cell extracts. This implies that intrinsic characteristics do determine susceptibility to proteolysis (18). For example, it is generally held that the smaller a protein, the more resistant it is to proteolysis. A frequently observed manifestation of this is the disappearance of higher M_r bands, but not lower M_r bands, from samples analysed by the SDS-PAGE technique (see Section 3). The factors that influence the proteolytic susceptibility of proteins to proteolytic attack are discussed in the Chapter 10. Moreover, subtle changes in structure can lead to dramatic alterations in proteolytic susceptibility. For example, the phosphorylation of the troponin inhibitory subunit makes it vulnerable to degradation by calpain (19). Phytochrome is interconvertible between a far-red-absorbing form (P_{fr}) and a red-absorbing form (P_r), and the former is considerably more resistant to proteolysis during purification (20). Glycosylation can also provide protection against proteolysis *in vitro* (21,22).

Extraction of proteins can involve the removal of components that normally protect parts of the structure from proteolytic attack, or which otherwise maintain the protein in a resistant conformation. Thus, the isolation of membrane proteins is often prone to problems of proteolysis. Individual components of multimeric proteins may also have vulnerable sites which are hidden until the components are dissociated. Proteins are probably most resistant in their native state (susceptibility to proteolysis can be used to assess the degree of folding of proteins, see Chapter 10), and any perturbation of this native structure during an extraction procedure increases the likelihood of a proteolytic problem.

2.2 The influence of other molecules on susceptibility to proteolysis

Any small molecules which either directly or indirectly affect a protein's conformation will influence the degree to which it is a target for proteolytic attack. The binding of molecules such as substrates and cofactors tends to maintain a protein in a more resistant conformation, although some effector molecules may have the opposite effect (18). Molecules which have a general effect on protein structure will indirectly affect resistance to proteolysis. These include reagents such as glycerol and DMSO, which stabilize protein conformation and reducing agents which prevent the oxidation of thiol groups and thus maintain native protein structures, although the latter may have a destabilizing effect by reducing disulphide bridges.

2.3 Properties of endogenous proteinases

Clearly, proteolysis and its attendant problems will depend on the nature of the proteolytic enzymes present in a sample. Proteolysis by exopeptidases *in vitro* has been difficult to detect, which may explain why it is not considered to represent a major problem, and precautions directed specifically against exoproteolytic attack are rarely taken. However, mass spectrometry may indicate many more examples of *C*- or *N*-terminal trimming. An understanding of the properties of endoproteinases and exoproteinases is essential to the design of effective strategies to counter these problems, even when the troublesome proteinase is unknown. A detailed account of proteinase properties is in Chapter 4.

3 Identification of proteolysis as a problem

3.1 Changes in protein properties

Proteolytic problems can result from the hydrolysis of a single peptide bond in a protein molecule at one extreme to extensive degradation involving the hydrolysis of many bonds at the other. The ease with which unwanted proteolysis is identified will depend not only on the extent of the proteolysis but also on the

Table 2 Symptoms of a proteolytic problem

Symptom	Comments
Absence of a specific protein	Lack of expected activity or immunoreactive protein. Genuine absence can be confirmed only if the most stringent precautions against proteolysis are taken
Poor yields or loss of activity	Purification of proteins may prove impossible unless protective measures are adopted. To improve yields, consider alternative purification methods or sources of material. Loss of activity on storage is frequently caused by proteolysis. Trace amounts of proteinase can still cause problems in otherwise homogeneous preparations of protein
Changes in activity, e.g. specificity, pH dependence	Result from limited proteolysis which may be difficult to detect. If it occurs rapidly upon sample preparation, physiological forms of proteins may never be detected. The properties observed may be misleading. Apparent changes in activity during physiological or developmental processes could be due to variations in endogenous proteinases
Structural changes revealed by electrophoresis	See Table 3
Discrepancies between reported properties of specific proteins	Proteolysis may account for differences in the patterns of multiple forms separated by gel filtration, isoelectric focusing, etc., in isoenzyme patterns, in terminal amino acid data

Table 3 Symptoms on gel electrophoresis indicative of a proteolytic problem

Poor resolution (smearing) of bands and high background staining along gel tracks
Loss of higher M_r bands and increase in lower M_r material, especially at the dye front
Discrepancy in the apparent M_r of a specific protein between crude and purified preparations and the absence of specific bands from crude samples
Discrepancy between the M_r values reported by different laboratories for the same protein, or discrepancy in the M_r values observed for the same protein purified by alternative procedures
Stepwise decrease in apparent M_r during storage
Microheterogeneity, multiple bands in 'pure preparations' of protein
'Nicking', the appearance of two lower M_r bands concurrently with the loss of a single higher M_r band. The aggregate size of the smaller polypeptides is equivalent to that of the larger protein
Variable numbers of subunits in preparations of multifunctional proteins

type of study being undertaken. In studies in which proteins are analysed primarily by measuring activity, for example during enzyme purification, proteolysis may go undetected, as limited proteolysis does not always affect the catalytic properties of an enzyme. Proteolysis will be more obvious when physical properties are analysed. The most frequent symptoms of unwanted proteolysis *in vitro* are given in *Table 2*.

3.1.1 Detection by electrophoresis

SDS-PAGE is one of the analytical procedures used to study total cell proteins, products of translation *in vitro* (labelled proteins detected by autoradiography), including radiolabelling (*in vivo* or *in vitro*) of selected proteins (e.g. phosphorylation, protein kinase assays), immunoprecipitation of radiolabelled proteins, Western analysis (proteins stained with specific antibody after blotting on to nitrocellulose) and activity staining (detecting enzyme activity). Many proteolytic problems come to light during SDS-PAGE (*Table 3*). Differences in reported isoenzyme patterns obtained by electrophoresis using starch gels or cellulose acetate may also be caused by proteolysis during sample preparation.

3.1.2 Problems arising during electrophoresis

Although electrophoretic analysis is a powerful method for diagnosing proteolytic problems, (*Table 4*) it can itself create proteolytic artefacts. Endogenous proteinases may not necessarily be the only cause of proteolysis, however, as chemical hydrolysis in the presence of SDS has been reported (24,25). The symptoms of electrophoresis-related proteolysis will be the same as those described in *Table 3* and symptoms such as the smearing of bands and the differential loss of high-M_r proteins may be accentuated. Problems resulting from electrophoresis can be differentiated from those arising during sample preparation by their independence of either sample storage or preparation time.

Table 4 Causes of proteolytic problems arising during electrophoresis

Denaturing conditions such as those used in SDS-PAGE unfold proteins and make them more vulnerable to proteolytic attack
Some proteinases are not completely inactivated by standard denaturing conditions
Many proteinases remain active in mild denaturing conditions (e.g. mixing with SDS and reducing agent but not heating). This is a necessary requirement for the success of the Cleveland technique for peptide mapping (22) and the ease with which proteinases can be detected in gels after SDS-PAGE (23).
Endogenous inhibitor–proteinase complexes will be dissociated.
Reducing agents activate cysteine proteinases

3.1.3 Detection by mass spectrometry

Two mass spectrometric methods have been developed that are applicable to protein characterisation: ESI-MS (26) and MALDI-TOF (27). The former can give more accurate masses of intact proteins, and a resolution of 0.05% is readily attained. This equates to 1 Da in 5000, which is substantially higher than the resolution needed to observe the loss of one (or more than one) amino acid from one terminal or the other. MALDI-TOF mass spectrometry does not enjoy the same resolution as ESI-MS, but for small peptides can still indicate the loss of amino acids. The analysis of a limit peptide map by MALDI-TOF can pinpoint such changes with considerable confidence, however (see Chapter 7). (For a recent example of the identification of an unexpected, exoproteolytic cleavage by trypsin identified by ESI-MS, see ref. 28.)

3.2 Mimicking an effect with added proteinases

If novel forms of proteins have arisen during sample preparation it may be possible to mimic the process by adding exogenous proteinases. However, if protein interconversion does occur, this does not necessarily confirm that proteolysis *in vitro* by endogenous proteinases was involved, since proteolysis could have occurred inside the cells as a normal physiological process. It will, however, identify whether protein forms could be proteolytically related.

3.3 Checking samples for proteinase activity

The presence of contaminating proteinases in otherwise pure preparations of proteins might be detectable by conventional proteinase assays. A variety of proteinase assay techniques are available, details of which can be found in Chapter 3. The complete absence of a proteinase may be difficult to prove and, since trace amounts of proteinases can have damaging effects, very sensitive detection methods may be required. Pringle (13) described how as little as a one-millionth part of the bacterial proteinase subtilisin mixed with serum albumin gave extensive breakdown of the latter during denaturation with 1% SDS and 1% mercaptoethanol at 25 °C. The presence of proteinases that are active during electrophoresis can be revealed by adding radiolabelled protein to samples of interest just prior to electrophoresis and checking for degradation of radio-labelled protein by autoradiography.

4 Inhibition of proteinases

4.1 Outline of approaches for reducing proteinase activity

For the isolation and analysis of any protein, protocols should be designed so that the activity of the endogenous proteolytic enzymes is minimized and, if possible, eliminated completely. This requires procedures that reduce the levels of proteinases in the initial samples and allow the rapid removal of those proteolytic enzymes that remain; these aspects are discussed in Section 5. Whatever the procedures adopted, immediate and total removal of proteinases will be impossible, and so it is essential that precautions are taken to prevent those proteolytic enzymes still present from having an effect on the sample proteins. A successful strategy involves (i) the selection of conditions which will suppress proteolysis by active proteinases, (ii) the inactivation of proteinases by denaturation and (iii) the suppression of proteinase activity by inhibitors.

4.2 Suppression of endogenous proteinase activity

Table 5 lists a number of points which should be considered when selecting appropriate conditions for the extraction of proteins. It must always be borne in mind that none of these remove the proteinase, they only prevent the enzymes present from displaying activity. A change in conditions during preparation or analysis could activate a cryptic proteinase and cause proteolytic problems at later stages of a preparation. This may happen, for example, as a result of a change of pH, as in acidification, or when a sample is denatured prior to electrophoretic analysis. There have been a number of reports of so-called 'firmly-bound' or 'sticky' proteinases which remained tightly attached to proteins throughout purification, but whose presence was not detected until purification was complete. Such proteinases may only become apparent as a result of a proteolytic attack on the protein of interest and may not be detectable by any other means, for example as stained bands on SDS-PAGE analysis.

4.3 Preventing proteolysis by denaturation

If proteins of interest are not required in an active form, as in the case of total cell protein analysis by SDS-PAGE, the most direct means of preventing proteinase activity is to use strongly denaturing conditions either during or immediately following cell disruption. Protein extraction can be carried out in the presence of urea, SDS or guanidinium hydrochloride. However, proteinases are often more resistant to denaturation than other proteins, and mild denaturation conditions may accelerate proteolysis by exposing sites on target proteins without inactivating the proteinases. Proteinase inhibitors should be included whenever mild denaturation conditions are used, for example if samples are mixed

Table 5 The avoidance of proteolysis during sample preparation

Factor	Comment
pH and buffers	Use a buffer with a pH either above or below the optimum for proteinase activity. It is almost certain that there is no range of pH at which all proteolytic enzymes could be considered to be inactive, but above neutral pH the non-specific and highly active enzymes of the lysosomes or vacuoles will be minimally active. If, as in yeasts, the cellular pH is normally lower than this, a high concentration of buffer should be used to make certain that the required pH is obtained when the cell contents are released. A neutral pH favours interactions between proteinases and any endogenous inhibitors that could provide some degree of protection. Some stabilizing effects have been noted with particular buffers, for example phosphate (41)
Low temperature	Proteinase activity is minimized by working throughout at low temperature. Extended procedures should be carried out in the cold and samples stored frozen
Time	The shorter the preparation time, the less opportunity for proteolysis to occur. Lengthy procedures such as overnight dialysis should not be undertaken without the inclusion of proteinase inhibitors in the dialysis buffer. In some systems, for example yeasts and some filamentous fungi, prolonged incubation of extracts results in proteinase activation because endogenous inhibitors are degraded
Stabilizing agents	Protein stabilizing reagents can provide protection against proteolysis. The inclusion of glycerol (15–35% v/v) or DMSO (10% v/v) may be useful during preparation and storage. Reducing agents such as DTT (0.1–1 mM) or mercaptoethanol (1–10 mM) ensure that free thiol groups are not oxidized but will activate cysteine proteinases. The reagents could affect the activity of the protein of interest and reducing agents will diminish the effectiveness of some proteinase inhibitors
Exogenous protein	Proteins such as BSA (1–5 g/l) may provide protection by offering an alternative substrate to endogenous proteinases. Additional steps in protein preparation may have to be included in order to remove this exogenous protein later
Effectors	Low molecular weight effectors such as substrates, substrate analogues and cofactors can help to maintain a protein in a stable conformation and prevent proteolytic attack. Not all substrates or cofactors will necessarily be effective. Each substrate and cofactor should be tested as part of general screening process for protective agents. Any protective effect has to be balanced against the cost of using expensive chemicals in a large-scale preparation
Activators omitted	Potential proteinase activators such as divalent cations (these can activate and/or stabilize the enzyme) should be excluded from the extraction buffer
Denaturation	See Section 4.3
Proteinase inhibitors	See Section 4.4

with reducing agent and SDS but not boiled. If the samples are heated as part of the denaturation process (this should ensure the inactivation of most proteinases) this should be carried out immediately after addition of denaturing agents. Incubation at 100 °C for 2–3 min is sufficient for the preparation of samples for electrophoretic analysis. To ensure rapid temperature equilibration when samples are placed in a boiling-water bath, sample volumes should be as small as possible; slow warming will encourage proteolysis. Some proteinases can survive fairly harsh denaturation conditions and their activity will be restored when denaturants are removed. Denaturation may not, therefore, provide a permanent solution. Even heating samples with SDS and mercaptoethanol has not always eliminated proteolytic problems associated with SDS-PAGE analysis. Other methods of sample preparation or alternative electrophoretic procedures should be investigated in these circumstances.

Two examples of the analysis of plant proteins illustrate some relevant points. The proteins of petunia anthers are highly susceptible to proteolysis, and high-M_r proteins are degraded within a few minutes of homogenization in buffered solutions (30). Even the presence of 1% SDS in the homogenization buffer failed to prevent a complete loss of these proteins in SDS-PAGE analysis. The inclusion of proteinase inhibitors (PMSF, benzamidine) did not eradicate the problem. An alternative extraction method involving denaturation with TCA (trichloroacetic acid) was devised (*Protocol 1*).

Protocol 1

Plant protein extraction involving denaturation with trichloroacetic acid[a]

Equipment and reagents

- Microcentrifuge
- TCA
- Acetone
- Buffer: 0.06 M TrisHCl, pH 6.8, 1% (w/v) SDS, 3 mM benzamidine

Method

1 Samples are homogenized in 10% (w/v) TCA and incubated at 4 °C for 30 min. The precipitated protein is collected in a microcentrifuge. It is essential that TCA is present during homogenization and not added afterwards.

2 The protein sediments are washed once in 5% (w/v) TCA and twice in pre-chilled (−20 °C) 90% (v/v) acetone.

3 The sediments are dried and solubilized by sonication in the buffer before electrophoretic analysis.

[a] Method from Wu and Wang (30).

Plant plasma membrane proteins are especially vulnerable to proteolysis, which causes poor resolution of their component polypeptides in SDS-PAGE.

Thorough denaturation of samples (heating with SDS and DTT) inactivates the proteinases responsible but has the undesirable effect of causing protein aggregation. This problem was resolved in two ways (31). Firstly, samples were not boiled prior to SDS-PAGE but proteinase inhibitors were included during homogenization, cell fractionation and electrophoresis. The inclusion of PMSF and chymostatin significantly improved resolution with SDS-PAGE, although the inhibitors were not totally effective at preventing proteolysis under denaturing conditions. When samples containing the inhibitors are incubated in 5% (w/v) SDS for 1 h at 20 °C, high-M_r polypeptides are still lost. Secondly, an alternative electrophoretic procedure, PAU-PAGE is employed. This utilizes the denaturing effects of phenol and urea and the intrinsic charge of proteins to separate polypeptides according to molecular weight. PAU-PAGE rapidly inactivates proteinases and solubilizes aqueous insoluble proteins. This procedure has been adapted for use with slab gels (32).

4.4 Use of proteinase inhibitors

4.4.1 General points

The inclusion of inhibitors in solutions during and following tissue disruption represents the most favoured answer to the problem of unwanted proteolysis.

By using appropriate inhibitors, proteinases can be kept inactive throughout a preparation. The variation among the proteinases present in different organisms, and in different tissues of the same organism, rules out a universal procedure, and reagents must be selected to suit the particular problem encountered. However, because there are only a small number of types of proteinase, there is considerable overlap in the ways in which proteolysis in different systems can be controlled.

This section considers the inhibitors available, how they are prepared, the limitations in their use and the points to be considered in choosing appropriate reagents. A list of those most widely used as protective agents is given in *Table 6*. Some other reagents which have been used to prevent proteolysis in a more limited number of studies include bacitracin (an inhibitor of bacterial cell wall synthesis), TAME (a trypsin substrate); Cbz-Phe; methylamine; and β-ammoniopropionitrile fumarate. The inclusion of serum has also been shown to have a protective effect, probably due to the presence of α_2-macroglobulin (see Chapter 5), but the use of serum has a distinct disadvantage since it involves the addition of proteins during procedures which are often designed to eliminate all proteins other than the one of interest. A perfect, general purpose proteinase inhibitor should:

- have broad specificity
- be inexpensive and effective at low concentrations
- be soluble in aqueous media, thus avoiding the use of excess amounts of organic solvents
- be stable under the conditions used
- be non-toxic and therefore easy to handle
- react rapidly and irreversibly with proteinases in a reaction unaffected by other components of the extraction buffer

Table 6 Proteinase inhibitors used as protective agents (see Appendix III)

Proteinase type	Inhibitors	Comments[a]
Aspartic	Pepstatin A	Reversible but binds very tightly. Limited solubility. The only aspartic proteinase inhibitor in widespread use
Cysteine	Iodoacetic acid, *N*-ethylmaleimide, PHMB	Not specific, may inactivate other proteins
	E64	Effective at micromolar concentrations.
Cysteine/serine	TLCK	Offensive smell. Limited specificity with serine proteinases (inhibits trypsin-like enzymes)
	TPCK	Limited solubility. Limited specificity with serine proteinases (inhibits chymotrypsin-like enzymes)
	PMSF	Limited solubility and unstable in aqueous solution. More inhibitor should be added during lengthy preparations. Salted out at high concentrations of ammonium sulphate. Cysteine proteinase inactivation reversed by reducing agents. Inhibition independent of enzyme specificity. Some side-effects reported (34)
	Leupeptin, antipain	Reversible inhibitors effective at micromolar concentrations. Limited specificity with serine proteinases (inhibit trypsin-like enzymes)
	Chymostatin	Reversible inhibitor. Limited solubility but effective at micromolar concentrations. Limited specificity with serine proteinases (inhibits chymotrypsin-like enzymes)
Serine	DIPF	Highly toxic, not recommended for routine use. Inactivation independent of enzyme specificity. Inhibits other enzymes besides proteinases (serine esterases)
	Benzamidine, 4-aminobenzamidine	Reversible inhibitors of trypsin-like enzymes
	Elastatinal	Reversible inhibitor, not yet widely tested. Limited specificity (inhibits elastase-like enzymes)
	SBTI Aprotinin (Trasylol)	Reversible polypeptide inhibitors. Other polypeptide inhibitors are available but these are the two most widely used
Metallo-	EDTA, EGTA, 1,10 phenanthroline	Chelating agents, reversible. May also destabilise some serine proteinases or metal-stabilized enzymes.
	Phosphoramidon	Reversible inhibitor effective at micromolar concentrations. May have limited specificity. Not widely used
Calpains	Cysteine proteinase inhibitors	Only effective when enzyme is in an active form, i.e. calcium must be present
	Chelating agents	Reversible, removes calcium ions

[a]Unless otherwise indicated, proteinases are inhibited irreversibly.

- be specific for proteinases with no side-effects on other proteins
- be easy to remove from the purified protein, with low molecular weight being an advantage.

No such cheap, easily handled, potent and universal inhibitor exists, so any strategy will be based on specific targeting of a single problem proteinase, or a cocktail of inhibitors targeted to all classes of proteinase. The majority of the inhibitors fulfil most of the requirements, but for protection against all types of proteinase a number of inhibitors will be required.

4.4.1.1 Screening inhibitors of endogenous proteinase activity

A knowledge of the properties of proteinases present in the source material, especially of their mechanistic type, allows selection of appropriate inhibitors. However, not all of the endogenous proteolytic enzymes may be known and in many cases details of their properties will be incomplete. If there is insufficient information available, then screening of individual reagents or combinations of reagents will be necessary. This can be carried out in the following ways.

Proteinase activity in the sample material can be assayed using a general proteinase substrate to establish which reagents inhibit this activity most effectively. Ideally, this should be conducted on a small scale to avoid wasting material and expensive inhibitors. Colorimetric assays using nitroanilide derivatives of amino acid and peptides as substrates can be carried out conveniently in microtitre plates and reactions followed with an ELISA reader. This provides a quick way of screening a number of different inhibitor/sample combinations. A range of such substrates is available (see Chapter 3). Another type of screening method described recently involved the use of a diffusion plate assay using casein as a substrate (33). This identified PMSF and chymostatin as reagents to protect against proteinases in various plant cell fractions. Assays involving exogenous substrates have the advantage of being relatively easy and quick to undertake, but suffer from the drawback that only some of the proteolytic enzymes may be detected. The screening may not help to identify inhibitors of enzymes of restricted specificity responsible for limited proteolysis.

A more appropriate screening may require the analysis of changes to endogenous proteins during sample incubation and storage by assaying the activity of particular marker enzymes in cell extracts, or by electrophoretic analysis involving direct staining or, if antibody is available, Western blotting. Should changes in enzyme activity or band pattern occur during sample preparation or storage, proteinase inhibitors can be screened for their ability to prevent these changes. The monitoring of other endogenous proteins may provide a better model for proteins of interest, and it may be possible to identify inhibitors of proteinases responsible for limited proteolysis. The disadvantage is that it takes longer and the analysis is likely to be more complicated than direct assays of proteinase activity.

In many instances, a persistent proteolytic problem may only be resolved by direct screening of inhibitors on changes to the actual protein(s) of interest. This may be necessary if a previously uncharacterized minor proteinase is responsible for the problem. This approach could be limited by the amount of material available, and laborious when the protective effects can only be checked after a multistep preparation. However, the solution will, in this case, be one which is guaranteed to work, while there is no assurance that strategies predicted from indirect screening procedures will be successful for all proteins.

The experience of other workers who have used the same source of material to isolate and analyse proteins or another source to isolate or analyse the same protein(s) can be very helpful. Past papers should be read thoroughly, as useful information relating to the control of proteolysis is not always given prominence.

4.4.2 Inhibitor preparation

In the following section the most important factors concerning the preparation of inhibitors are discussed. Specific details can be found in *Table 6*, and further information on inhibitors is available in Chapter 5 and Appendices II and III.

(a) Inhibitor solubility and stability. Some reagents are not very soluble in aqueous media and stock solutions must be prepared in organic solvents (see *Table 7*). Whenever these are added to cell extracts or protein preparations, they should be mixed well to avoid localized high concentrations of solvent. Solvents which are likely to alter the properties of the proteins of interest should not be used.

Table 7 Suggestions for inhibitor cocktails for protein isolation

Tissue	Inhibitor (final concentration)[a]	Stock[b]
Animal tissues[c]	PMSF (1 mM)	0.2 M in methanol[d]
	EDTA (1 mM)	0.1 M in aqueous solution
	Benzamidine (1 mM)	0.1 M in aqueous solution
	Leupeptin (10 μg/ml)	1 mg/ml in aqueous solution
	Pepstatin (10 μg/ml)	5 mg/ml in methanol
	SBTI or aprotinin (1 μg/ml)	0.1 mg/ml in aqueous solution
Plant tissues[c]	PMSF (1 mM)	0.2 M in methanol[d]
	Chymostatin (20 μg/ml)	1 mg/ml in DMSO
	EDTA (l mM)	0.1 M in aqueous solution
Protozoa	PMSF (1 mM)	0.2 M in methanol[d]
	Leupeptin (10 μg/ml)	1 mg/ml in aqueous solution
Slime moulds[c]	PMSF (1 mM)	0.2 M in methanol[d]
	Leupeptin (10 μg/ml)	1 mg/ml in aqueous solution
	Benzamidine (1 mM)	0.1 M in aqueous solution
	TLCK (0.1 mM)	10 mM in 1 mM HCl
Yeast and fungi	PMSF (1 mM)	0.2 M in methanol[d]
	Phenanthroline (5 mM)	1 M in ethanol
	Pepstatin (15 μg/ml)	5 mg/ml in methanol
Bacteria	PMSF (1 mM)	0.1 M in methanol[d]
	EDTA (1 mM)	0.1 M in aqueous solution

[a] These are probably the minimal requirements; alterations, including additions of other inhibitors, may be necessary in specific instances.

[b] More details of inhibitor stock solution preparation can be found in Appendix III.

[c] If cysteine proteinases are thought to be the source of problems, the use of E64 should be checked. A concentration 1–10 μg/ml ought to be effective, and stock solutions can be prepared in aqueous solution at 1 mg/ml. E64 has not been widely tested, hence it has not yet been included in any of the recommended cocktails.

[d] Dilute in buffer immediately prior to use. Repeat additions should be made during sample preparation.

(b) Reagent concentration. *Table 7* provides recommended working concentrations for particular inhibitors. In practice, there is considerable variation in the concentrations used by different workers and in some instances the amounts of inhibitor appear to be unnecessarily high. The need for a high concentration should be checked carefully whenever expensive inhibitors such as pepstatin, antipain, leupeptin or E64c are considered. All of these are effective at micromolar concentrations, which helps offset their high cost. Stock solutions prepared in organic solvents should be as concentrated as possible to avoid the addition of excess solvent to the sample.

(c) Safety. Any reagent which inhibits enzyme activity is potentially harmful and should be treated with due care. Special precautions are definitely required with some of the reagents listed in *Table 6*. Dip-F is a highly toxic nerve gas which can be absorbed by the skin and its routine use is not recommended. PMSF is toxic and should also be treated with caution. Gloves should be worn when handling both solid and solutions. TLCK has a strong and offensive odour which is very easily transferred to equipment, glassware, door handles, light switches, clothing, skin, etc., if insufficient care is taken. The smell may linger for many hours, and careless handling of TLCK can be considered 'anti-social'. This can be avoided by wearing gloves which should be removed immediately after using the reagent. Inhibitors such as AEBSF (Pefabloc) offer almost as effective inhibition as PMSF and Dip-F, but are less toxic and more stable than the older analogues.

4.4.3 Specific recommendations for inhibitor use

The following information is based on a survey of the usage of proteinase inhibitors in studies reported in the recent literature. No one inhibitor or combination of inhibitors is preferred for all systems. Some workers favour the inclusion of inhibitors of all four types of proteinase, although this is rarely justified on the basis of known effects of proteolysis on the proteins of interest. Others are more selective and base their choice on factors discussed in this chapter. The compositions of the inhibitor cocktails in *Table 7* are formulated on the basis of these reports and are offered as a guide rather than a set of absolute recommendations. Some manufacturers now offer pre-prepared cocktails of proteinase inhibitors, sometimes in tablet form, that can provide a straightforward, if expensive solution to proteinase problems (see Core List of Suppliers).

(a) Animal tissues. PMSF and EDTA are the most frequently used inhibitors in work with animal samples. The aqueous instability of PMSF and its poor solubility are not considered to be sufficient drawbacks to limit its use. Protection against serine proteinases is frequently reinforced by the inclusion of more selective inhibitors such as benzamidine, TPCK, TLCK, leupeptin, chymostatin and antipain (the last four will also deal with cysteine proteinases) and by the use of soybean trypsin inhibitor or aprotinin. EDTA, and less frequently 1,10 phenathroline or EGTA inhibit metalloproteinases and metal-activated proteinases. There are relatively few accounts of problems involving aspartic proteinases, and the incorporation of pepstatin in a cocktail designed to counter a broad spectrum of proteinases is normally a sufficient safeguard.

(b) Plant tissues. Most of the well-characterized plant proteinases are of the serine and cysteine types. In general, protective measures involving inhibitors are directed against serine proteinases through the use of PMSF in combination with another inhibitor such as chymostatin or leupeptin. The last two reagents also prevent problems from cysteine proteinases. EDTA is also used frequently.

(c) Protozoa and slime moulds. In certain respects, the proteolytic systems of these organisms resemble those of higher animals. Many species have cysteine proteinases present in lysosome-like organelles, but these are a greater nuisance than their counterparts in higher animals as they are not inactivated above neutral pH, and so can be active during both protein isolation and electrophoretic procedures. Protection against proteolysis usually requires the use of cysteine proteinase inhibitors such as leupeptin. It may also be necessary to avoid using reducing agents which will activate cysteine proteinases. In preparations of *Trichomonas vaginalis*, for example, the inclusion of DTT caused a marked loss of adenylate kinase activity (35). PMSF has also been used and its inclusion is advisable.

(d) Fungi, including yeasts. The proteolytic system of baker's yeast (*Saccharomyces cerevisiae*) has been the subject of detailed investigations and proteolytic problems encountered with this organism have received close attention (13,14). In yeast, the major vacuolar aspartic and serine proteinases are regarded as the most likely causes of proteolytic problems. They are countered by the inclusion of pepstatin and PMSF in inhibitor cocktails. Yeast cells also contain other types of proteinase and additional inhibitors, for example 1,10-phenanthroline, may be required. Aspartic and serine proteinases are also the major proteinases of most other fungi and inhibitor cocktails similar to those in use with yeast are suitable for fungal protein preparations.

(e) Bacteria. The predominant proteinases in most bacteria are of the serine and metallo- types and precautions normally involve the inclusion of PMSF and EDTA in the extraction buffer.

Illustrations of the use of inhibitors are provided by *Figures 1* and *2*. Extensive proteolysis can occur during electrophoretic analysis of proteins of the protozoan *T. vaginalis*, especially if samples are not completely denatured. If heat treatment is not included, proteins of higher M_r are not apparent (*Figure 1*). The loss is not affected by sample storage time, but results directly from the electrophoretic analysis.

Trichomonads have high levels of cysteine proteinases and the problem is cured by the inclusion of leupeptin, a potent inhibitor of these enzymes. Another cysteine proteinase inhibitor, antipain, is much less effective.

Figure 2 shows the use to which proteinase inhibitors have been put in the analysis of rat myelin proteins. Inhibitors were screened to solve the problem of

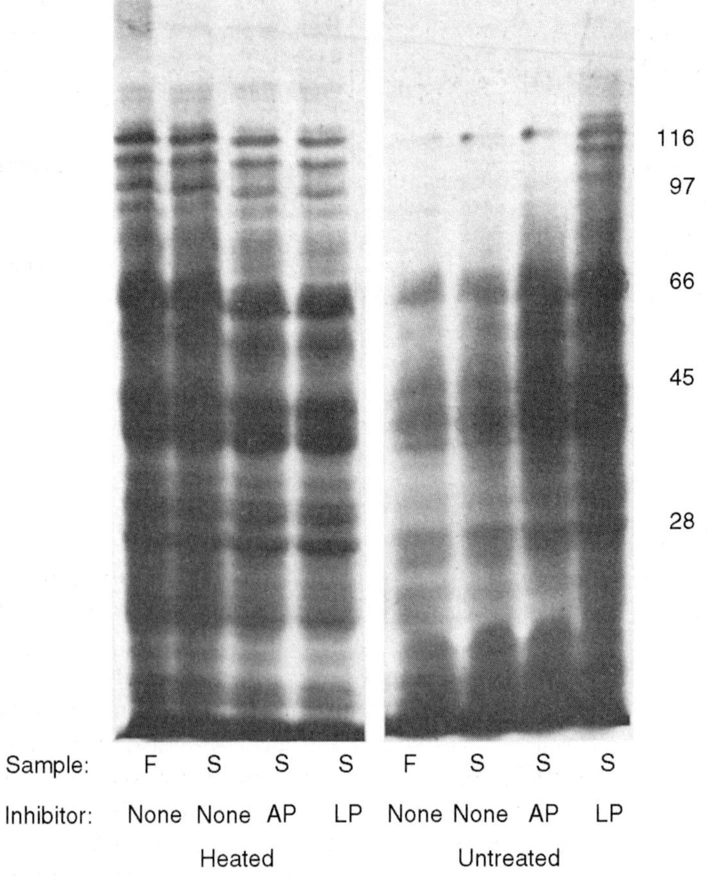

Figure 1 The protection of trichomonad proteins by leupeptin during electrophoresis. Frozen pellets of *Trichomonas vaginalis* were lysed by treatment with 0.25 M sucrose containing 0.25% Triton X-100 and proteinase inhibitors, as indicated. The samples were then stored frozen for 2 weeks (S) or used fresh (F). Electrophoretic analysis was carried out by mixing samples with an equal volume of electrophoresis sample buffer: 0.0625 M Tris-HCl, pH 6.8, 2% (w/v) SDS, 5% (v/v) mercaptoethanol, 10% (v/v) glycerol and 0.002% bromophenol blue. Some samples were heat-treated by incubation in a boiling-water bath for 2 mm. Samples were applied to a 7.5% (w/v) acrylamide gel which was run overnight at 5 mA. Proteins were stained with Coomassie Brilliant Blue. None, no inhibitor added; AP, plus antipain (0.1 mg/ml); LP, plus leupeptin (0.1 mg/ml). The positions of standard proteins (molecular sizes in kilodaltons) are indicated.

the formation of a lower-molecular-weight active form of 2′,3′-cyclic nucleotide 3′-phosphodiesterase (CNPase, Wolfgram protein) which appears if central nervous system myelin is left for a few hours at room temperature (*Figure 2a*). Proteolysis was the suspected cause, as the extra band was absent when freshly prepared myelin was analysed. Inclusioin of TLCK in the homogenization buffer prevented the appearance of the extra band. None of the other inhibitors tested was effective. TLCK is now included routinely in preparations of this protein.

Analysis of a second myelin protein, in this case one from the peripheral nervous system, showed that the use of TLCK does not provide a universal answer to myelin-associated proteolytic problems. The 170 kDa protein is detected by immunoblotting using specific antiserum and, in addition to a major 170 kDa band, a series of faster (lower M_r) bands are apparent, a symptom characteristic of stepwise proteolytic degradation (*Figure 2b*). Proteinase inhibitors, including TLCK, were again screened but no single agent afforded complete protection. The problem was solved by using a cocktail of inhibitors. The 170 kDa protein is probably especially vulnerable to proteolytic attack because of its size and needs far more stringent protection than the CNPase despite the similarity of the source material.

5 Removal of proteinases

5.1 Choice of starting material

5.1.1 Physiological and tissue-specific effects

The proteinase levels in tissues and cells vary enormously. For example, in animals, levels of lysosomal cathepsins are high in liver, spleen, kidney and macrophages. In many microorganisms, proteinases are inducible and are synthesized only if protein is available as a nutrient, although this applies more to the regulation of extracellular enzymes, which should not cause problems with studies on cellular proteins if well-washed cell preparations are used. Proteinase production may be subject to carbon, nitrogen or sulphur repression and can often be suppressed by appropriate supplements to the medium such as glucose, ammonia and sulphate. In some microbial systems, proteinase levels are elevated during starvation and sporulation. Thus, proteolytic problems may result from the use of stationary-phase cultures but are avoided by using log-phase cells. If the harvesting of cells takes a long time, this may mimic starvation and stimulate proteinase production. Regulation also applies to animal tissues. Well-fed animals have lower liver proteinase levels than starved animals.

5.1.2 The use of mutants

For the isolation of protein from microorganisms, mutants deficient in one or more proteinase offer a particularly effective means of eliminating some proteolytic problems. Mutants have been isolated in studies concerned with the physiological function of proteinases, but their availability can be of considerable value in studies of other proteins from the same organism. This is exemplified by the yeast proteolytic system for which mutants lacking one of four proteinases (*yscA*, *yscB*, *yscD* and *yscF*), two aminopeptidases, two dipeptidyl aminopeptidases and three carboxypeptidases have been isolated (36). Strains lacking more than one enzyme have also been constructed. This approach is normally limited to those organisms for which proteinase-deficient mutants are already available. The isolation of proteinase-deficient strains can be a time-consuming and tedious

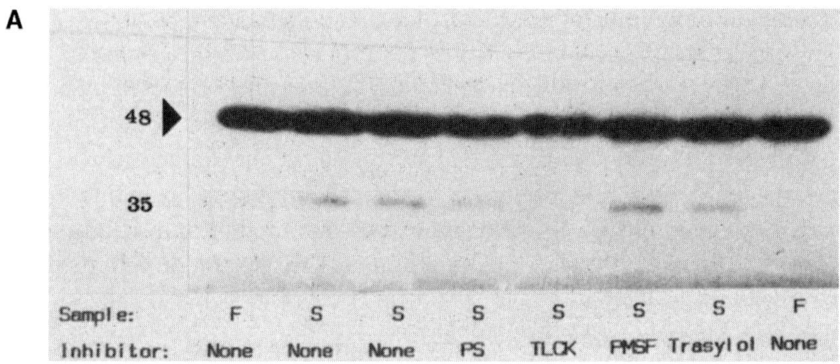

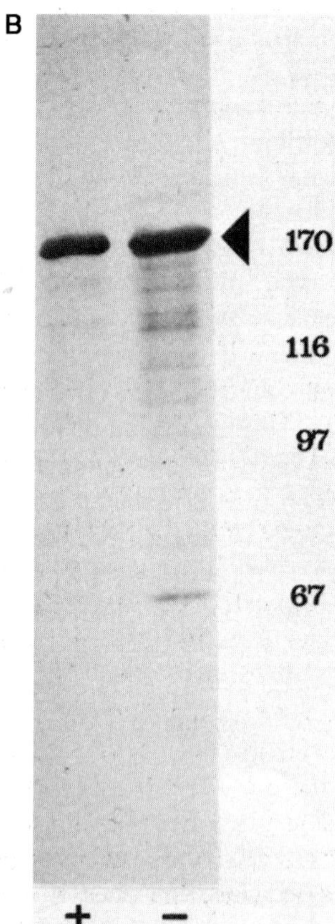

process and could not be routinely incorporated into a study directed towards the analysis of other cellular proteins. Neither can it be assumed that any of the mutants isolated would actually lack the proteinase responsible for a particular artefact. When extracellular proteinases pose problems, as they can do in the preparation of secreted proteins, selection of mutants may be easier. For example, a proteinase-negative mutant of the fungus *Aspergillus ficum* was selected on casein medium and used to obtain an electrophoretically homogeneous α-amylase (37).

Figure 2 The protection of myelin proteins against proteolysis. (A) Central nervous system myelin was prepared from homogenized rat brain and used either fresh (F) or after storage for up to 6 h at room temperature (S). For some preparations, proteinase inhibitors were added to the homogenization buffer. Samples of myelin were subjected to SDS-PAGE, the proteins blotted on to nitrocellulose filters and stained for 2′,3′-cyclic nucleotide 3′-phosphodiesterase (CNPase) activity. The inhibitors used were pepstatin (PS) (0.1 mg/ml), 1 mM TLCK, 0.5 mM PMSF or Trasylol (200 U/ml). Note that the activity of CNPase is extremely resistant to denaturation and survives boiling with SDS and DTT for 2.5 min, electrophoresis and blotting. Native CNPase is 48 kDa and the proteolytic product is 35 kDa. (B) The protection of a myelin protein by a cocktail of proteinase inhibitors. Peripheral nervous system myelin was prepared from homogenized rat sciatic nerve with (+) or without (–) the addition of a cocktail of proteinase inhibitors to the homogenization buffer. The cocktail contained 1.2 mM PMSF, 1 mM TLCK, 10 mM benzamidine, 1 mM EDTA, 1 mM EGTA, leupeptin (10 μg/ml), antipain (10 μg/ml), chymostatin (5 μg/ml) and pepstatin A (15 μg/ml). Samples of myelin were subjected to SDS-PAGE, the proteins blotted on to nitrocellulose and immunostained using an anti-rat 170-kDa protein antiserum. The position of the 170 kDa protein is indicated. The molecular sizes, determined from standards, are in kilodaltons. All data courtesy of C. S. Gillespie.

This enzyme (M_r 88 000) is raw-starch absorbable, but is easily converted to a raw-starch non-digesting enzyme (M_r 54 000) by proteolysis which removes the raw-starch affinity site. In the mutant, this conversion does not occur.

Low yields of proteins encoded by cloned foreign genes can result from degradation *in vivo* by host proteolytic enzymes which regard such proteins as abnormal. *Escherichia coli* strains with altered proteinase activity are being used to overcome this problem (38). The strains include *lon*$^-$ mutants which lack the ATP-dependent proteinase La, and *htpR* mutants in which heat-shock gene regulation is defective (the lon protease is a heat-shock protein). An alternative procedure involves the introduction of the bacteriophage T4 pin gene which encodes a protein that prevents the degradation of abnormal proteins. It appears to be effective against protease La but does not eliminate some of the other proteinases involved (39).

The following example provides an interesting illustration of how manipulation of a cloned gene itself overcame an *in vitro* proteolytic problem. It also reinforces the importance that the structure of a protein has in determining its susceptibility to proteolysis. Expression of the hepatitis B virus surface antigen gene *P31* in transformed yeast results in the accumulation of glycosylated gene products, GP37 and GP34. During their extraction, a portion of each of these products becomes digested by yeast proteinases to form smaller species devoid of polyalbumin receptor (PAR) activity. To overcome this problem, a modified *P31* gene has been constructed, the product of which can be isolated with PAR activity (40). The modification involved the deletion of a coding sequence for six amino acids (Ser44–Thr49) which removed one of the most susceptible sites of action of trypsin-like enzymes (Arg48-Thr49) and created a proteinase-resistant gene product. This strategy was complemented by the use of proteinase-deficient mutants as hosts. Use of this approach is clearly very limited.

5.2 Cell disruption and fractionation

The distribution of cellular proteolytic enzymes among different compartments means that careful disruption of the cells followed by subcellular fractionation can yield samples which are low in endogenous proteolytic activity. Tissues should be disrupted so as to avoid breakage of proteinase-rich vacuoles and lysosomes. This requires the use of gentle techniques (homogenization, freeze–thaw in liquid nitrogen and the omission of detergents such as Triton X-100 which dissolve vacuolar/lysosomal membranes). To prevent osmotic rupture of organelles, samples should be homogenized in isotonic solution (e.g. 0.25 M sucrose). Digitonin (0.8 mg/ml) can be used to perforate the plasma membrane without the lysis of organelles. After homogenization, the vacuolar fraction can be removed by differential centrifugation. Following a low-speed spin (5 min at 800 **g**) to remove unbroken cells and nuclei, a further centrifugation for 10 min at 12 000 **g** in a standard preparative centrifuge should be sufficient to obtain a supernatant depleted of vacuolar enzymes. Removal of the vacuolar fraction should be carried out as rapidly as possible, and samples kept at 4 °C throughout to minimize rupture. The degree of success can be checked by assaying for a lysosomal enzyme such as cathepsin D or cathepsin L (see Chapter 3), which should have been removed.

For preparations involving delicate proteins, different methods of cell breakage should be checked to find one that gives the least contamination with lysosomal hydrolases. It may be necessary to sacrifice yield to avoid excessive breakage of lysosomes. For example, in a recent study of actin-binding proteins in *Dictyostelium discoideum*, 80% cell lysis by homogenization was found to be optimal to avoid proteolysis (41). Note that preparations of other subcellular fractions such as mitochondria may be contaminated with vacuoles and special precautions may be needed to counter proteolysis caused by vacuolar enzymes.

Breakage of cell walls for the preparation of fungal and plant proteins may necessitate the use of harsh conditions in which vacuolar disruption is unavoidable. In some situations this can be avoided by preparing sphaeroplasts or protoplasts in which the cell wall is removed by digestion with enzymes. The cells can then be broken carefully by isoosmotic lysis. Preparation of yeast sphaeroplasts allows the isolation of cytosolic and mitochondrial fractions which are essentially free of vacuolar proteinases (42). The preparation of sphaeroplasts and protoplasts may be considered too laborious and expensive for large-scale procedures.

Autolysis, in which cell breakage is achieved by the action of endogenous enzymes, is not recommended. With yeast, this involves a long incubation period during which proteolysis can occur and activation of proteinases take place as endogenous inhibitors are digested.

A further caution is necessary: vacuoles and lysosomes are not the sole compartments for proteolytic enzymes. Removing a known source of proteinases does not always guarantee that problems are eliminated.

5.3 Selective removal of proteinases during purification

Protein purification schemes are designed to optimize the recovery of the protein of interest at the expense of all other proteins. Whenever proteolysis is a problem, some thought should be given to the inclusion of specific steps to remove proteinases, even though the individual step has little effect on the overall purity of the protein concerned. If proteinases have been purified from the same cells or tissue, refer to published purification schemes and select methods which would exploit differences in properties between proteinases and protein. Affinity techniques, normally used in proteinase purification, can be adopted to remove the latter selectively from other preparations. The affinity adsorbents are immobilized proteinase inhibitors or substrates. Amersham Pharmacia Biotech supply agarose-soybean trypsin inhibitor (AG-SBI™), benzamidine-Sepharose 6B:agarose-haemoglobin (AGHEM™), lysine-Sepharose 4B and arginine-Sepharose 4B. Sigma supply soybean trypsin inhibitor-agarose, lima bean trypsin inhibitor-agarose, benzamidine-Sepharose 6B, benzamidine-agarose, Aprotinin-agarose, Aprotinin-acrylic beads and Pepstatin A-agarose.

The choice of affinity medium will depend on the type of proteinase(s) present in the sample. The high cost of the commercial affinity adsorbents may limit the scale on which this procedure can be used. Laboratory methods are available for the preparation of these and other adsorbents (see Chapter 2 and ref. 15 for examples).

Acknowledgements

The authors thanks to Stewart Gillespie for allowing the inclusion of his data on myelin proteins.

References

1. Garimorth, K., Welte, T., and Doppler, W. (1999). *Exp. Cell Res.* **246**, 148.
2. Vellekamp, G., Sihag, R. K., and Deutscher, M. P. (1985). *J. Biol. Chem.* **260**, 9843.
3. Puri, N. K., Cardamone, M., Crivelli, E., and Traeger, J. C. (1993). *Protein Exp. Purif.* **4**, 164.
4. Weidner, M. F., Grenier, D. and Mayrand, D. (1996). *Oral Microbiol. Immunol.* **11**, 103.
5. Potts, A. J., Croall, D. E., and Hemler, M. E. (1994). *Exp. Cell Res.* **212**, 2.
6. Shome, B., Parlow, A. F., Liu, W. K., Nahm, H. S. Wen, T., and Ward, D. N. (1988). *J. Protein. Chem.* **7**, 325.
7. Kapcala, L.P. and Albers, J. M. (1988). *Peptides* **9**, 437.
8. Oliver, C. N. and Stadtman, E. R. (1983). *Proc. Natl. Acad. Sci. USA* **80**, 2156.
9. Martinez, J. H., Solano, F and Lonazo, J. A. (1989). *Cell Biochem. Funct.* **7**, 21.
10. Berrens, L. and van Rijswijk-Verbeek, L. (1975). *Int. Arch. Allergy Appl. Immunol.* **49**, 632.
11. Samuels, D. S. and Shimizu, N. (1994). *Mol. Biol. Rep.* **19**, 99.
12. Hassel, M., Klenk, G., and Frohme, M. (1996). *Anal. Biochem.* **242**, 274.
13. Pringle, J. R. (1975). In *Methods in cell biology* (ed. Prescott, D. M.), Vol. 12, p.149. Academic Press, London.

14. Pringle.J. R. (1979). In *Limited proteolysis in microorganisms* (ed. Cohen, G. N. and Holzer, H.), p. 191. US Government Printing Office, Washington, DC.
15. Beynon, R. (1988). In *Methods in molecular biology* (ed. Walker, J. N.), p. 1. Humana Press. Clifton.
16. Beynon, R. J and Oliver, S. (1996). In *Protein purification protocols* (ed. Doonan, S.), p. 81. Humana Press, Clifton.
17. Scott, R. E., Lam, K. S., and Gaucher, G. M. (1986). *Can. J. Microbiol.* **32**, 167.
18. Goldberg, A. L. and Dice, J. F. (1974). *Annu. Rev. Biochem.* **43**, 835.
19. Toyo-Oka, T. (1982). *Biochem. Biophys. Res. Commun.* **107**, 44.
20. Vierstra, R. D. and Quail, P. H. (1983). *Biochemistry* **22**, 2498.
21. Werner, R. G., Noe, W., Kopp, K., and Schluter, M. (1998). *Arzneimittelforschung* **48**, 870.
22. Ladenheim, R. G., Seidah, N. G., and Rougeon, F. (1991). *Eur. J. Biochem.* **198**, 535.
23. Hames, B. D. (1981). In *Gel electrophoresis of proteins: a practical approach* (ed. Hames, B. D. and Rickwood, D.), p. 219. IRL Press, Oxford.
24. Cookson, E. J. and Beynon, R. J. (1987). *Biosci. Rep.* **7**, 209.
25. Rittenhouse, J. and Marcus, F. (1984). *Anal. Biochem.*, **138**, 442.
26. Burlingame, A. L., Boyd, R., K., and Gaskell, S. J. (1998). Anal. Chem. **70**, 647R.
27. Chaurand, P., Luetzenkirchen, F., and Spengler, B. (1999). *J. Am. Soc. Mass. Spectrom.* **10**, 91.
28. Wu, C., Robertson, D. H. L., Hubbard, S. J. Gaskell, S. J., and Beynon, R. J. (1999). *J. Biol. Chem.* **274**, 1108.
29. Lockwood, B. C, North. M. J., Scott, K .I., Bremner, A. F., and Coombs, G. H. (1987). *Mol. Biochem. Parasitol.* **24**, 89..
30. Wu, F.-S. and Wang, M.-Y. (1984). *Anal. Biochem.* **139**, 100.
31. Gallagher, S. R. and Leonard, R. T. (1987). *Plant Physiol.* **83**, 265.
32. Gallagher, S. R. and Leonard, R. T. (1987). *Anal. Biochem.* **162**, 350.
33. Gallagher, S. R. Carroll, E. J., and Leonard, R. T. (1986). *Plant Physiol.* **81**, 869.
34. Meyers Hutchins, B. L. and Frazier, W. A. (1984). *J. Biol. Chem.* **259**, 4379..
35. DeClerck, P. J. and Muller, M. (1987). *Comp. Biochem. Physiol.* **88B**, 575.
36. Suarez Rendueles, P. and Wolf, D. H. (1988). *FEMS Microbiol. Rev.* **54**, 17.
37. Hayashida.S. and Teramoto, Y. (1986). *Appl. Environ. Microbiol.* **52**, 1068.
38. Kane, J. F. and Rartley, D. L. (1988). *Trends Biotechnol.* **6**, 95.
39. Skompski, K., Tomaschewski, J., Riger, W., and Simon, L. D. (1988). *J. Bacteriol.* **170**, 3016.
40. Itoh, Y. and Fujisawa, Y. (1986). *Biochem. Biophys. Res. Commun.* **141**, 942.
41. Hock, R. S. and Condeelis, J. S. (1987). *J. Biol. Chem.* **262**, 394.
41. Schwenke, J., Canut, H., and Flores, A. (1983). *FEBS Lett.* **156**, 274.

Chapter 10
Proteolysis of native proteins as a structural probe

Simon Hubbard
Department of Biomolecular Sciences, UMIST, PO Box 88, Manchester, M60 1QD UK

Robert J. Beynon
Department of Veterinary Preclinical Sciences, University of Liverpool, PO Box 147, Liverpool L69 3BX, UK

1 Introduction

A common use of proteinases *in vitro* is the degradation of proteins to limit peptide mixtures in which all susceptible bonds have been cleaved (Chapter 8). To facilitate this process, the substrate would usually be first denatured, by heat, urea, guanidinium hydrochloride, by precipitation with trichloroacetic acid or similar treatments. As a result of the destruction of the higher-order structure in the substrate, all susceptible (i.e. in accordance with the primary cleavage specificity) bonds will be attacked rapidly and with rates that, compared with native state digestions, are approximately similar and influenced only by local changes in subsite occupancy; in this way, the peptide mixture limit is rapidly generated. The need for the prior denaturation step implicitly assumes that retention of native structure will slow or modify the proteolytic attack, and this is indeed the case. The role of three-dimensional structure of a protein substrate in modification of extent and rapidity of proteolysis is incompletely understood, but a recurrent observation is that proteolysis is more limited, both in the extent and in the rate of hydrolysis. This restriction of proteolytic attack on native proteins also underpins most proteolytic interactions of physiological relevance, and thus, the study of limited proteolysis is also relevant to processes *in vivo*.

Limited proteolysis is a powerful tool for the study of protein structure. The range of applications of this technique is broad, from preparation of protein derivatives in which a specific function has been disabled or excised, to probes of flexibility mediated by ligands, or during the process of protein folding. It is not appropriate in this chapter to give comprehensive lists of examples, but *Table 1* includes most types of application, together with some specific examples, and key references or reviews. Because limited proteolysis typically requires the

Table 1 Uses of limited proteolysis as a probe of protein structure

Use	Conditions	Examples
Surface scanning	Typically low [E]:[S] (e.g. 1:50 to 1:100) Non-denaturing conditions	V8-proteinase incubation with a lipase from *Pseudomonas aeruginosa* gives surfaces sites(confirmed via multiple alignment to homologue with known structure). Information is incorporated into a three-dimensional model (1)
Identification of inter-domain links	Typically low [E]:[S] (e.g. 1:50 to 1:100) Non-denaturing conditions	Domain assignments made to pancreatic lipase via limited proteolysis experiments with chymotrypsin on the horse, pig and human forms of the lipase (2). The locations of the colipase-binding and active site domains were assigned
Monitoring of conformational change due to ligand binding	Low [E]:[S] (e.g. 1:50 – 1:100) although higher enzyme concentrations may be necessary Non-denaturing conditions	Avidin proteolysis by proteinase K is dependent on the presence of the natural ligand, biotin (3). The holoenzyme is not measurably nicked while the protease cleaves specifically at two bonds in a loop above the ligand-binding site in the apo-protein
Probing partly folded states	Higher [E]:[S] to ensure cleavage takes place faster than unfolding. Partially denaturing conditions such as acid, urea, guanidinium hydrochloride, trifluoroethanol or high temperature	Experiments on ribonuclease A using heat and guanidinium chloride demonstrated the unfolded segments around residue 25 and 31–33 and that significant structure was retained by three core regions, even up to 60 °C (4–8)

interaction of two proteins in their native state, the behaviour of both must be taken into consideration. Whereas protocols for generation of limit peptide maps by digestion of denatured proteins are straightforward (Chapters 7 and 8), no such generic protocols can be presented for limited proteolysis of native proteins. Experimentation is warranted and indeed, is necessary, although there are guiding principles that can be helpful.

There are different reasons why one might wish to treat a native protein with a proteinase:

- to generate defined products that retain some or all properties of the parent proteins, or which gain new properties
- to measure the rate of digestion as a parameter defined by the structure of the target protein
- to monitor the rate and/or route of limited proteolysis as a function of conformational changes caused by, for example, ligand binding or oligomerization
- to gain insights into the domain structure of a protein, by cleavage of more digestible regions between such domains.

In some applications, the requirement of the digestion is to maximize the amount of a particular digestion product that appears as a transient product on the route of digestion, and thus, the reaction should be optimized for maximal yield. In other applications, it is the rate of digestion, measured precisely, that is the key datum derived from the experiment. In both instances, knowledge of the factors that underpin the interaction can substantially enhance the value of the experiment.

2 Factors influencing susceptibility

In order to exploit limited proteolysis as a technique for probing protein structure, a good understanding of the various factors that underpin it is required. The structural basis of primary sequence specificity (such as the preference of trypsin-like enzymes to cleave *C*-terminal to the basic amino acids lysine and arginine) is reasonably well understood. The active site of the enzyme contains subsites, which are binding pockets formed by the enzyme that are complementary in shape and chemistry to amino acids in the substrate molecule. For example, in some proteases, the main subsite is the S_1 subsite, which recognizes the P_1 side chain of the substrate polypeptide located immediately before the peptide bond to be cleaved. For example, in trypsin this pocket is deep and has an acidic aspartate side-chain at its base, which forms a strong ionic bond to lysine and arginine side-chains (9). Other enzymes possess strong preferences at the S_1' site (immediately after the scissile peptide bond) and also additional preferences removed from the bond under attack, such as subtilisin which prefers hydrophobic side chains at P_4' in addition to its P_1' preferences (10).

However, when a native protein is attacked by a proteinase, under typical reaction conditions, not all the potential hydrolytic sites in a folded, native protein structure are cleaved and, therefore, the proteolysis must be limited by higher-order structural phenomena. In many cases, only one or two bonds are cleaved over a few hours' incubation and, sometimes, no digestion can be measured over the limited time scale of the experiment. This defines the sequence–structure paradigm of limited proteolysis and several studies have been undertaken to discover those structural features that are responsible for limiting the reaction.

2.1 Molecular recognition and limited proteolysis

Early studies considered the locations of limited proteolytic sites in proteins of known three-dimensional structure and concluded that they occur at 'hinges and fringes' (11). Further works attempted to quantify this in terms of parameters such as accessibility to a probe rolled over the surface of the protein (12). Fontana and co-workers placed more emphasis on the flexibility of the protein chain being the prime determinant after studies on the hydrolysis of thermolysin by subtilisin (13,14) as well as on other systems (15). This conclusion is supported by structural analyses and loop modelling studies that identified local unfolding as a prerequisite for the limited proteolytic reaction (16,17). Indeed, the expected degree of local unfolding required is upwards of 10 or 12 amino acids around the scissile bond.

The primary specificity of the protease has consequences for the theoretical understanding of limited proteolysis and its structural determinants. We may make a crude distinction into two classes: narrow specificity proteases (e.g. trypsin, V8-proteinase and endopeptidase Arg C) which cut after only one or two of the 20 commonly occurring amino acids and broader specificity proteases that cleave after many residue types (e.g. subtilisin, thermolysin and proteinase K). Since the broader specificity enzymes can cleave after a wider range of amino acids, often including hydrophobic ones, the structural factors determining the cleavage might be expected to be more pronounced. This is because a narrow specificity enzyme like trypsin cuts only after two amino acids that are already expected to be at the protein surface because of their charged state. Hence, from a practical point of view, the potential for information gain is lower, and from a theoretical point of view, the analysis and prediction might be harder. This is why broader specificity enzymes are often chosen for conformational probing experiments.

2.2 Prediction of nicksites

There is considerable value in a prediction method that could identify sites of limited proteolysis in a known protein. It would allow rational design of the protein in order to increase or decrease its overall proteolytic susceptibility that is typically correlated with its thermal stability (18). There have been few attempts to do this. Simple correlations with loops and β-turns have been considered (19,20), but for restricted protein subsets.

A more general approach has been developed to permit the prediction of sites of limited proteolysis in any protein of known structure (21,22). If this task proved impossible, then it would demonstrate that limited proteolysis is not a reliable tool for probing protein structure. The approach also allows us to evaluate the differing contributions of the underlying structural factors that determine nicksites. The analysis and prediction algorithm is briefly summarized here.

The different conformational parameters used in the analysis, and the rationale for their inclusion, are described in *Table 2*. Building on earlier work (16), these conformational parameters were calculated for every protein structure in a data set of proteins known to be cut by a range of serine proteases.

The data set was split into proteins for which the proteolysis data were primarily for narrow specificity proteinases and those cut by broad specificity enzymes. After normalization, the calculated parameters were then combined using a series of smoothing windows and weights for each parameter, thus:

$$N_x(i) = \frac{\sum_{i-(h_x-1)}^{i+h_x} F_x(i)}{s_x} \quad \text{and} \quad P_x(i) = \frac{\sum_{x=1}^{x=n_f} w_x N_x(i)}{n_f}$$

where s_x = window length for feature x, h_x = half-window length, $F_x(i)$ = normalized feature score at residue i, $N_x(i)$ = normalized, smoothed feature score at i,

Table 2 Conformational parameters and their relationship to limited proteolysis

Parameter	Reason for inclusion	Method/technical details
Solvent accessibility	Nicksites already known to be generally at surface-exposed sites	A probe rolled around the protein exterior assigns areas in Å^2 to each atom which are summed over each residue (23)
Protrusion index	Nicksites would be expected to protrude from the protein surface to enable accommodation into the enzyme active site	Each amino acid is assigned a score from 0 to 9, depending on its protrusion index calculated from a set of similar equimomental ellipsoids with origins at the protein centre of mass (24)
Temperature factors	Nicksites are found at flexible regions of the protein, as characterized by temperature factors or B-values from X-ray crystallographic determinations	The mean residue temperature factor is calculated by summing and averaging individual atomic values
Ooi numbers	Nicksites would be expected to be at weakly packed regions of the structure, which are able to unfold locally more easily	Each amino acid is assigned an Ooi number score which is simply the number of α-carbon centres within a fixed radius (25). Scores are normalized and subtracted from unity so that higher scores are more favourable
Secondary structure	Nicksites are rarely found in regular secondary structure and there are geometric and energetic reasons for their exclusion from helix and sheet	Each amino acid is assigned to one of the three secondary structure states: α-helix, β-strand or coil (26). Each state is assigned a score, with an additional penalty for disulfide bonds
Hydrogen bonding	Nicksites would be expected to be at regions of the structure which are not overly pinned down by interactions with the bulk of the protein such as hydrogen bond	The number of non-local hydrogen bonds is calculated from residues within a fixed window about each amino acid to those outside it. Values are normalized and subtracted from unity so that higher scores are more favourable

w_x = weighting value for feature x, n_f = number of features and $P_x(i)$ = prediction score for residue i.

A schematic of the prediction algorithm is illustrated in *Figure 1*. The method allows a Monte Carlo optimization procedure to optimize the weights and window sizes, driven by three criteria:

- limited proteolytic sites should be distinguished from other sites that do not contain the correct residues flanking the scissile bond
- limited proteolytic sites should be distinguished from other potential sites that are not cut, at least under the same conditions
- the mean ranking of nicksites should be as low as possible.

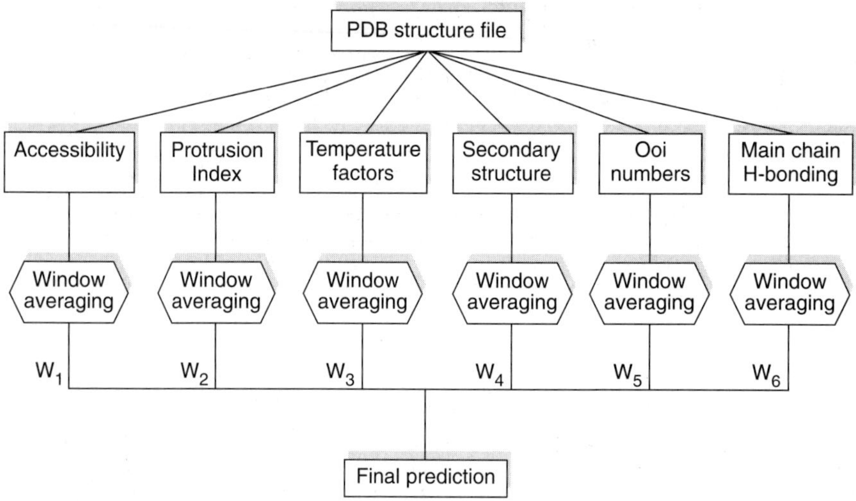

Figure 1 Schematic flowchart of the NICKPRED prediction algorithm.

The data obtained through this process allow a direct comparison of the relative importance of the different conformational parameters in determining limited proteolysis. The optimal weights and windows obtained from an optimization on the narrow specificity data set revealed that most parameters were strongly linked with the exception of protrusion index. Similarly, in agreement with the loop modelling studies, smoothing window lengths were generally in excess of 10 or 12 residues. This suggests that isolated residues in possession of high accessibility/flexibility are not good predictors. Rather, an extended segment is required.

The optimized algorithm is able to identify correctly a nicksite residue as the highest scoring residue for all of the 12 proteins contained in the narrow-specificity data set. This demonstrates that the underlying principles are valid and that the corollary is true, that limited proteolysis can be used reliably as a structural probe. When the same parameters were applied to the broad specificity data set only two from eight were correctly identified. Furthermore, the prediction could not be drastically improved by re-optimization against the broader data set. However, prediction of the broader specificity sites is clearly a more challenging problem. Indeed, on close examination the predictions for the broad specificity sites were quite good. In all cases but one, the nicksites were located at high-scoring 'peaks' in the prediction profiles and were usually within two or three amino acid positions of the highest-scoring residue. This shows that the structural principles still hold and that subtleties of specificity and steric fit must dictate the final precise site of cleavage

Some example prediction profiles are illustrated in *Figure 2*. Generally, all the nicksites are located at the highest peak in the profile, or are the highest-scoring residues which match the sequence specificity.

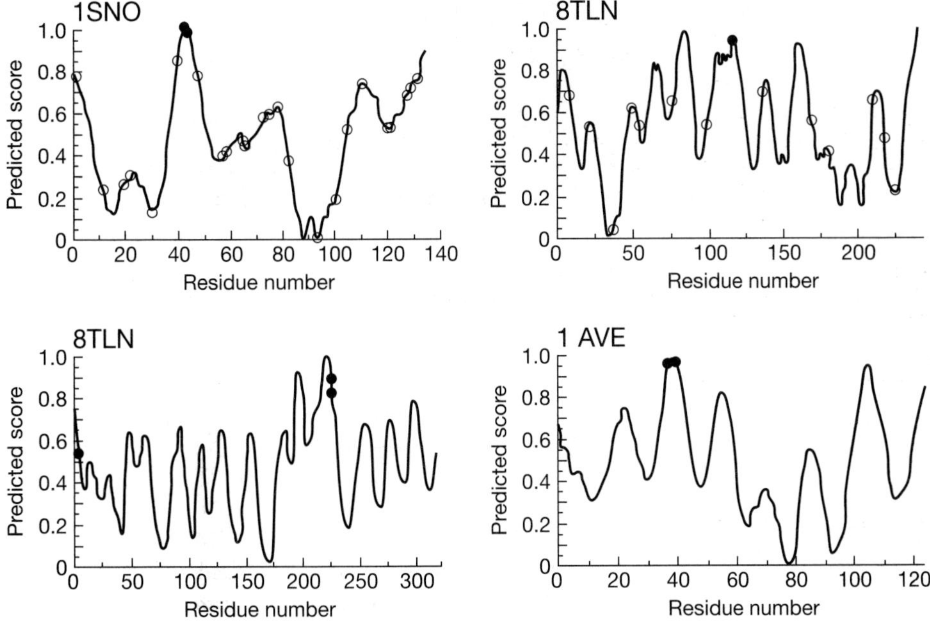

Figure 2 Examples of prediction profiles from the NICKPRED prediction program. Profiles are shown for two proteins cut by trypsin, staphylococcal nuclease (Databank code, 1SNO) and elastase (3EST), and the subtilisin-nicked protein thermolysin (8TLN) and avidin (1AVE) cut by proteinase K. In all plots, the profile score is plotted against the residue number along the protein chain. Limited proteolytic sites are marked by a filled circle on the profile, and for the trypsin-cleaved proteins, all other lysine and arginine positions are indicated by an open circle. For clarity, these putative sites were not marked on the broader protease-cleaved profiles.

2.3 A tool to aid in prediction of sites of limited proteolysis

The prediction program, NICKPRED, has been made available via the internet and can be reached via the URL: http://sjh.bi.umist.ac.uk/nickpred.html (*Figure 3*). Users may obtain predictions from structures in the current release of the Brookhaven Databank (27) or may submit their own models/structures in PDB format. Results are returned via e-mail. The output options of the software allow the user access to the raw, normalized conformational parameter scores, as well as the processed prediction scores on a residue-by-residue basis and collated statistics on putative and non-putative nicksites. Furthermore, a simple Rasmol output script can be sent via e-mail which may be saved and quickly run which colours the structure by prediction score (from 0 to 1, 1 being the most highly predicted). A sample prediction profile for the protein elastase is shown in *Figure 4*, along with the Rasmol image showing the prediction score along the protein backbone (*Figure 5*).

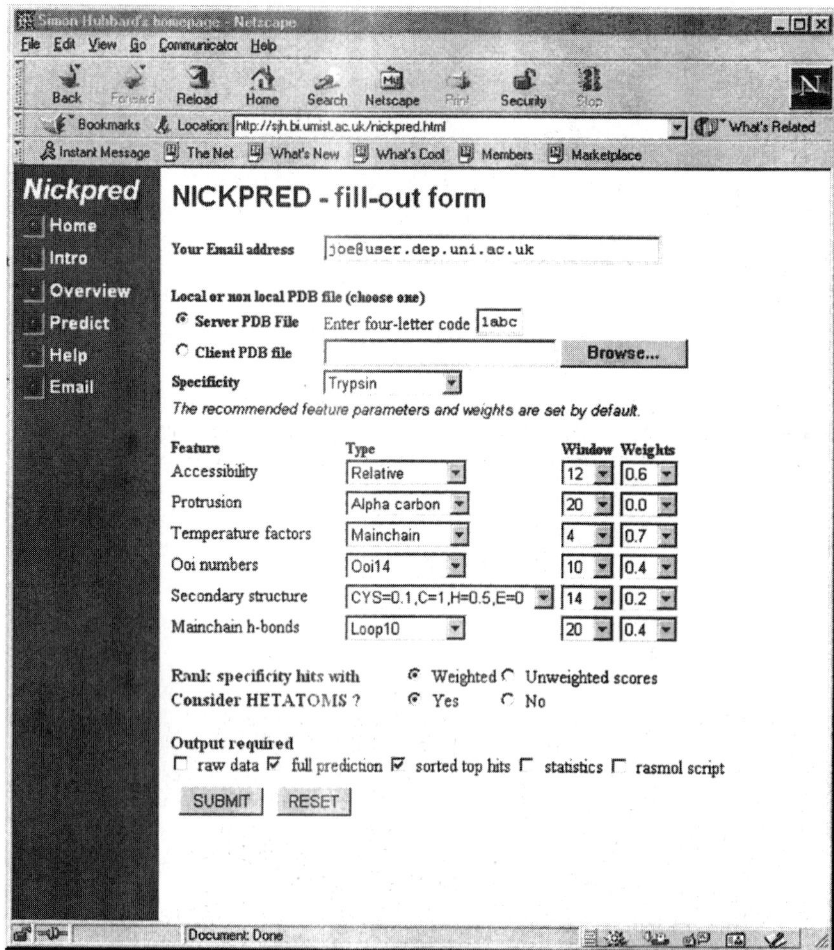

Figure 3 The NICKPRED Web server submission page. Users may enter a Brookhaven four-letter code identifier or upload one of their own PDB format files from their local computer. Users may control which proteinase is to be considered as a cleavage agent, as well as the conformational parameters to use as predictors. The optimized weights and windows are used as defaults. Varying output formats may be specified, which are returned via electronic mail to the user.

3 Experimental considerations

Many factors determine the nature of a specific protein–proteinase interaction. The interaction is so dependent upon the particular molecular recognition events intrinsic to the process that exact guidelines are impossible. Nevertheless, some general guiding principles can usually be applied. In particular, the potential manipulation of limited proteolysis experiments is greatly enhanced if the kinetics are understood.

To a first approximation, proteolytic reactions, especially those that are

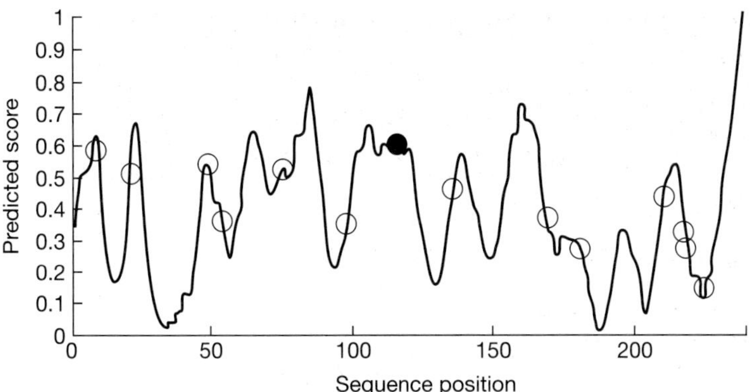

Figure 4 Example NICKPRED prediction profile calculated using default parameters for the protein elastase. All putative tryptic sites are circled with the single, true nicksite indicated by the filled circle. It is the highest scoring lysine or arginine residue in the protein.

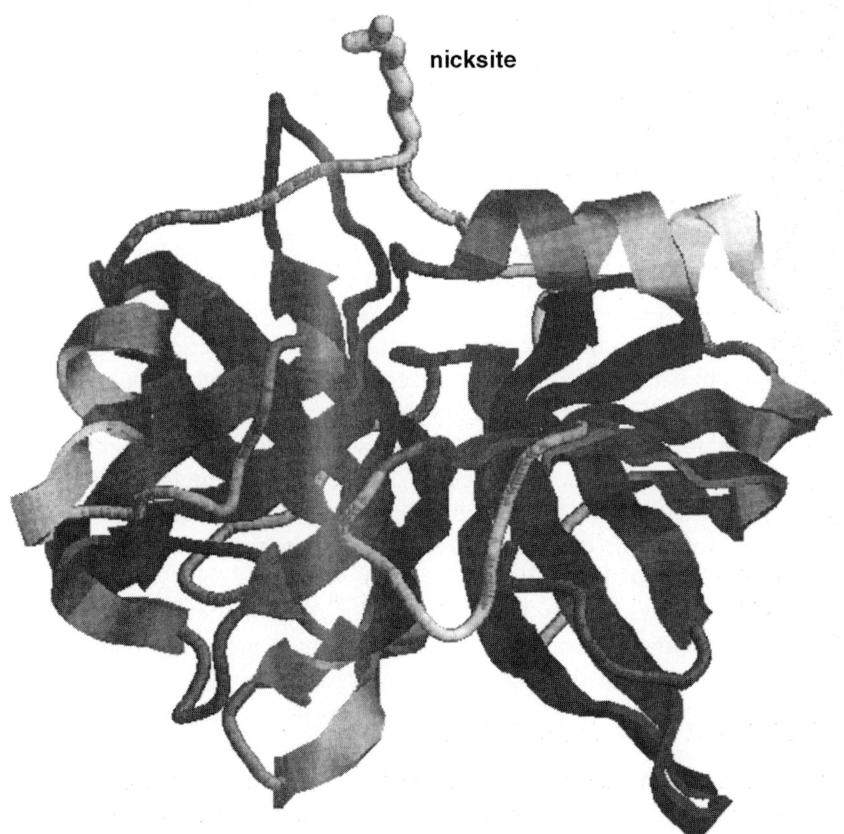

Figure 5 Rasmol image of elastase coloured by NICKPRED prediction. Regions of the protein predicted to be more susceptible to proteolysis are white, and those least susceptible are black (this range is normally colour-coded from red to blue). The true nicksite lies in a long loop region at the top of the protein that is marked 'nicksite' and which is at the arginine residue with the backbone included.

limited to a single cleavage site, can be construed as pseudo first-order, because the concentration of the proteinase does not alter throughout the reaction and because the interaction between proteinase and substrate is stochastic. Moreover, this simple model dictates that the rate constant for proteolysis is linearly dependent upon the concentration of proteinase in the digestion but of course, is independent of the concentration of the substrate. The first-order rate constant and concentration of proteinase in the system can therefore be used to yield a second-order rate constant which expresses the dependence of the rate of the reaction on the intrinsic susceptibility of the substrate and the amount of proteinase.

For a fixed amount of proteinase, the amount of substrate that is digested in a given time is therefore solely dependent on the amount of starting material. If the reaction monitoring method, or post-proteolysis analysis requires a minimal amount of material, it is necessary to increase the concentration or volume of starting material and adjust the proteinase concentration. This requires knowledge of the pseudo second-order rate constant, which can be estimated from published data, or from prior experimentation, to ensure that the required amount of a product is generated in a specific reaction time.

If it is necessary to monitor the reaction, then judicious calculation of the proteinase concentration can guarantee that the time course is represented for informative data points, over which the substrate decays and the product is generated. The practical outcome of this is that it is possible to design the digestion conditions to produce the required effect over a suitable sampling time.

There is one further important consideration. The attack of a proteinase upon a native protein can be seen as the sum of a series of 1 to n hydrolytic reactions which proceed at different rates ($k_1 \ldots k_n$). Sometimes these reactions occur independently and in parallel, and in other circumstances a second proteolytic reaction may have to follow the first (the first cleavage makes the second possible). Modelling of such reaction schemes can provide some insight into the behaviour of the system (see later). In particular, the transient accumulation of a specific intermediate will be strongly influenced by the relative magnitudes of the rate constants for all of the cleavage reactions. Optimal generation of a specific intermediate may require that the reaction be stopped at a specific stage in the progress of the digestion (the time at which the reaction is stopped will depend on other parameters, such as the absolute amount of proteinase that is added). Moreover, the ratio of the participating rate constants might be such that it is impossible to obtain more than a few percentage of any particular intermediate. In such cases, altering the digestion period or the amount of proteinase are totally ineffective as strategies to enhance yields of an intermediate.

3.1 Choice of proteinase

3.1.1 Specificity

The primary specificity of a proteolytic enzyme is usually defined in terms of the P_1 and P_1' residues. Although this primary specificity may define the bond that is eventually cut in a native protein, it fails to consider the structural factors that make one susceptible bond more readily digested than another less susceptible bond that has an equivalent primary sequence (see above). Since highly polar, charged residues are predominantly located on the surface of proteins, proteinases with specificities for such residues might be the most obvious choice. However, there are many examples of systems where the sole cleavage is at a hydrophobic residue that might not have been anticipated. The practical outcome of these considerations is that it is rarely possible to predict which proteinase might be most informative. Experimentation is warranted and exploratory studies using a range of proteinases (with different primary specificities) can indicate unexpected and valuable cleavage reactions (3). It may be necessary to use high proteinase concentrations, and care must therefore be taken in the subsequent inactivation of the proteinase before analysis of the proteolytic reaction.

Appendix II lists the proteinases that are commercially available, and includes information on their primary specificity.

3.1.2. Ease of inhibition or removal of the proteinase

It may be necessary to arrest the proteolytic reaction at a particular stage in the digestion process. Simplistically, it might be expected that the reaction could be stopped by rapid denaturation of the proteinase. However, this may not be possible as the denaturing treatment would also denature the substrate protein and destroy the biological properties of the substrate. Note that denaturation can also effect a rather more rapid unfolding of the substrate than the proteinase, with the outcome that the substrate becomes exhaustively digested during a transient period of high activity.

There is a need, therefore, for specific methods of inactivation or removal of the proteinase. These two strategies have markedly different outcomes. Inactivation will disable the proteinase, but it will remain in the reaction mixture. If the inactivation is attained through addition of a reversible inhibitor, subsequent dilution or desalting steps can allow the enzyme to regain activity. Even if the inhibitor is irreversible, sufficient inhibitor must be added to achieve virtually complete inactivation of the proteinase in a short time frame, so that proteolysis does not extend beyond the stage required (Chapter 5).

Removal of proteinases from a digestion mixture must also be rapid in order that the reaction can be terminated abruptly. This is especially true if the kinetics of proteolysis are being monitored. The most effective method is the simultaneous absorption and inhibition of the proteinases onto an immobilized inhibitor. For example, trypsin and chymotrypsin are effectively inhibited and bound by soybean inhibitor, lima bean inhibitor or ovomucoid, when covalently attached to agarose (all of these are commercially available, for example, from Sigma). The reaction mixture can be passed through a small bed of inhibitor gel, or the gel can be mixed with the digestion reaction mixture, and the supernatant recovered after centrifugation. Other absorption methods might also work, but removal of the proteinase by ion exchange chromatography, for example, might not cause inhibition of the proteinase and thus activity will continue until the ion exchanger is separated from the digestion products.

3.1.3 Stability

Most native-state proteolytic reactions are conducted in non-denaturing solvents, and the physicochemical stability of the enzyme is only an issue where denaturants such as urea or guanidinium hydrochloride are included in the solution, or when the digestion is conducted at elevated temperature. If such conditions are necessary, some experimentation is warranted. In our experience, many proteinases are remarkably tolerant to moderate concentrations of urea. The effect of the urea is to diminish the rate of hydrolysis of unfolded substrates, but the pattern of digestion of unfolded substrates remains unchanged (our unpublished data).

If the proteinase undergoes reversible or irreversible denaturation, the effective decline in proteinase concentration will make the rate of reaction appear to diminish (remember that the first-order rate constant for digestion is proportional to the proteinase concentration). When the rate of inactivation of the proteinase occurs at rates similar to the rate of digestion, the kinetics will not obey simple first-order behaviour and data processing to recover rate constants may be more difficult, or even impossible. Some knowledge of the rate of denaturation of the proteinase will allow the reaction conditions to be adjusted, such that the proteolytic reaction can be analysed before there is significant loss ($> 10\%$) of proteinase activity.

Some proteinases can autolyse (digest themselves). Since this is another example of proteolysis of a native protein, the rate of autolysis will vary in proportion to the proteinase concentration. Experiments in which the proteinase concentration is high, such as those implied here, will therefore be vulnerable to autolytic losses. This will manifest itself in the same way as any other type of irreversible inactivation of the proteinase and the progress of the reaction will deviate from simple first-order behaviour. The solution conditions can be adjusted to minimize autolytic losses—high concentrations of substrate protein, or the inclusion of stabilizing molecules that prevent/slow autolysis. A well-known example is the marked stabilizing effect of calcium ions on trypsin and some of the bacterial M4 metalloendopeptidases, such as thermolysin (28). It follows that buffers used with these proteinases should not contain chelating agents.

3.1.4 Catalytic mechanism of the proteinase

There are circumstances in which the catalytic mechanism of the proteinase might impose unacceptable restrictions on the reaction conditions. For example cysteine proteinases require that a relatively high concentration of a reducing agent (such as 2-mercaptoethanol, or DTT) be maintained in the reaction mixture. However, by the same chemistry, the reductant could reduce disulphide bonds within the substrate, and alter susceptibility to digestion. Calcium-activated or calcium-stabilized proteinases can require high concentration of calcium ions in the reaction. However, this precludes use of buffers for which the calcium salts are of low solubility (such as phosphate or carbonate).

A less likely outcome derives from the ability of many proteases to catalyse peptide synthesis, through transpeptidation, a reversal of the normal hydrolytic reaction (see Chapter 11). In normal circumstances, this is insignificant, but when the proteinase and substrate protein concentrations are both high, and particularly if the water activity is reduced, by inclusion of an organic co-solvent or by high concentrations of salts, such synthetic reactions are more likely. These can manifest themselves as higher mass products, whether monitored by SDS-PAGE or MS.

3.2 Ratio of proteinase to substrate

The rate of hydrolysis of even the most susceptible of peptide bonds can be considerably slower than that seen with denatured substrates. Typical digestions to generate limit peptide mixtures (see Chapters 7 and 8) use a proteinase–substrate ratio of 1:50 or 1:100. In contrast, native state digestions can often require equal weights of substrate and proteinases, and in some instances the molar ratio of proteinase to substrate can exceed unity by a considerable margin (the authors have used molar ratios of proteinase–substrate of 10:1) This high concentration of proteinase can cause problems during later stages in monitoring of the reaction. The careless use of such high proteinase concentrations will also not endear one to colleagues using denaturing systems in the laboratory. Even a small degree of cross-contamination with such strong proteinase solutions has the potential to cause unwanted proteolysis in other work. For example, the authors have encountered trace levels of trypsin activity in SDS-PAGE sample buffer.

The key outcomes of such high proteinase concentrations are the need to ensure that the proteinases are of sufficient purity, and to guarantee that the proteinase can be effectively stopped. In one example from the authors' laboratories, chymotryptic digestion of a single bond in glutamate dehydrogenase increased the susceptibility of the product to trypsin by at least two orders of magnitude. The dramatically enhanced susceptibility to trypsin meant that trace amounts of contaminating trypsin in the chymotrypsin preparation attacked the partly proteolysed product (29).

3.3 Solution conditions

Native state proteolytic reactions require reaction conditions that are compatible with two native proteins: the proteinase and the substrate.

The pH of the digestion mixture is able to affect the catalytic activity of the proteinase through active-site ionizations, and the structure of the proteinase and the substrate. The optimal pH for the proteinase may not be the same as that for maximum stability or relevant functional state of the substrate. However, the pH optima of most proteinases are broad, and adequate digestion may be obtained at suboptimal pH values. A higher rate of digestion can also be obtained by increasing the concentration of proteinase, although the demands of post-proteolysis analysis may set limits on this tactic to regain adequate rates.

The choice of buffer ions may also be significant. For example, phosphate is a ligand of many proteins, and phosphate buffers might therefore alter proteolytic susceptibility of the substrate proteins through specific interactions with the substrate. Almost all of the organic buffers should be treated with some caution, since they may have specific effects. In this category are buffers such as acetate, malate and citrate. Buffers such as phosphate can precipitate metal ions (a good example is the formation of insoluble calcium phosphate) and citrate is an effective metal ion chelator, thereby reducing effective metal ion concentrations in the digestion system.

Ionic strength is a generic solution property that is often ignored, yet it is capable of eliciting significant effects on rates of digestion. Higher ionic strengths will generally either protect the substrate, or will be without effect; it is rare for increased ionic strength to enhance the rate of digestion.

A satisfactory strategy for the design of buffers, and for control of ionic strength is to use the buffer species at a low concentration, commensurate with control of the relatively small changes in proton concentration that will occur during the digestion (typically less than 1 mM). However, the ionic strength is then maintained at a relatively high value (such as 150 mM) by addition of a monovalent salt such as NaCl or KCl. The monovalent ions shield the proteinase and substrate from buffer ions, thus diminishing buffer effects and providing a controlled ionic environment. This is particularly important when a series of buffers are used to cover a pH range; although the pH may be varied smoothly, the ionic strength will vary to a greater extent, especially if some buffer species are polyvalent. A pH curve should be prepared such that the ionic strength is held constant by monovalent ions, and at a value that is significantly higher than the ionic strength due to buffer species alone. Buffer recipes, thermodynamically corrected, are readily calculated through the web site http://www.bi.umist.ac.uk/buffers.html (30).

3.4 Determination of site of proteolysis

It is highly likely that most proteolytic experiments require, as an outcome, information on the site(s) of digestion. There are several strategies for acquisition of this information.

3.4.1. SDS-PAGE

Although the mass accuracy of SDS-PAGE is poor, there are some circumstances in which the sizes of the products, coupled with knowledge of the restricted specificity of the proteinase can be enough. For example, cleavage with endoproteinase LysC will reduce the options for digestion to relatively few sites, and in a small protein a single site of cleavage might be readily identified by the sizes of the products.

3.4.2 Edman degradation

If there is enough material, it is possible to take a product band, separated on SDS-PAGE and blotted onto PVDF membranes, and obtain several cycles of Edman degradation to define the site of cleavage. The authors usually prefer to let the sequenator run for four or five cycles to obtain an unambiguous answer. Although this defines the *N*-terminal of one of the product peptides it does not define the *C*-terminus, and although the size of the product on SDS-PAGE can give some clues, it is important to be aware that some unexpected cleavages might not manifest themselves this way (3).

3.4.3 Mass spectrometry

Accurate mass is an exceptionally powerful piece of information in proteolysis studies. Electrospray MS has a resolution of ≈ 1 Da in 10 000 Da, which, given an average amino acid mass of 110 Da, is more than adequate to define the site of cleavage of a moderately sized protein, especially when coupled with knowledge of the primary specificity of the proteinase. For example, the authors have used electrospray MS to identify an unusual tryptic-catalysed exoproteolytic cleavage at a *C*-terminal Arg-Glu bond in a 19 kDa protein (31). Although MALDI-TOF MS is less effective with larger proteins, it can give precise mass determination of smaller peptide products (see Chapter 7), and can be used to identify smaller products, or to monitor later stages of proteolysis. SDS-PAGE is not very useful for such small peptides.

If two sections of a polypeptide chain are linked by disulphide bonds, then a bond can be cleaved with no change in mass on non-reducing SDS-PAGE gels. With electrospray MS, the product will have increased in mass by 18 Da, a mass shift that is readily observed.

Finally, MS-MS can be used to provide sequence information on small peptides, and thus aid in the identification of cleavage sites. However, for most circumstances, in which the primary sequence of the substrate is known, this will not be necessary.

3.5 Strategies for limited proteolysis experiments

The key factor in implementing the experimental design for a limited proteolysis experiment is the need to ensure that proteolysis only occurs within the course of the reaction. The design of the experiment must ensure that the proteinase only acts during the incubation period. If kinetic data are required, it is also essential that the activity of the proteinase does not change throughout the reaction, and thus, autolysis or inactivation must be prevented.

In the absence of any preliminary data, it will be necessary to conduct some exploratory experiments. *Protocol 1* outlines the planning behind these preliminary experiments. Once the preliminary data are available, more detailed protocols can be designed, giving due consideration to the factors summarized in *Table 3*.

Table 3 Checklist for limited proteolysis studies

Solution conditions	pH	Must be compatible with both proteinase and substrate
	Ionic strength	Maintain reasonable concentration of monovalent salt relative to buffer species and fix ionic strength
	Buffer species	Avoid difficult buffers—chelation, protein binding, formation of insoluble inorganic salts, etc.
	Temperature	Compatible with both proteinase and substrate
Reactants	Substrate concentration	Half-life is independent of substrate concentration
	Proteinase concentration	First-order rate constant for digestion ($0.693/t_{1/2}$) is proportional to proteinase concentration
Stopping the reaction	Denaturation	Make sure denaturation step is near-instantaneous
	Inhibition	Use high concentrations of fast inhibitors. Irreversible inhibitors are preferred
Analysis of products	SDS-PAGE	Ensure proteinase cannot be reactivated and digest material in sample buffer.
	Chromatography	Ensure material is fully dissolved before analysis. Use rapid, quantitative separation protocol.
	Activity	Make sure that proteolysed substrate is stable, thermodynamically or to reactivated proteinase

Protocol 1

Design of a native-state proteolytic reaction

Equipment and reagents

- Protein sample
- Proteinase
- Suitable buffer
- Trichloroacetic acid
- SDS-PAGE apparatus

Method

1 Decide which proteinase to use, using the NICKPRED algorithm if a three-dimensional structure or homologue, or possibly, a built model is available, a priori knowledge of the target site, published data or by quick survey experiments.

2 If the protein is pure, incubate the substrate (5 μg) with a predefined proteinase, or a range of proteinases (trypsin, chymotrypsin, thermolysin, proteinase K, subtilisin), at each of two substrate–proteinase weight ratios of 50:1 and 5:1, for 1 h at 20–30 °C. The buffer should be carefully chosen, and should be at or near neutral pH values, of low buffer species concentration (20 mM) but at higher ionic strength (100 mM). If thermolysin, trypsin or chymotrypsin are used, include calcium ions at 2 mM

Protocol 1 continued

to stabilize the proteinase, which precludes the use of a buffers that precipitate calcium ions, such as phosphate.

3 Stop the reaction by addition of TCA to a final concentration of 5% (w/v) and allow the samples to stand at 4 °C for 15 min.

4 Collect the denatured, precipitated proteins by centrifugation in a bench microcentrifuge for 5 min and wash the pellets three times carefully with acetone to remove residual TCA. Analyse the precipitated proteins by SDS-PAGE, using whatever detection method (dye or silver staining, Western blotting) is appropriate to monitor for the loss of starting material and the appearance of any products.[a]

5 From the analysis, estimate the approximate second-order rate constant for the reaction. If A_t% of the activity/material has been digested/inactivated in t min, the first-order rate constant (/min) is calculated by $k = -\ln(A_t/100)/t$

6 Calculate a pseudo second-order rate constant (/min,/μM) by dividing the pseudo first-order rate constant by the proteinase concentration E.[b] This will then allow a strategic design of further proteolysis experiments, in which the time-course can be predicted, and which can be manipulated by control of proteinase concentration or reaction time. Again, note that the time course of the reaction is independent of the amount of starting material, unless more complex concentration-dependent behaviours are in evidence.

7 Refer to *Table 3* for a checklist of issues to be considered in the design of proteolytic experiments.

[a] If the substrate has a biological or catalytic activity that is readily measured, you might choose to monitor this process in place of SDS-PAGE analysis. Be aware however, that a protein can be digested with no immediately apparent change in catalytic activity. Also, since this approach is unlikely to include a denaturation step, the proteinase must be inactivated or inhibited, or the assay of biological activity conducted over a short time (less than 1%) of the total digestion time. In any circumstances, include a blank in which the substrate is incubated in the absence of the proteinase.

[b] Example: if 40% of the starting material/activity remains after 60 min digestion with 2 μM proteinase, the first order-rate constant is given by $k = -\ln(0.4)/60 = 0.015$/min and the second-order rate constant is therefore 0.015/2 = 0.0075/min.μM

4 Analysis of limited proteolysis data and simulations

In addition to cleaving proteins at vulnerable or functionally relevant sites and simply analysing the products, it is possible to monitor the kinetics of proteolysis, and to derive information on the rates of digestion of individual sites. The requirements of this approach are for accurate measurement of the extent of digestion, monitored as loss of starting material, or as appearance of products, or both. Second, the points must be distributed over a time-frame that defines a significant proportion of the reaction. In exploratory studies, there is some advantage in setting sampling time points in an expanding series, e.g. 2, 5, 10, 20, 60 and 120 min, such that both fast and slow processes have a reasonable chance of being represented in the sampling profile.

4.1 Obtaining quantitative data

The requirement for high-quality quantitative data is simple: the measured variable must be a smooth and preferably linear function of the amount of material (substrate or product) that is present. The linearity of this relationship is best checked experimentally, using a standard curve of substrate, or of a proteolytic reaction in which the proteolysis has been completely halted at different times by the addition of an inhibitor or removal of a proteinase. A gel or series of HPLC traces should be prepared in which a range of material has been applied and the response (band intensity, peak height or peak area) measured. The linearity of the relationship should then be assessed. Some curvature is acceptable, provided that the unknown samples are interpolated on a well-constructed standard curve. The range should encompass the amounts of material that will be used routinely.

4.1.1 Densitometry

Many proteolytic reactions will be monitored by SDS-PAGE and visualized with a protein dye such as Coomassie Brilliant Blue or Fast Green. These give a reasonably linear response, over 0–10 μg of protein in a standard mini-gel. Colloidal Coomassie Brilliant Blue, or some other recently introduced stains can also yield linear responses at higher sensitivities. Silver staining can be more problematic and high backgrounds, variable colours of stained product, and non-linearity conspire to eliminate it as a quantitative measure of digestion in most applications.

To recover quantitative data, the gel must be scanned. Dedicated gel scanners will give excellent results and are supported by software for accurate recovery of the amount of material in each band on the gel. However, the authors have found that a standard scanner for a desktop computer will work well. Current models are full colour (allowing sensitivity to be optimized at the wavelength most appropriate to the dye) and high resolution (300–600 dpi optical resolution, giving a spot size of about 40 μm). Some scanners proclaim higher resolutions, but often these use software interpolation to 'fill in' values between two optically distinct dots and do not recover additional biological information from the gel. Optical resolution is the critical parameter and 300 dpi should be the minimum requirement. Lower resolutions can lead to loss of information (*Figure 6*).

Most scanners operate in reflectance mode only, in which they measure the amount of attenuation of a reflected signal. However, for best results, the scanner should be capable of working in transmittance mode, whereby the light source is shone through the gel, and attenuation of the transmitted beam is measured. The image acquired can be saved in a number of formats. Some file formats can result in loss of data. For example, 8-bit greyscale mode will rep-

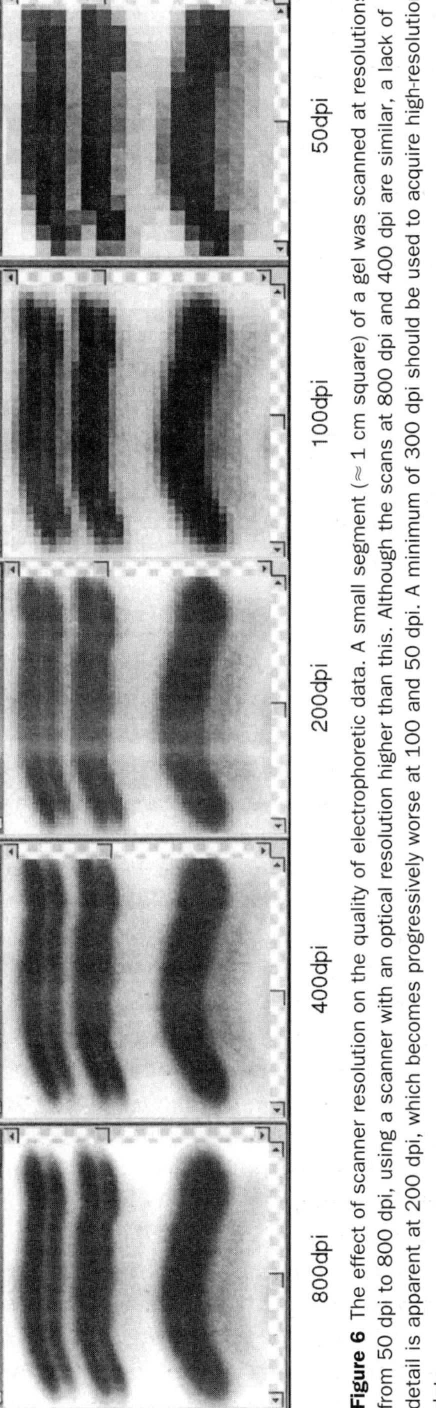

Figure 6 The effect of scanner resolution on the quality of electrophoretic data. A small segment ($\approx$ 1 cm square) of a gel was scanned at resolutions from 50 dpi to 800 dpi, using a scanner with an optical resolution higher than this. Although the scans at 800 dpi and 400 dpi are similar, a lack of detail is apparent at 200 dpi, which becomes progressively worse at 100 and 50 dpi. A minimum of 300 dpi should be used to acquire high-resolution data.

resent the image as one of 255 different levels of grey; if the gel is lightly stained, the dynamic range might be as little as 25% of that greyscale range, giving a total of only about 50 levels for quantification. Colour modes may be preferable.

Although disk storage is rarely a critical issue these days, bear in mind that high-resolution scans of gels can create large data files. For example, a mini-gel, scanned in 8-bit greyscale at 300 dpi will create an uncompressed image of ≈ 1MB, which will fit on a standard high density floppy disk. A 24-bit colour image of the same gel, at the same resolution, will create a file that is three times larger. The two outcomes of creating such large image files are the time taken to process them and the problem of storing them on standard floppy disks. Some image formats can be compressed, using a range of algorithms but some of these compression methods are 'lossy' (data is discarded from the image to create a highly compressed file) and the original cannot be reconstructed when decompressed: either avoid compression, or use a loss-free compression algorithm (e.g. TIFF, LZW compression).

Image processing software might allow one to resample the image at a lower resolution, say from 300 dpi to 150 dpi. The process of averaging adjacent pixels to yield the new image may distort the data, so the gel should really be rescanned at the new resolution. Similarly, it is not possible to increase the resolution by software, as the new pixels are interpolated and no additional data is added. Any change in scanning requirements, such as greyscale/colour, change in resolution, or adjustments to brightness or contrast should be made at the data-acquisition stage, not in the subsequent image-processing steps

The intensities of different bands can be assessed using simple software, or more complex packages that allow for distorted bands or gels. A popular freeware program which allows quite sophisticated analysis of well behaved gels is NIH-IMAGE (http://rsb.info.nih.gov/nih-image/), which was written for the Macintosh, and the port of this program to the PC by Scion (http://www.sciончcorp.com/). A more sophisticated commercial package used to acquire band intensity data is the 1-d Quantifier package from Phoretix (http://www.phoretix.com) (*Figure 7*).

4.1.2 Chromatography

Column chromatography, using modern rigid matrices and good pumping systems can effect separations so quickly that on-line monitoring of the digestion of the reaction might be possible. The reaction mixture can be analysed by nondenaturing separation technologies, such as ion exchange chromatography or by reversed-phase chromatography, in which the components of the reaction mixture are likely to be partially or fully denatured. The usual caveat applies—denaturation of the reaction mixture might affect the substrate more than the proteinase and, thus, there could be some adventitious proteolysis during the early phase of the analytical step. There are two strategies for limited proteolytic reactions analysed by chromatography. In the first, the reaction is completed, and serial samples are extracted and the proteinase is inactivated. The serial chromatography runs can then be conducted at leisure. The second strategy

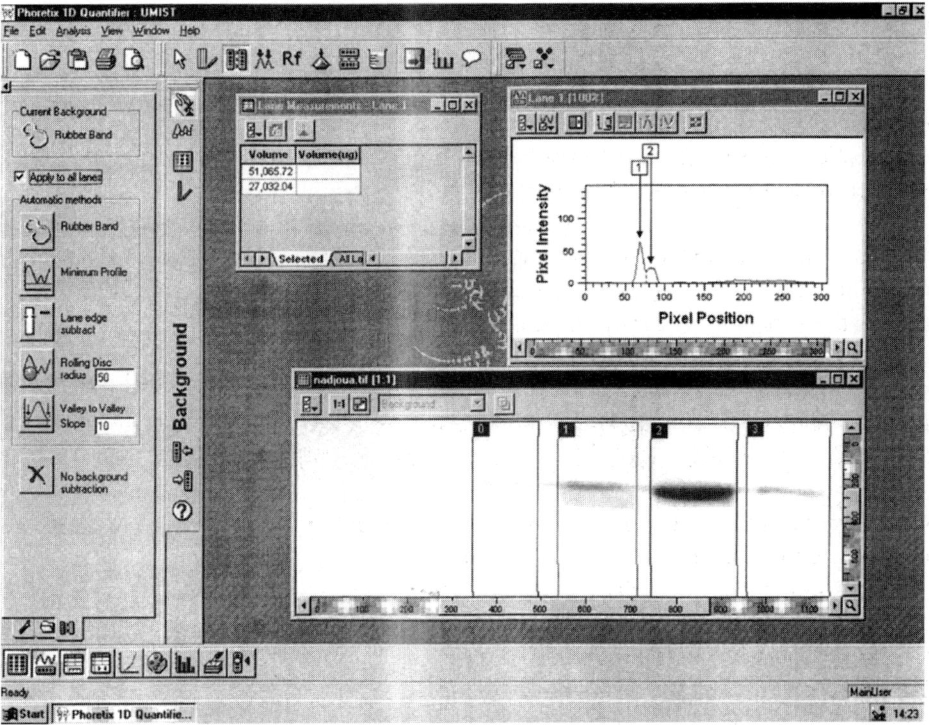

Figure 7 Gel densitometry package for quantification of band intensities. The package (1d-Quantifier, Phoretix) can be used for more sophisticated functions than illustrated here, and includes the ability to correct for gel or band distortions. The screenshot indicates the range of methods that can be used for background subtraction and simple quantification of two merged peaks.

requires chromatography of sufficiently high speed that the chromatographic analyses can be conducted between sampling intervals. If this is the preferred strategy, sufficient time must also be allowed for regeneration of the column after, for example, a gradient of salt or non-polar solvent. If there is a mismatch between the chromatographic time and the digestion reaction, it is preferable to extend the latter by the use of lower amounts of proteinase. Speeding up the chromatography by increasing the flow rate can lead to altered separations or additional wear on pumps and seals.

The quantification should be based on peak area, or in the case of sharp, symmetrical peaks (such as are seen in a well-designed reversed HPLC separation) peak height. In either instance, it is worth the effort to confirm that the measured parameter (area or height) is genuinely a linear function of the amount of material that is applied.

In one study, the authors used repeated sampling of a reaction in an autosampler vial to monitor the reaction. The sacrifice of precise temperature control (although relatively stable in an air-conditioned room) was more than offset by the convenience of rapid and automatic accumulation of a progress curve for several reactions running in parallel, in adjacent autosampler vials that were sampled successively, and for which no operator intervention was needed.

4.1.3 Mass spectrometry

The increasing use of ESI or MALDI-TOF MS has facilitated the definition of sites of proteolysis, as these techniques provide a mass resolution that is much greater than the residue masses of any amino acid. Less clear, however, is the ease with which the size of the mass peaks can be taken to represent absolute amounts of substrate or product, since different ions are detected with different efficiencies. If an internal standard, comprising an isotopically heavier variant of the substrate or product is included, then quantitative information is recoverable; however, this will not be an option in most experimental systems. At present, the primary use of MS analysis is in the identification of cleavage sites, not in the quantification of progress curves.

4.1.4. Other methods

There are many other methods that can be used to monitor a proteolytic reaction, most of which will be specific for a particular substrate. Extensive proteolysis can bring about a substantial change in the environment of internal aromatic residues, which can be measured as a change in absorbance spectrum, or in the fluorescence of tryptophan residues. Small changes in absorbance spectra can sometimes be isolated from the UV spectrum by analysis of the fourth derivative of the absorbance curve. The substrate may possess other chromophoric reporters, such as pyridoxal 5′ phosphate, that are in a protected environment in the native proteins, such that proteolytic attack can lead to a change in fluorescence.

Under extreme conditions, the liberation of a fluorophoric reporter can be measured as a change in the polarization of fluorescence. In the protein-bound state, the fluorophore will be tumbling more slowly in solution and thus will not be capable of depolarizing a plane-polarized fluorescent beam. When liberated, the higher rate of rotation of the reporter molecule will bring about greater depolarization. Thus, the anisotropy of the signal should decay over time.

The main advantage of such methods is that they allow continuous monitoring of the reaction, and direct extraction of the rate constants from a data set that is much more finely grained than that obtained from, for example, serial sampling for SDS-PAGE. Unfortunately, such changes are often a consequence of extensive proteolysis, and thus, can rarely be used to monitor early proteolytic events. The exception to this is where the first cleavage is rate limiting, and thus the proteolysis fits the 'all or none' model. Under such circumstances, the change in a conformationally sensitive optical property will reflect the rate of initial hydrolysis.

4.2 Analysing the data by non-linear curve fitting

The data set derived from a simple proteolytic reaction will be a set of (t, A_t) data pairs to which will be fitted an equation describing the behaviour. All first-order reactions are described by equations that are non-linear. This does not mean that the line of A_t as a function of t is curved, but that the independent variable has a non-linear relationship to the parameters of the equation—in this instance, A_t (at any one t value) does not vary linearly according to the value of the rate constant k. In contrast, the equation is linear with respect to the other parameter, A_0, since A_t is a linear function of A_0, at a given t.

The practical outcome of the non-linear behaviour of the rate equations is that simple methods of analysis to obtain the first-order rate constant that are based on a linear transformation of the data set (a plot of $\log_{10}(A_t)$ or $\log_e(A_t)$ versus t, for example) will yield parameter estimates that are incorrect and biased. The only expectation to this outcome is when the data are perfect, in which case all analytical methods yield the same result, and if the data were known to be prefect, then the rate constant could be calculated from a pair of (A_t, t) data. In reality, the acquisition of At data by densitometry, chromatography or enzyme assay will contain significant error and therefore a method of derivation of the rate constants must be found that treats the error properly and avoids bias in the parameter estimates.

Undoubtedly, the preferred method is to use a program for non-linear curve fitting. These packages fit the theoretical (A_t, t) line (corresponding to a particular set of values for the parameters) to the untransformed data, calculate an error sum (usually, the sum of the squares of the residuals) and calculate a goodness of fit. This process is then repeated with new parameter estimates until the fit is deemed adequate. Thus, the process is iterative, starting from initial parameters guesses and refining those values until an exit condition is satisfied.

There are a number of algorithms and programs that perform such an analysis. Some packages require that the operator makes the initial guesses of parameter values, while others use simplifications or linear transformations to create the initial guess. It could be argued that the user should make the initial guess, since they should understand the function well enough to make reasonable estimates. Moreover, the process of making initial guesses can be very instructive and can illustrate the effect of varying any parameter, and of the sensitivity of the function to variation in a particular parameter.

Key features that are desirable in a non-linear curve-fitting package include the ability to display the data and function for any set of parameter values, the ability to constrain parameter estimates (for example, a rate constant cannot be less than zero), the ability to enter new functions, with reasonable numbers of parameters, estimates of the certainty of the parameter estimates and the ability to plot 95% confidence limits for the fitted curve. Additional features that are helpful include covariance analysis to indicate the interdependence of the different parameters, the ability to plot residuals to look for inappropriate models and 'housekeeping' tasks such as comprehensive data and graphics export/import and macro capabilities. *Figure 8* is an illustration of one package that provides both sophisticated analyses and flexible graphics generation (FigP, Biosoft, http://www.biosoft.com). However, for simpler analyses, the 'Solver' function within Microsoft Excel (http://www.microsoft.com) provides a powerful non-linear optimizer that can be valuable for initial exploration of the dataset.

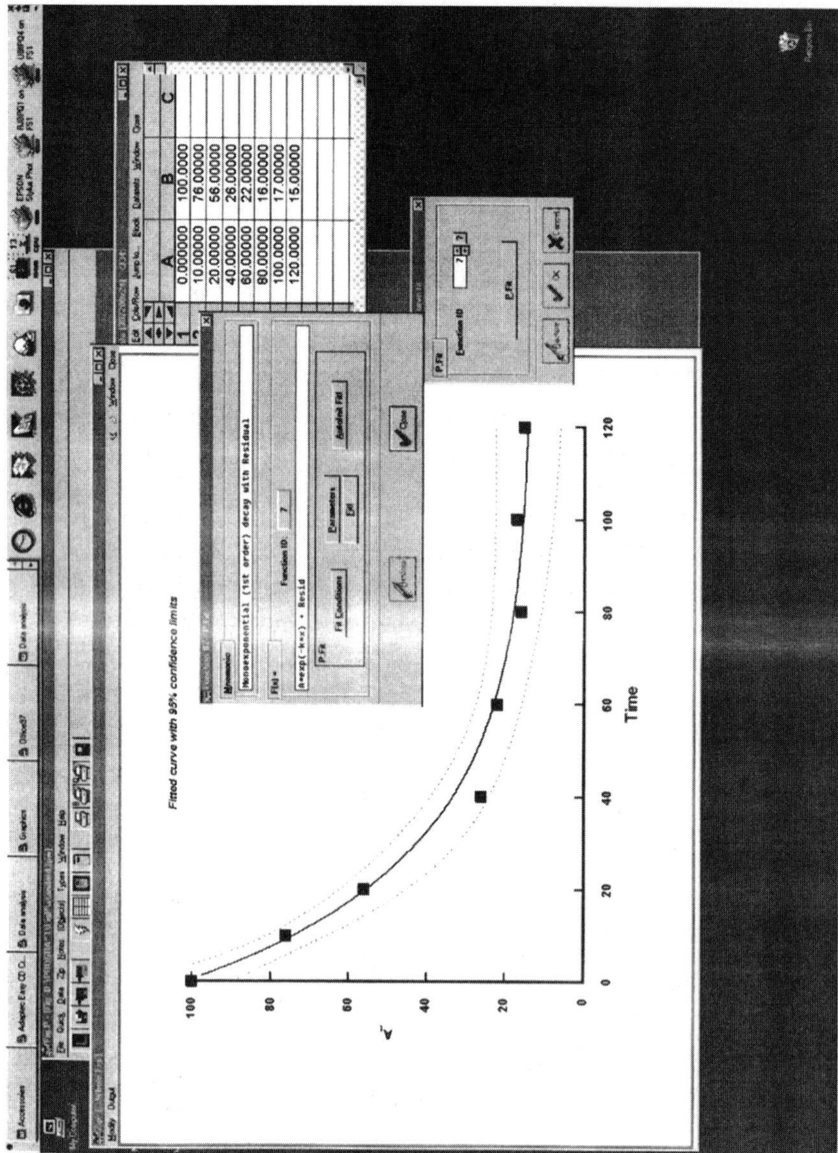

Figure 8 Analysis of monoexponential decay data using non-linear curve fitting. The program (FigP, Biosoft) includes a powerful non-linear curve fitting tool that generates complex statistical analyses of the data, and which can produce confidence limits on the fitted function (as shown here). The program also offers comprehensive control of the appearance of the final fitted data set.

4.3 Example reaction schemes

4.3.1 Simple first-order digestion

A single proteolytic cleavage, in which neither substrate nor product interferes with the proteolytic reaction, should obey pseudo first-order kinetics. For example, the cleavage of a protein ab into fragments a and b.

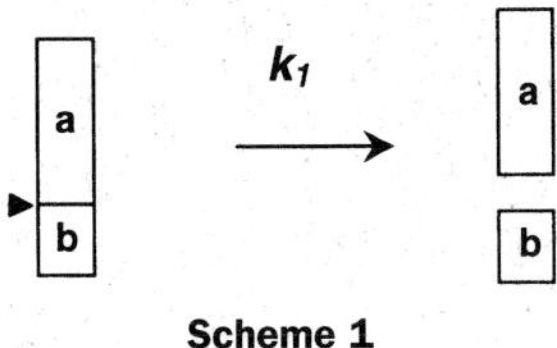

Scheme 1

The general equation describing all types of simple, first order, proteolytic reactions is:

$$A_t = A_f + (A_f - A_0)\cdot\exp(-k\cdot t)$$

Where A_t is the measured property at time t, A_0 defines the property at the start of the reaction, A_f the property at the end, and k is the rate constant for proteolysis. For example, a simple inactivation has $A_0 = 100\%$, or the true measured activity, and $A_f = 0$. Thus, at time zero, $A_t = A_f + (A_0 - A_f)\cdot 1 = A_0$, and at time t, $A_t = A_f + (A_0 - A_f)\cdot 0 = A_f$. In other systems, A_f might be non-zero, for example, where a proportion of substrate is indigestible, (see for example, ref. 3) or where the proteolysed form of the substrate has a residual activity. Unless

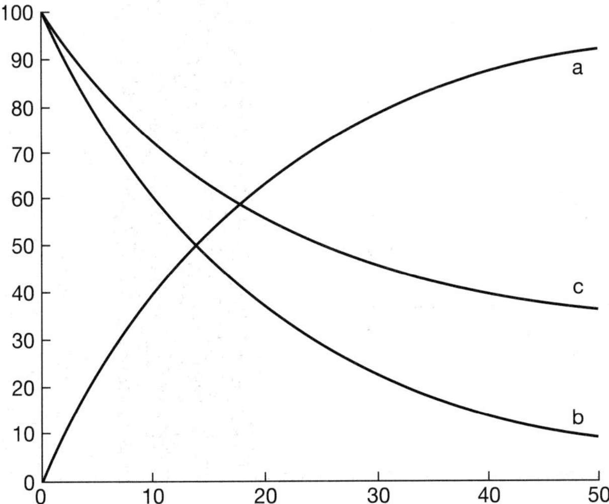

Figure 9 Different behaviours of first-order reactions. For all three simulations, the same equation has been used. The three behaviours are (a) where a property is acquired, (b) where a property is lost completely, and (c) where the product retains in part that property. In order to discriminate between the last two behaviours, it is necessary to allow the reaction to run for several half-lives.

there are good reasons to assume that A_f will genuinely be zero, data analysis should aim to recover a value for this parameter. In other circumstances, A_f will be larger than A_0 (for example, in proteolytic activation of a precursor of an enzyme). Examples of all three types of behaviour are seen in *Figure 9*.

4.3.2. Sequential first-order reactions

This reaction scheme is more complex, but is commonly met. It invokes the situation where the product of the first cleavage (ab) is the substrate for a second proteolytic attack. Indeed, most proteolytic reactions *in vitro* (and *in vivo*) are amenable to this analysis. If k_1 is much greater than k_2, B accumulates; If k_2 is much greater than k_1, then little of ab is detectable.

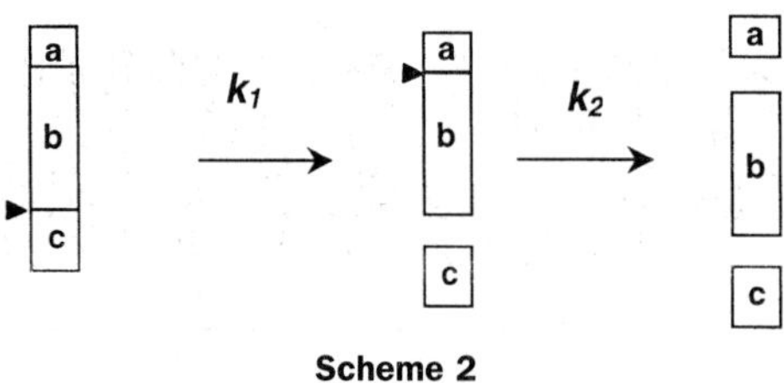

Scheme 2

This simple model is strictly sequential. The analytical solution of the rate equations for A and for B are relatively straightforward. If A = abc, and B = ab

$$A_t = A_0 \cdot \exp(-k \cdot t)$$

$$B_t = \frac{k_1}{(k_1 - k_2)} \cdot (\exp(-k_1 \cdot t) + \exp(-k_2 \cdot t))$$

The loss of A(abc) is simple first order, and is analogous to the situation in the previous section. At the beginning of the reaction, no B(ab) is present, and at the end of the reaction, all B has been consumed through conversion to peptides a, b and c. Thus, B has a transient existence during the reaction (*Figure 10*), and the appearance and subsequent disappearance of B are under the control of the two first-order rate constants k_1 and k_2. This, incidentally, is why it can prove impossible to increase the yield of an intermediate B. Although the amount of starting material A can be increased, if k_2 is larger than k_1, and the time at which B is maximal can be calculated, a large proportion of B will never be accumulated. The only strategy to enhance the yield of B is to find a means to reduce the rate of the second reaction (for example, by addition of a protective ligand that inhibits the B to C reaction).

4.4 Simulations and modelling

As proteolytic reaction schemes become more complex, exact analytical solutions are progressively harder to derive. An alternative strategy for analysis of such data, based on the simplifying assumption that each reaction is strictly first order, is to use a general modelling package that can find solutions to the

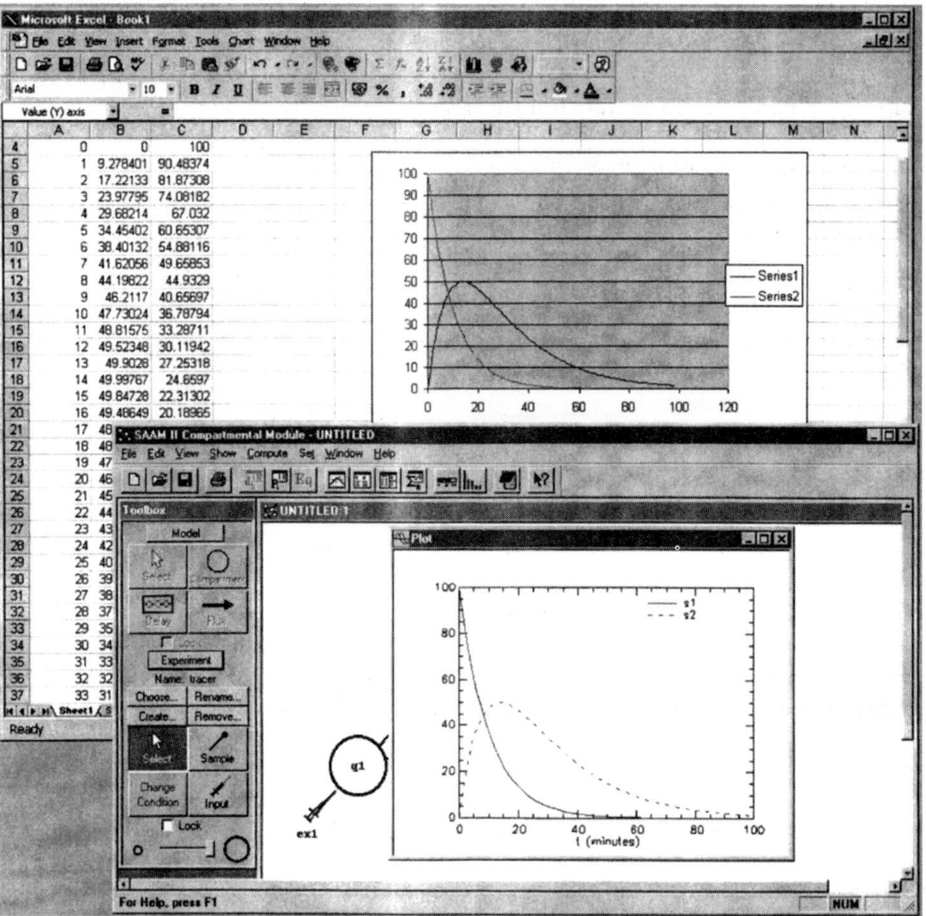

Figure 10 Sequential proteolytic reactions. The screenshot is of a simple, sequential proteolytic reaction A?B?C where the concentrations of A and B are plotted as a function of time. The reaction was modelled numerically in Excel (rear window) and simulated as a compartmental model in SAAM II (front window).The transient appearance of B is evident in both displays.

proteolytic reactions, and which can fit such models to sets of experimental data. Several such packages exist, but the authors have experimented with the package SAAM II (http://www.saam.com/), which facilitates both simulation and fitting of models to data sets. The model is easily described using icons for pools (substrate, products) and arrows that join pools represent the reactions, each of which is assumed to be first order, with a transfer coefficient that is the same as the first-order rate constant. SAAM II then creates the system of differential equations that defines the model from the model structure that has been drawn on-screen. The model is subsequently used as a template for specific experiments, in which starting pool sizes and transfer coefficients are defined and specific pools are sampled to monitor their size during the reaction. Finally, the

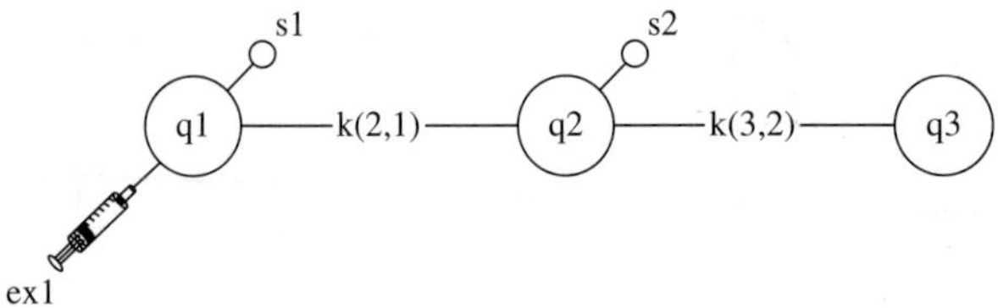

Figure 11 Compartmental model (SAAM II) of a simple, two-cleavage, sequential proteolytic system.

simulation is run for a defined time and the data on all pool sizes is generated as a data file and as graphical output. More complex models can be defined, for example, to model the addition of a protective ligand half way through the reaction.

Simple models are readily constructed. For example, the two-stage sequential proteolysis described in Section 4.3.2 is readily designed (*Figure 11*) and then used both as a modelling tool and as a data analysis tool.

The packages can allow for alternative routes of proteolysis and quite complex models can be described. Consider, for example, a two-stage sequential proteolysis whereby a protein abc is cleaved into three products a, b and c. The first cleavage can occur between segments a and b, or b and c. Thus, the proteolysis can take one of two routes, where site 1 (ab) is hydrolysed before site 2 (bc), or vice versa. There are four rate constants that define the cleavage reactions.

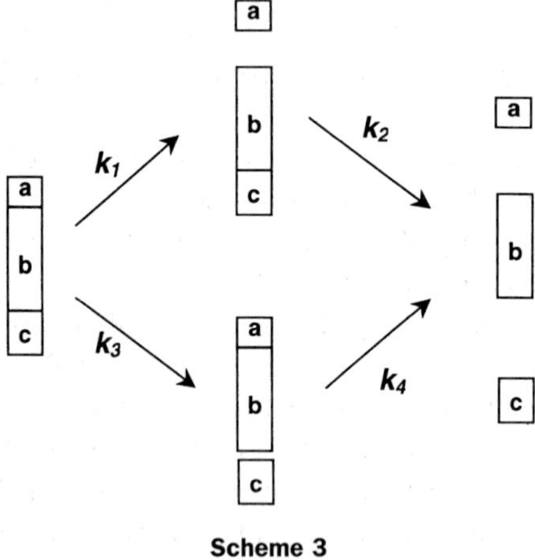

Scheme 3

This model is sufficiently general to encompass all possible behaviours. Setting any of the rate constants to zero creates a simpler model; for example, setting k_3 to zero (irrespective of the value of k_4) will simplify the model to a sequential pair of first-order reactions, as in Section 4.3.2. However, the most complex scenario is where all four rate constants are finite and non-zero.

The description of a model of this process with SAAM II is a graphical one. Each of the pools represents one of the reactants (abc, ab, bc, a,b,c), and the arrows between pools represent the transfer coefficients (or first order rate

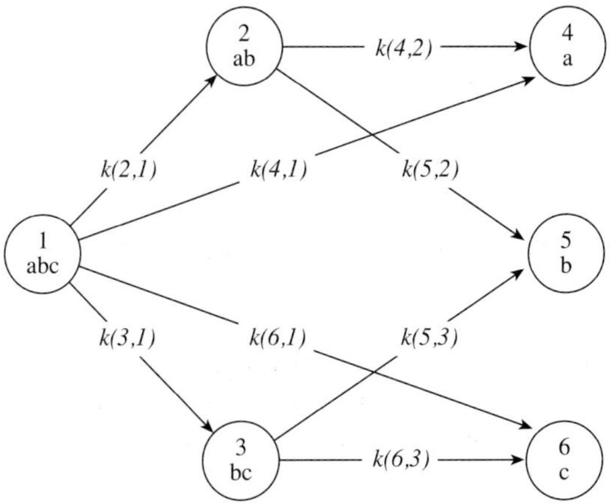

Figure 12 Compartmental model (SAAM II) of a two-cleavage, random-order proteolytic system. Each pool represents one of the reactants (abc, ab, bc, a,b,c) and the arrows represent the transfer coefficients between each pool (unidirectional because proteolysis is considered to be irreversible). The nomenclature of the transfer coefficients is that the parentheses indicate the pool to which materials, followed by the pool from which material flows. Thus, *k*(6,1) is the first-order rate constant between pool 1 (abc) and pool 6 (c). This is the same reaction that generates ab, and thus the model can be simplified by setting equalities such as *k*(6,1) = *k*(2,1).

constants) for each pool (*Figure 12*). The analytical solution can be predicted to be quite complex, but the whole reaction scheme can be represented in SAAM II as outlined in *Figure 11*. The graphical output from this program is crude, but nonetheless informative (*Figure 13*).

Moreover, data on the size of the different pools can be entered into the program and the rate constants for the different stages of the reaction can be derived by non-linear curve fitting. The models quickly become complex as new reactions are added. *Figure 14* includes a SAAM II model for a three-site random-order proteolysis scheme in which the pools are now abcd, abc, bcd, ab, bc, cd, a,b,c and d. However, such a simulation is easy to set up and can be used to explore the behaviour of progressively more complex systems.

The models inevitably look more complex than simple reaction schemes because each proteolysis reaction is represented by two arrows, as the formation of both products must be specified. As an example, the attack on hirudin by V8 protease has been modelled; this is a study for which very good quality data is available on several intermediates over time (33). As can be seen in *Figure 15*, the model might seem to be a reasonable interpretation of the proteolytic process, but the best fit of the model to the total data set reveals that there are some incongruities and that there may be additional components to the model that cannot be measured experimentally. Of course, in the study referenced (33) the objective was not the derivation of a complete kinetic model of the proteolysis, and the inability to fit the model to the data set in no way invalidates the published analysis. However, the ability to generate, simply and flexibly, such

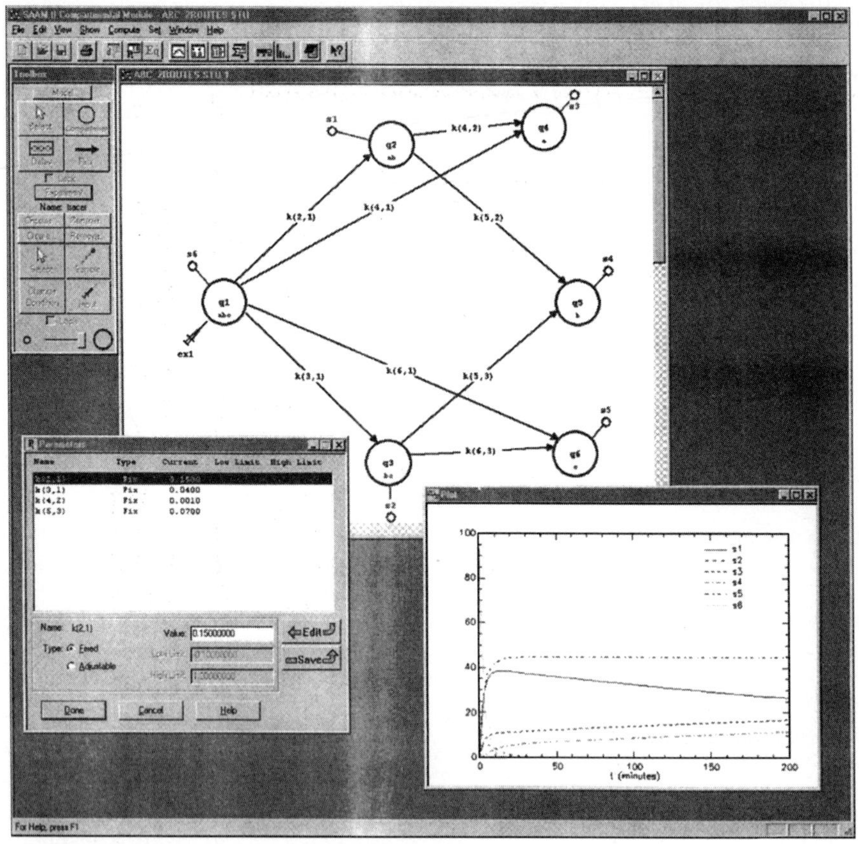

Figure 13 Simulation of the model described in *Figure 12*. For a set of rate constants (top left window), and the model described in the top right window, the graphical output (bottom left window) describes the behaviour in time of the five pools that are sampled (s1 to s5 on the model).

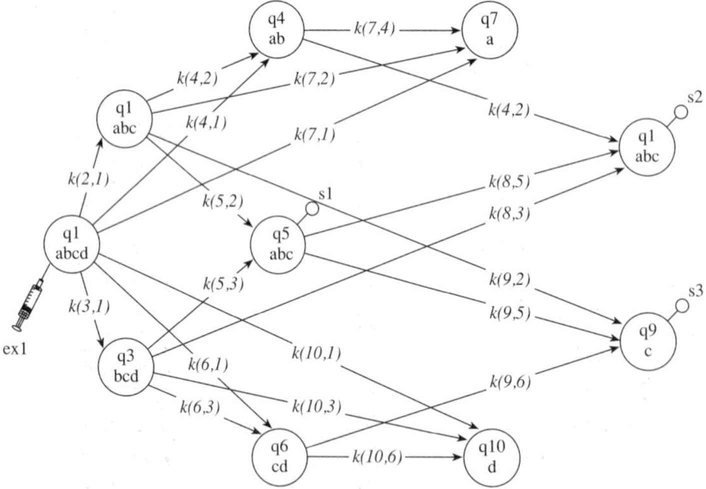

Figure 14 Complex, multiple-cleavage, random-order proteolysis schemes. Compartmental analysis can be extended to more complex schemes. The upper panel is that of a random order three-cleavage system. The lower panel is a compartmental model for the hydrolysis of human growth hormone by V8 protease (32).

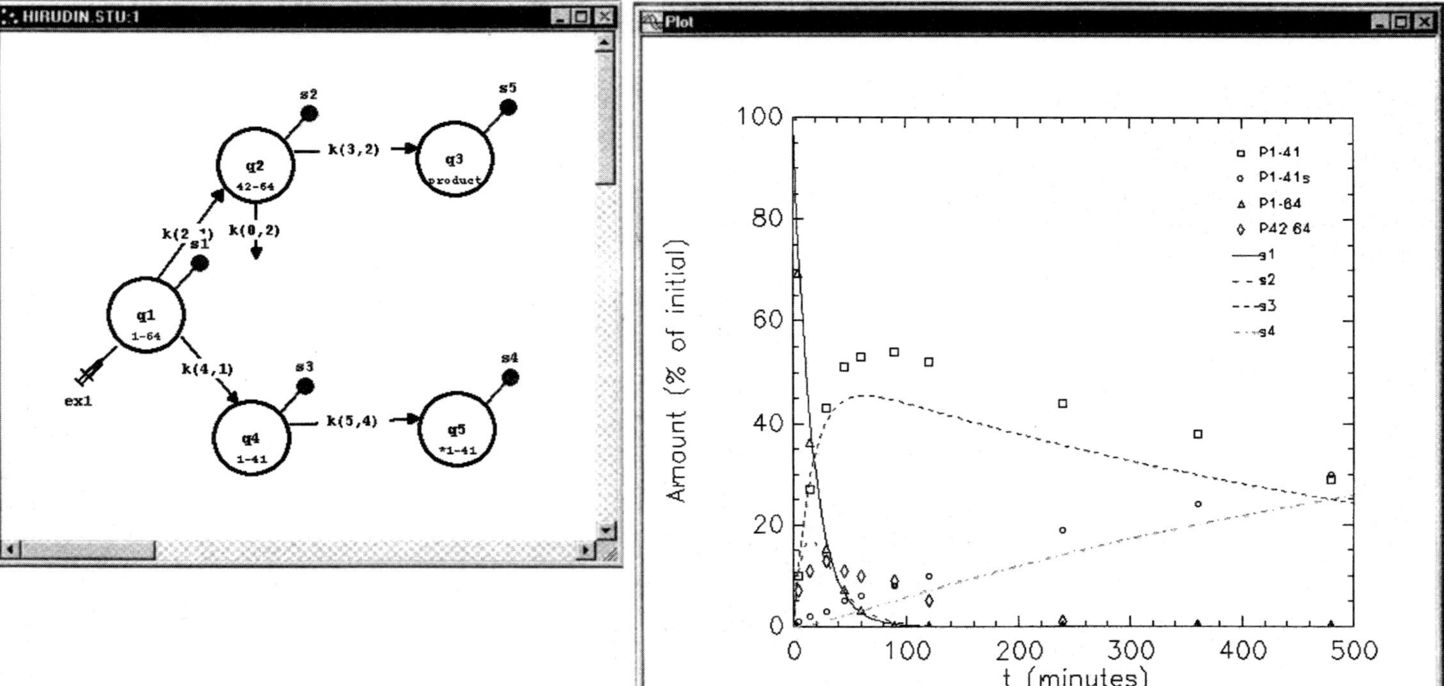

Figure 15 Analysis of complex proteolytic schemes by limited proteolysis. The model shown on the left is the compartmental description of the proteolysis of hirudin by V8 protease, using the scheme proposed by the authors of this work (33). The panel on the right shows the best fit of the model to the data set, as extracted roughly from the published figure. Although the model seems reasonable, there are discrepancies between the data and the simulation, suggesting that there may be other routes of proteolysis and other compartments.

reaction schemes and then run simulations indicates the power of the compartmental modelling approach.

References

1. Jaeger, K-E., Ransac, S., Kock, H. B., Ferrato, F., and Dijkstra, B. W. (1993). *FEBS Lett.* **332**, 143.
2. Abousalham, A., Chaillan, C., Kerfelec, B., and Foglizzo, E. (1992). *Protein Eng.* **5**, 105.
3. Ellison, D., Hinton, J., Hubbard, S. J., and Beynon, R. J. (1995). *Protein Sci.* **4**, 1337.
4. Rupley, J. A. and Scheraga, H. A. (1963). *Biochemistry* **2**, 421.
5. Ooi, T., Rupley, J. A., and Scheraga, H. A. (1964). *Biochemistry* **2**, 432.
6. Arnold, U., Ricknagel, K. P., Schierhorn, A., and Ulbrich-Hofman, R. (1996). *Eur. J. Biochem.* **237**, 862.
7. Yang, H. J. and Tsuo, C. L. (1995). *Biochem. J.* **305**, 379.
8. Polverino de Laureto, P., Scaramella, E., de Filippis, V., Vruix, M., Rico, M., and Fontana, A. (1997). *Protein Sci.* **6**, 1.
9. Ruhlmann, A., Kukla, D., Schwager, P., Bartels, K., and Huber, R. (1973). *J. Mol. Biol.* **77**, 417.
10. Gron, H., Meldal, M., and Breddam, K. (1992). *Biochemistry* **31**, 6011.
11. Neurath, H. (1980). In *Protein folding* (ed. Jaenicke, R.), p. 501. Elsevier, Amsterdam.
12. Novotny, J. and Bruccoleri, R. E. (1987). *FEBS Lett.* **211**, 185.
13. Vita, C., Dalzoppo, D., and Fontana, A. (1985). *Biochemistry* **24**, 1798.
14. Fontana, A., Fassina, G., Vita, C., Dalzoppo, D., Zamai, M., and Zambonin, M. (1986). *Biochemistry* **25**, 1847.
15. Fontana, A. (1989). In: *Highlights of modern biochemistry* (ed. Kotyk, A., Skoda, J., Paces, V., and Kostka, V.), p. 1711. VSP International Science Publ., Zeist.
16. Hubbard, S. J., Campbell, S. F., and Thornton, J. M. (1991). *J. Mol. Biol.* **220**, 507.
17. Hubbard, S. J., Eisenmenger, F., and Thornton, J. M. (1994). *Protein Sci.* **3**, 757.
18. Parcell, D. A. and Sauer, R. T. (1989). *J. Biol. Chem.* **264**, 7590.
19. Bek, E. and Berry, R. (1990). *Biochemistry* **29**, 178.
20. Monsalve, R. I., Menendez-Arias, L., Lopez-Otin, C., and Rodriguez, R. (1990). *FEBS Lett.* **263**, 209.
21. Hubbard, S. J., Thornton, J. M., and Campbell, S. F. (1992). *Faraday Discuss.* **93**, 13.
22. Hubbard, S. J., Beynon, R J., and Thornton, J. M. (1998). *Protein Eng.* **11**, 349.
23. Lee, B. and Richards, F. M. (1977). *J. Mol. Biol.* **55**, 379–400.
24. Taylor, W. R., Thornton, J. M., and Turnell, W. G. (1983). *J. Mol. Graph.* **1**, 30.
25. Nishikawa, K. and Ooi, T. (1986). *J. Biochem. (Tokyo)* **100**, 1043.
26. Kabsch, W. and Sander, C. (1993). *Biopolymers* **22**, 2577.
27. Bernstein, F. C., Koetzle, T. F., Williams, G. J. B., Meyer Jr, E. F., Brice, M. D. *et al.* (1977). *J. Mol. Biol.* **112**, 532.
28. Beynon, R. J. and Beaumont, A. (2001). *The thermolysins. Protein profiles.* Academic Press, London (in press).
29. Place G. A. and Beynon, R. J. (1983). *Biochim. Biophys. Acta* **747**, 26.
30. Beynon, R. J and Easterby, J. S. (1996) *Buffers for pH control – the basics.* Bios Scientific Publ., Oxford.
31. Wu, C., Robertson, D. H., Hubbard, S. J., Gaskell S. J., and Beynon R. J. (1999). *J. Biol. Chem.* **274**, 1108.
32. De Laureto, P., Toma, S., Tonon G., and Fontana, A. (1995). *Int. J. Peptide Protein Res.* **45**, 200.
33. Vindigni, A., De Fillipis, V, Zanotti, G., Visco, C, Orsini, G., and Fontana, A. (1994). *Eur. J. Biochem.* **226**, 323.

Chapter 11
Proteases in peptide synthesis

Volker Kasche
Technical University at Hamburg-Harburg, Biotechnologie II, Denickestrasse 15, D–21071 Hamburg, Germany

1 Introduction

The basis for the classification of proteases as enzymes is their biological function as hydrolases. As catalysts, the proteases must also catalyse the reverse reaction, i.e. the formation of a peptide bond (1,2). For some time this was treated as a curiosity rather than a property that could be used to gain more knowledge of the catalytic function of proteases, or for peptide synthesis *in vitro*. Recently, their potential use for the latter has led to a marked increase in studies on the synthetic function of proteases.

Proteases can catalyse the synthesis of peptides in an equilibrium-controlled or a kinetically controlled process (3). The different processes are shown schematically in *Figure 1*. Their time dependence and dependence on enzyme properties, temperature and pH are given in *Figure 2*. In the kinetically controlled process (I–III), an activated substrate (peptide or amino acid) is used and the enzyme catalyses the transfer of the acyl group to a nucleophile (amino acid or peptide) in reaction I. In this reaction, the protease acts as a transferase. Competing with

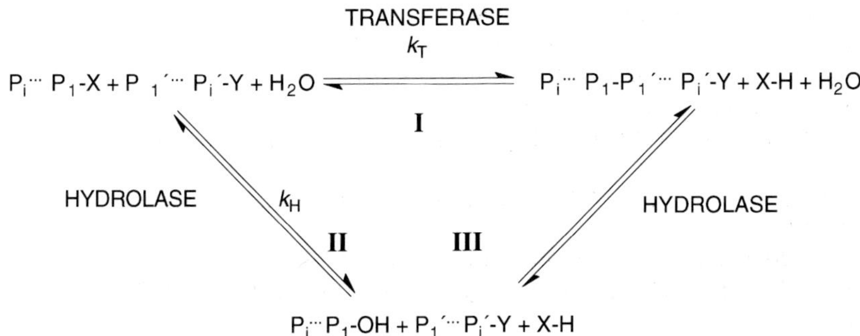

Figure 1 Protease-catalysed kinetically controlled (I + II + III) and equilibrium-controlled (III) synthesis of peptides $P_i \cdots P_1–P_{1'} \cdots P_{i'}$. The protease acts as a transferase in (I) and as a hydrolase in (II, III): k_T, k_H, apparent second-order transferase and hydrolase rate constant of the enzyme used; P_i, P_i', amino acid residues; -X, group used to activate the carboxyl end of peptide $P_i \cdots P_1$-; -Y, group used to protect the carboxyl end of peptide (amino acid) $P_i' \cdots P_i'$-.

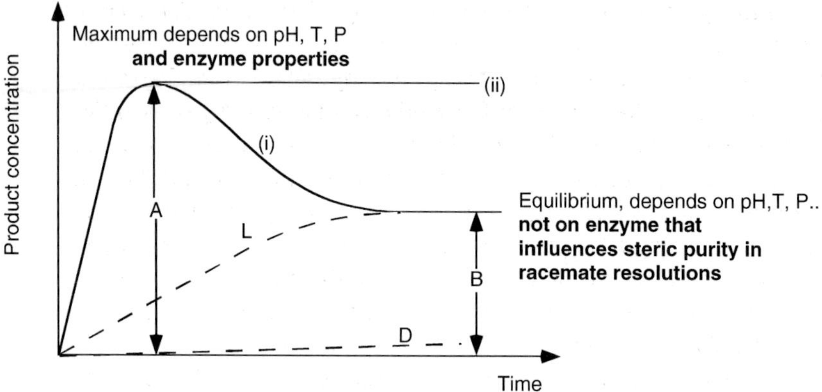

Figure 2 Time dependence in kinetically (solid line) and equilibrium (dashed line) controlled peptide synthesis. For the kinetically controlled synthesis two cases are considered: (i) synthesis rate of the same order of magnitude as the rate of hydrolysis of product peptide, and (ii) synthesis rate >> hydrolysis rate. For the equilibrium-controlled synthesis the time dependence for the condensation of a racemic mixture to a L-amino acid (peptide) using a L-specific protease is given.

this reaction is the hydrolysis of the activated substrate (reaction II). The product peptide is also a substrate for the protease, and can be hydrolysed in reaction III. The product yield depends on the apparent ratio of the transferase to hydrolase rate constant $(k_T/k_H)_{app}$, and the rate with which the product peptide is hydrolysed. When the ratio $(k_T/k_H)_{app}$ is large, the product -P_1-P'_1- can be formed in high yields even when the thermodynamically favourable products are -P_1-OH- and -P'_1. . ., i.e. the equilibrium favours hydrolysis.

This formation of non-equilibrium concentration of peptides (and other condensation products) requires activated substrates. Contrary to the equilibrium-controlled process, the product yield depends on the properties of the enzyme used $(k_T/k_H)_{app}$ and the substrate specificity. The enzymes used to catalyse the synthesis of peptides (*in vivo* transferases and ligases with a transferase function as amino acid-tRNA synthetase) may also catalyse the hydrolysis reactions (II or III in *Figure 1*). Examples here are hydrolytic steps in proof-reading mechanisms —amino acid activation or DNA-synthesis (4,5). The ratio of the apparent transferase to hydrolase rate constant for these transferases can be estimated to be $> 10^6$ from the few experimental data published here (3). Such high ratios are necessary to minimize the energy waste involved in the acyl or peptidyl transfer to H_2O (≈ 55 M) at the low concentrations ($< 10^{-3}$ M) of peptides *in vivo*. Only proteases that form covalent acyl-enzyme intermediates can be used as catalysts in kinetically controlled peptide synthesis. They have values of $(k_T/k_H)_{app}$ in the range 10^2–10^4 (3).

In an equilibrium-controlled process reaction (reaction II or III in *Figure 1*) the protease is used only to increase the rate at which the equilibrium is established. The enzyme cannot influence this equilibrium.

The stereospecificity of a protease gives information on the enantiomer (L or D) that is preferentially converted in the enzyme-catalysed reaction. The stereospecificity of an enzyme may differ in the S_1 and S_1'-binding subsites. Recently, it has been shown that enzymes are not strictly stereospecific. To describe this, a quantitative measure for the stereospecificity—the stereoselectivity or enantioselectivity—has been introduced. It is a ratio of the rate constants involved in the enzyme-catalysed reaction for the L- to the D-enantiomer (Section 2.1.3). As a ratio of rate constants, it should depend on temperature (6). This has also recently been shown to apply for proteases (7). Proteases can therefore also be used for equilibrium or kinetically controlled synthesis of peptides with both L- and D-amino acids and racemate resolutions. In this case, the steric purity of the products depends on the stereoselectivity of the protease used.

Since the first edition of this book the use of proteases in peptide synthesis has been extensively covered in books and reviews (8–11). This reflects the continued interest in evaluating the potential of proteases as catalysts for peptide synthesis as an alternative or complement to chemical peptide synthesis. Different proteases are now used for the synthesis of peptide antibiotics and hormones, neuropeptides, peptides used as sweeteners such as aspartame, and modifications of recombinant proteins such as in the production of human insulin (12).

To evaluate the further potential of proteases as catalysts in peptide synthesis, the quantitative analysis of the yield-controlling factors, their P_1 and P'_1 specificity, and stereoselectivity of their S_1 and S_1' binding subsites, not covered in existing reviews, will be stressed in this chapter.

2 Enzyme properties influencing the product yield and steric purity in protease catalysed peptide synthesis

2.1 Kinetically controlled synthesis

2.1.1 Mechanism

Studies on the kinetically controlled synthesis of condensation products such as oligo-saccharides (13), peptides (14) and β-lactam antibiotics (3) using hydrolases as catalysts have presented kinetic evidence that the nucleophile NH must be bound to the acyl- (or glycosyl-) enzyme before it can deacylate this intermediate. The reactions determining the yield of the condensation product from the acyl-enzyme are shown in *Figure 3*. This mechanism also applies for other condensation products. Therefore, a more general designation will be used here for the activated substrate (AB) and the nucleophile (NH) than that in *Figure 1*.

2.1.2 Product yield

The peptide formed in the kinetically controlled synthesis is a substrate for the hydrolase. The hydrolysis of this product influences the maximum product yield when the hydrolysis rate is of the same order of magnitude or larger than the

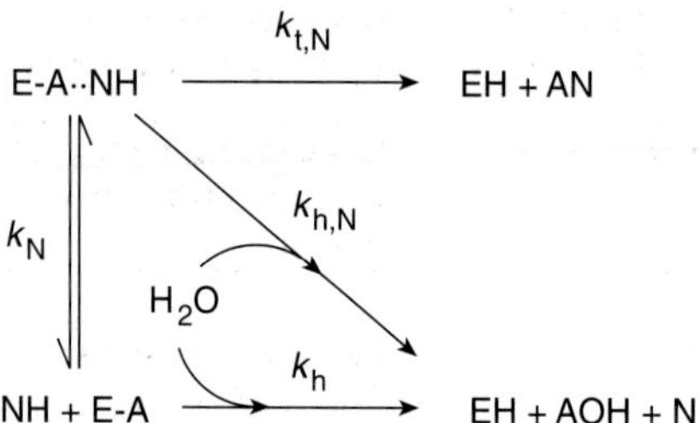

Figure 3 Mechanism of kinetically controlled synthesis of a peptide AN from the acyl-enzyme E-A, involving the binding of the nucleophile NH (amino acid or peptide). It can only deacylate the acyl-enzyme when it has been bound to it. The formation of the peptide competes with the deacylation of the acyl-enzyme by water (hydrolysis).

rate of the synthesis of the condensation product AN. Thus, two cases must be considered.

(i) The hydrolysis rate of the condensation product is not negligible when compared with the synthesis rate. In this case the product concentration is maximal when its rate of hydrolysis equals the synthesis rate

(ii) The hydrolysis rate of AN is negligible compared with the synthesis rate.

In both cases, the apparent ratio of the rate of condensation (AN) to hydrolysis product (AOH) formation can be derived from *Figure 3*. When the rate of equilibration of E–A, NH and E–A··NH is much faster than the other reaction rates the following relation is derived (3)

$$(k_T/k_H)_{app} = k_{t,N}/(k_h K_N + k_{h,N}[NH]) \qquad 1$$

The quantity $(k_T/k_H)_{app}$ can easily be determined from the ratio of the initial rates for the formation of the transferase, v_{AN}, and hydrolase, v_{AOH} product:

$$(v_{AN}/v_{AOH}) = (k_T/k_H)_{app}([NH]/[H_2O]) \qquad 2$$

In other studies, a partition constant p has been used as a measure for the apparent hydrolase to transferase activity of the proteinase (9). The relation between p and $(k_T/k_H)_{app}$ is

$$(k_T/k_H)_{app} = [H_2O]/p \qquad 3$$

In Case i (*Table 1*) the condensation product concentration is maximal when its rate of hydrolysis equals the rate of synthesis of AN. Expressions for the maximum yield of condensation product are given in *Table 1* (3). In contrast to the maximal product yield in equilibrium-controlled peptide synthesis, the maximum product yield in kinetically controlled peptide synthesis thus depends on $(k_T/k_H)_{app}$ and the specificity constants, i.e. the properties of the enzyme.

In Case ii (*Table 1*) the hydrolysis of AN can be neglected. When $[NH]_0 >> [AN]_{max}$ the nucleophile content and $(k_T/k_H)_{app}$ is practically constant during the synthesis. For this condition, the expression for $[AN]_{max}$ given in *Table 1* can be

Table 1 Expressions for the maximum yield of the condensation product AN in the kinetically-controlled synthesis. The constants are defined in *Figure 1* and Equation 2; subscript '0' denotes initial concentrations

	Maximum concentration of product peptide $[AN]_{max}$	**Has been shown to apply for the kinetically controlled synthesis of**
Case i Rate of product hydrolysis ≈ rate of product synthesis	$\left(\dfrac{\left(\frac{k_T}{k_H}\right)_{app}[NH]}{[H_2O]+\left(\frac{k_T}{k_H}\right)_{app}[NH]}\right)\dfrac{\left(\frac{k_{cat}}{K_m}\right)_{AB,\,NH}}{\left(\frac{k_{cat}}{K_m}\right)_{AN,\,NH}}[AB]$	β-Lactam antibiotics Oligosaccharides
Case ii Rate of product hydrolysis << rate of product synthesis	$\dfrac{\left(\frac{k_T}{k_H}\right)_{app}[AB]_0+[H_2O]\ln\left(1-\frac{[AN]_{max}}{[NH]_0}\right)}{\left(\frac{k_T}{k_H}\right)_{app}}$ for $[NH]_0>>[AB]_0$ $\dfrac{\left(\frac{k_T}{k_H}\right)_{app}[AB]_0[NH]_0}{\left(\frac{k_T}{k_H}\right)_{app}[NH]_0+[H_2O]}$	Peptides with proteases whose esterase activity >> amidase activity

derived. In contrast to Case i, the maximum yields do not depend on the ratio of the substrate and condensation product specificity constants. In both cases, the maximum does not depend on the enzyme content. The latter only influences the time required to reach the maximum (3).

Whether a specific synthesis can be described by Case i or Case ii in *Table 1* cannot generally be stated. It has to be determined from the time course of [AN] in the kinetically controlled synthesis (*Figure 2*). Generally, Case ii applies for peptide synthesis using activated esters and proteases with large ratios of the ester to amide hydrolysis rate constants.

The relations for $[AN]_{max}$ given in *Table 1* can be used for rational selection of the optimal enzyme for the kinetically controlled synthesis of a given peptide bond based on experimental values on $(k_T/k_H)_{app}$ and specificity constants; or optimization of the maximal yield for a given peptide bond with one enzyme.

2.1.3 Steric purity

The steric purity is determined by the stereospecificity of the enzyme. This qualitative property gives information on which of the D- or L-enantiomers that are preferred by the enzyme. The stereoselectivity or enantioselectivity, E, has been introduced as a quantitative measure for the stereospecificity of enzymes. It is defined as the ratio of rate constants for the enzyme catalysed reactions with the L- or D-enantiomer, respectively (6). For the kinetically controlled synthesis shown in *Figure 1* the stereoselectivity E_S is defined as

$$E_S = \frac{\left(\frac{k_T}{k_H}\right)_{app.\,L}}{\left(\frac{k_T}{k_H}\right)_{app.\,D}} \tag{4}$$

From *Figure 1* follows that it is a measure for the stereoselectivity in the S_1'-binding subsite. For the hydrolysis of a substrate the stereoselectivity E_H is defined as

$$E_H = \frac{\left(\frac{k_{cat}}{K_m}\right)_L}{\left(\frac{k_{cat}}{K_m}\right)_D} \tag{5}$$

Depending on whether the chiral centre is located in the P_1 or P_1' part of the substrate it is a measure for the stereoselectivity in the S_1 or S_1'-binding subsite. For a given activated substrate and nucleophile E_S and E_H are properties of the enzyme. Kinetically controlled synthesis can be used for the kinetic resolution of a racemic nucleophile (15) or activated substrate. From *Table 1* and Equations 4 and 5 it follows that the steric purity of the product are influenced by both E_S and E_H.

2.2 Equilibrium-controlled peptide synthesis

2.2.1 Yield

The maximum concentration (yield) of a peptide AN formed in equilibrium-controlled synthesis is

$$[AN]_{max,e'} = K_{app} \frac{[AOH][NH]}{[H_2O]} \tag{6}$$

where K_{app} is the apparent equilibrium (association) constant. It can easily be determined from measurements of the total concentrations given in Equation 6. The enzyme used to catalyse the formation of AN cannot influence $[AN]_{max,e'}$ it only influences the rate with which the equilibrium is established. From Equation 6 it follows that the maximal peptide yield can be increased by the law of mass action, or by increasing K_{app}. The dependence on pH, temperature, ionic strength and solvent composition of K_{app} is given by the following relation (3):

$$K_{app} = K_{th} \frac{f_{UA} f_{UB}}{f_{UP} f_{H_2O}} \frac{\left(\frac{\Sigma\,[\text{peptide}]}{[\text{uncharged peptide}]}\right)}{\left(\frac{\Sigma\,[\text{acid}]}{[\text{uncharged acid}]}\right)\left(\frac{\Sigma\,[\text{base}]}{[\text{uncharged base}]}\right)} \tag{7}$$

where the sum Σ denotes the total concentration, i.e. the sum of all charged and uncharged forms of the product (reactants); K_{th} is the thermodynamic association constant for the formation of an uncharged peptide UP from the uncharged reactants UA (AOH) and UB (NH) containing the carboxyl and amino group from which the peptide bond is formed; and f is the activity coefficient. The quantity K_{th} cannot be determined directly, but is independent on solvent properties. The latter influences only the last two factors in Equation 7. They can be used for a quantitative analysis of the influence on pH and other system properties on K_{app} (see Sections 4.2, 4.3 and 4.5).

2.2.2 Steric purity

The following relation can be derived for K_{app} in equilibrium controlled peptide synthesis (16)

$$K_{app} = \frac{\left(\frac{k_{cat}}{K_m}\right)_{synth}}{\left(\frac{k_{cat}}{K_m}\right)_{hydr}} \qquad 8$$

where the subscripts synth and hydr denote the synthesis and hydrolysis of the peptide, respectively. From Equation 8 the following relation can be derived for the enantioselectivity in equilibrium-controlled peptide synthesis

$$E_{S,e} = \frac{\left(\frac{k_{cat}}{K_m}\right)_{synth,\,L}}{\left(\frac{k_{cat}}{K_m}\right)_{synth,\,D}} = \frac{K_{app,\,L}\left(\frac{k_{cat}}{K_m}\right)_{hydr,\,L}}{K_{app,\,D}\left(\frac{k_{cat}}{K_m}\right)_{hydr,\,D}} = \frac{K_{app,\,L}}{K_{app,\,D}} E_H \qquad 9$$

where the subscripts L and D denote the synthesis and hydrolysis of peptides with only L- enantiomer or with one D-enantiomer, respectively. From Equation 9, it follows that the steric purity, in contrast to the yield in equilibrium-controlled peptide synthesis, depends on E_H, i.e. the properties of the enzyme. No data are available for the ratio $(K_{app,D}/K_{app,L})$.

3 Selecting the optimal protease

3.1 Purity of the protease

The properties of the enzyme influence only the yield in kinetically controlled synthesis. The purity of proteases depends on the procedure used for the isolation of the enzyme from a homogenate or a medium (extracellular proteases) and the activation of zymogens. This can give rise to a heterogeneous preparation of enzymes with the same function but with different specificity constants and values of $(k_T/k_H)_{app}$ (17,18). Different bovine trypsin preparations have been found to vary with respect to relative amounts of α- and β-trypsin, and porcine trypsin preparations contain several active trypsin forms that can be separated by IEF or affinity chromatography (19, and own unpublished data). Recently, bovine α-chymotrypsin preparations with anionic chymotrypsins that were not present in older preparations have appeared (own unpublished data). This may be due to changes in the methods used to isolate the enzyme (18). Penicillin amidase preparations from different sources have also been shown to be heterogeneous (17,20). This may result in a poor reproducibility in product yields when using different enzyme preparations for the synthesis of the same product. A convenient method to analyse the heterogeneity of different enzyme preparations is IEF with substrate overlay techniques to detect active enzyme forms (17,20). In cases where the enzyme preparation is heterogeneous, it can be purified by affinity chromatography, provided that suitable affinity adsorbents are available (see Chapter 2), or by high resolution ion-exchange chromatography (17–20). Only homogeneous enzyme preparations should be used in fundamental studies.

3.2 P_1 and P_1' specificity

For the synthesis of a peptide bond P_1-P_1'- (Appendix I) it is necessary to use a protease with a high P_1'- or P_1-specificity for the amino acid that should be added to a given P_1 or P_1' amino acid. Data on the P_1-P_1' specificities of proteases, based on extensive studies on hydrolysis of different substrates, have been compiled (21–24). The P_1' specificity in peptide synthesis is given by the ratio $(k_T/k_H)_{app}$. For P_1' as a nucleophile in the kinetically controlled synthesis of P_1-P_1', $(k_T/k_H)_{app}$ increases with (k_{cat}/K_m) for substrates P_1-P_1', where P_1 is constant and P_1' varies (25). Thus, hydrolysis data could be used to determine the P_1'-specificity in the synthesis reaction. The substrates used in the studies on hydrolase P_1'-specificity have been mainly esters (-O-X in P_1', where X = CH_3, C_2H_5) or amides (with NH_2 or chromogenic amines as *p*-nitroanilides in P_1'). Consequently, they are of limited use for the determination of the specificities of amino acids or peptides in the P_1' (..P′) positions. However, data on limited proteolysis of peptides or proteins (P_1-P_1' specificity), $(k_T/k_H)_{app}$ and maximal yields in kinetically controlled synthesis of different peptides can be used as a measure of the P_1'-specificity. Data for the P_1, P_1-P_1' and P_1' specificities of different proteases that are commercially available and some proteases from extremely thermophilic organisms for which such data are available (*Table 2*) are given in *Figures 4* and *5* (21–23, 26–36). In *Figure 4* the data for the P_1 and P_1'-specificity, derived from separate studies on hydrolysis and synthesis, correlate well with the P_1-P_1'-specificity, derived from limited proteolysis of proteins and peptide hydrolysis. *Figure 4* can thus be used for a qualitative selection of the proteases that may be used for the synthesis of a specific peptide bond. From these, the optimal protease for kinetically controlled peptide synthesis can be selected. It is the one with the largest $(k_T/k_H)_{app}$ value (*Table 1* and *Figure 5*). The few existing data for proteases from extremely thermophilic organisms show that their P_1'-specificities are lower than for the best proteases from mesophilic organisms.

From the data in *Figures 4* and *5* it follows that the proteases that can be used for kinetically controlled synthesis have broader P_1' than P_1 specificities. For the endopeptidases, only P_1' amino acids with protected α-carboxylate groups can be used. These enzymes have a negative charge near the S_1' binding site that repels α- carboxylate groups and limits their use as exopeptidases. Unprotected P_1' amino acids can be used in kinetically controlled synthesis catalysed by the

Table 2 List of the proteases whose P_1- and P_1'-specificities are given in Fig. 4. Besides commercially available enzymes, proteases from extremely thermophilic organisms, an amidase for which such specificities have been determined, are included

Proteases forming covalent acyl-enzyme intermediates that can be used to catalyse both kinetically and equilibrium controlled peptide synthesis (serine-, cysteine-proteases)
1. EC 3.4.16.1 Serine type carboxypeptidase (former carboxypeptidase Y)
2. EC 3.4.21.1 Chymotrypsin
3. EC 3.4.21.4 Trypsin
4. EC 3.4.21.5 Thrombin
5. EC 3.4.21.6 Coagulation Factor Xa
6. EC 3.4.21.19 Glutamyl endopeptidase (former V 8 protease)
7. EC 3.4.21.26 Prolyl oligopeptidase
8. EC 3.4.21.36 Pancreatic elastase
9. EC 3.4.21.50 Lysil endopeptidase (former *Achromobacter* protease)
10. EC 3.4.21.62 Subtilisin
11. EC 3.4.21.64 Endopeptidase K (former Proteinase K)
12. EC 3.4.21.66 Thermitase
13. EC 3.4.21.68 t-Plasminogen activator
14. EC 3.4.21.73 u-Plasminogen activator (former Urokinase)
15. EC 3.4.21.93 Proprotein convertase
16. EC 3.4.22.2 Papain
17. EC 3.4.22.8 Clostripain
18. Non-classified serine protease from extremely thermophilic organisms (a) *Thermus* Rt41A (34) (b) *Thermococcus stetteri* (35) (c) *Streptomyces pactum* (36)
Proteases without covalent acyl-enzyme intermediates that can only be used for equilibrium controlled synthesis (aspartic and metallo-proteases):
19. EC 3.4.17.1 Carboxypeptidase A
20. EC 3.4.17.2 Carboxypeptidase B
21. EC 3.4.23.1 Pepsin[a]
22. EC 3.4.23.4 Chymosin
23. EC 3.4.23.15 Renin
24. EC 3.4.24.3 Microbial collagenase
25. EC 3.4.24.11 Neprilysin (former Enkephalinase)
26. EC 3.4.24.27 Thermolysin
Amidase forming covalent acyl-enzyme intermediates that can be used to catalyse both kinetically and equilibrium-controlled peptide synthesis
27. EC 3.5.1.11 Penicillin amidase

[a] Pepsin has recently been shown to perform kinetically controlled synthesis without a covalent acyl-enzyme intermediate. The rate of this reaction catalysed by pepsin is much slower than with enzymes that form covalent acyl-enzyme intermediates (37).

P'_1 / P_1	Ala	Arg	Asn	Asp	Cys	Gln	Glu	Gly	His	Ile
Ala	21		16			2	16	16	16	
Arg	3							3, 4, 10, 11		3, 5, 21
Asn	10, 16		2	10		10				
Asp	10			10, 16			21	10		
Cys										
Gln	10, 21	21	21	10			10		10, 11, 21	2
Glu	2, 10, 21		16, 21			16, 21			21	
Gly	16, 21	21	10, 21	21			10, 21	10		16, 21
His	2, 16, 21							2		2
Ile			2							
Leu						2, 10, 21	10, 16	10, 21	21	2, 21, 26
Lys	2 , 3,	10	16	3				2, 3, 16	3	3, 16
Met	2	25	2	10			10	2		2
Phe	2, 25, 26	10		21			2	2, 26		21
Pro	7									
Ser	2, 10, 16		10	10,21			21	10	10, 11, 16	21
Thr	2, 10, 21		2			10,21	2			
Trp	2							2, 10, 21		2, 10
Try			2, 10, 21					2, 10, 21		
Val	21	2	10, 21							21
P'_1-spec. from synthesis	1, 2, 8, 18, 17, 27	2, 3, 8, 16, 17, 18	1	3,27		1	3, 16, 27	1, 2, 16, 17, 24	1, 16, 17, 27	1, 16, 26

Figure 4 P_1–P'_1, P_1, and P'_1-specificity for some proteases. The data for P_1 specificity (last column) and P'_1 specificity [last row, (k_T/k_H)app $>$ 100 from *Figure 5*] were obtained from studies on substrate hydrolysis and protease-catalysed peptide synthesis, respectively. The

Leu	Lys	Met	Phe	Pro	Ser	Thr	Trp	Try	Val	P_1-spec. hydrolysis
16, 21	10		25		10		10		16, 21	9, 11, 21, 26
21					3	5			3, 10, 13	3, 4, 12, 14
16	2		10, 16, 21		2,10	2, 10, 21	21		2	
		16	21	21						6
21	2		10, 21		2				2, 21	
21	16, 21		21					21	21	6
26	10, 21		21, 26		10, 16	10			16, 21	16, 26
10, 16, 21	2		21		2				2, 21	16
21	21		16							8, 22
23, 26	2, 21				2, 16, 21	16		10, 11	10, 11, 21	2, 8, 10, 21, 23, 26
3,21	2, 3, 14, 16, 20				3	3		16	1, 3	3, 9, 16, 17
	2, 10		21		2				2	
2, 26		22	1,10, 21, 26	2	2, 3, 21	10, 16		1, 10, 21, 26	10, 21	2, 21, 25, 26
7	1								10	7, 26
	10		21		21	10, 21			21	
21, 26	2			1	2	21		16	2, 10, 21	
3, 10					26		21		2, 10, 21	2, 10
10,21	10		16		2, 10, 21	1, 2, 10, 11		2	21	2, 10, 16, 21
21	2					21		21	2, 21	8
1, 2, 3, 8, 16, 17, 26, 27	1, 3, 17, 18	1, 3, 8, 17, 18	1, 2, 3, 16, 17, 18, 26, 27	17	2, 17, 27	1, 3, 16	1, 17	17	1, 2, 3, 8, 16, 17, 26	

data for P_1–P_1'-specificity (first 20 rows, columns) were obtained from studies on limited proteolysis and hydrolysis of peptides. The blank spaces indicate lack of information. The proteases are given by the numbers in *Table 2*.

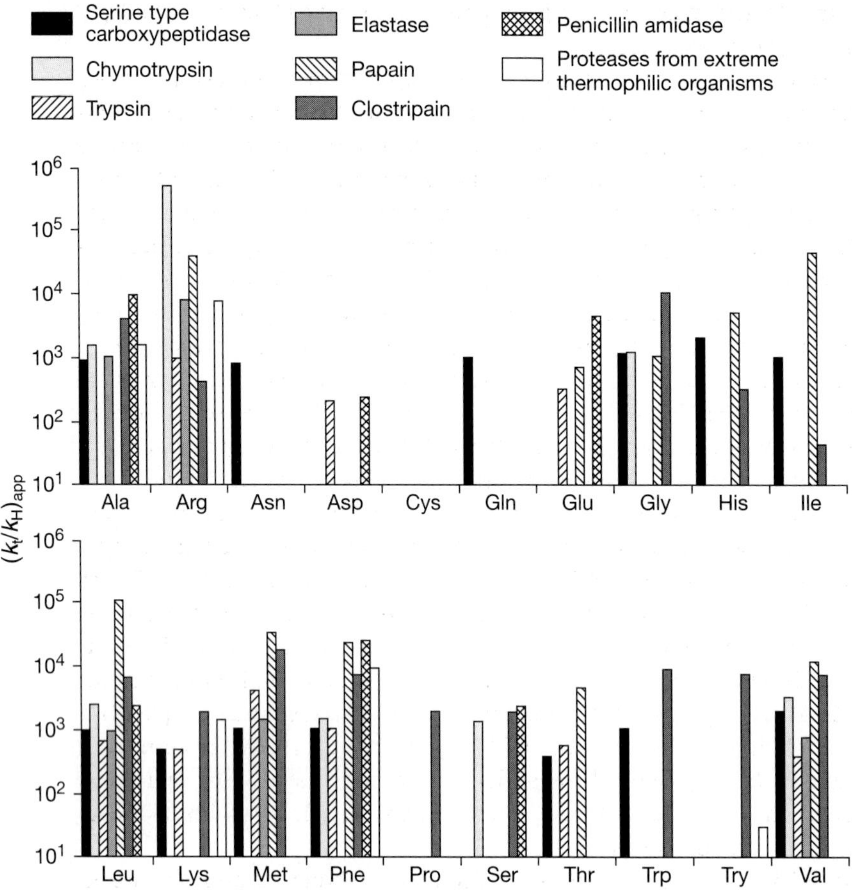

Figure 5 P_1'-specificity of serine and cysteine proteases, proteases from extreme thermophilic organisms and the amidase penicillin amidase expressed as $(k_T/k_H)_{app}$ for different L-amino acid amides or amino acids (penicillin amidase) used as nucleophiles. Conditions were as follows. Serine type carboxypeptidase Y (EC 3.4.16.1), P_1 = Ala, 25 °C, pH 9.5 (pH-Stat, 10% MeOH) (26); bovine α-chymotrypsin (EC 3.4.21.1) (27), pH 9.0, 25 °C, P_1 = Phe (pH-stat, 0.2 M NaCl) (20) or Tyr bicarbonate buffer, I = 0.2 M) (28,29); bovine trypsin (EC 3.4.21.4), P_1 = Arg, 25 °C, pH 8.0 (50 mM Tris-HCl, 5 mM $CaCl_2$, 0.5 M NaCl) (30); pancreas elastase (EC 3.4.21.36), P_1 = Leu, 25 °C, pH 9.0 (pH-stat, total ionic strength 0.2 M), (31); papain (EC 3.4.22.8), P_1 = Ala, 25 °C, pH 9.0 (pH-stat, 0.2 M KCl, 5 mM DTT) (32); clostripain (EC 3.4.22.8), P_1 = Arg, 25 °C, pH 7.3 (0.2 M Tris-HCl), (33); penicillin amidase (EC 3.5.1.11), P_1 = phenylacetyl and β-mandelyl (P_1' = His), 25 °C, pH 9.0 (bicarbonate buffer, I = 0.2 M) (28,29). Proteases from extreme thermophilic organisms: *Thermus* Rt41A, P_1 = Ala, P_1' = Tyr, 40 °C, pH 10 (50% DMF) (34); *Thermococcus stetteri*, P_1 = Arg, P_1' = Phe, 55 °C, pH 9 (0.2 M carbonate buffer) (35); *St. pactum*, *P1* = Arg, P_1' = Ala, Arg, Lys, and Phe, 40 °C, pH 9.0 (0.2 M carbonate buffer) (36).

proteases serine type carboxypeptidase (26) and papain (37, 38), and the amidase penicillin amidase (29). The last enzyme can also be classified as a protease (17). For the proteases, the yields are lower with unprotected than with protected amino acids. This does not apply for penicillin amidase that can be used to protect and deprotect amino groups of amino acids by adding or removing a phenylacetyl group (39).

The cysteine proteases have a preference for apolar P_1' amino acids and have, at similar nucleophile concentrations, $(k_T/k_H)_{app}$ values that are an order of magnitude larger than for the best serine proteases. Compared with the cysteine proteases, α-chymotrypsin and trypsin have a preference for basic P_1' amino acids. The exopeptidase serine-type carboxypeptidase (former carboxypeptidase Y) has the broadest P_1' specificity. For this enzyme the $(k_T/k_H)_{app}$ values are generally lower than for the other enzymes. Of the enzymes discussed here, only trypsin has been used for the synthesis of peptides with Asp or Glu in the P_1' position. Only carboxypeptidase Y could transfer acyl or peptidyl groups to cysteine. For proline in P_1', only clostripain has been shown to be a suitable catalyst. For endopeptidase K, $(k_T/k_H)_{app}$ (not shown in *Figure* 5) was found to be < 100 for Asp-, Arg-, Ser, Ala-, and Phe-NH_2 (29). For both equilibrium and kinetically controlled synthesis, enzymes with a high P_1-P_1' specificity for a desired peptide should be used as catalysts. For the kinetically controlled synthesis, the selection of the optimal enzyme depends on whether the synthesis can be described by Case i or ii conditions (Section 2.1.2, *Table 1*). Direct kinetic evidence that the kinetically controlled synthesis using these enzymes and an ester as the activated substrate can be described as a Case ii system (product hydrolysis rate $<<$ product synthesis rate) has been given for chymotrypsin (25,27,29), trypsin (30, 40), papain (32) and carboxypeptidase Y (26). The peptide yield should then increase with $(k_T/k_H)_{app}$ (*Table 1*). For experimental studies where $(k_T/k_H)_{app}$ and maximal yields have been determined under the same conditions, the observed yield equals the values calculated from *Table 1*, within the experimental error.

For kinetically controlled synthesis where the product hydrolysis rate is approximately equal to product synthesis rate (Case i), more enzyme properties than for Case ii influence the maximal product yield (Equation 1 and *Table 1*). To obtain maximal yield of a desired product as a function of enzyme, substrate activation and nucleophile content it is therefore, contrary to Case ii, not sufficient to use the activated substrate and enzyme with the highest value of $(k_T/k_H)_{app}$. This is demonstrated by the following experimental observations.

In the semisynthesis of the peptide antibiotics ampicillin and cephalexin, higher yields were obtained with D-phenylglycine-*O*-Me than with D-phenylglycine-*O*-Et or -NH_2, even when the latter are more specific substrates and have larger $(k_T/k_H)_{app}$ values for the enzyme used (*Escherichia coli* penicillin amidase) (*Table 1*). The selection of a suitable activated substrate should, in this case, also consider that the yield can be increased when the product can be continuously removed from the reaction mixture by precipitation, adsorption to product-selective adsorbents or extracted by suitable solvents. Such a process integration should always be considered in protease catalysed peptide synthesis on a larger scale (41).

In *Figure 6*, the maximal observed stereoselectivities the S_1 and S_1'-binding subsites of some proteases and penicillin amidase are given. For most proteases,

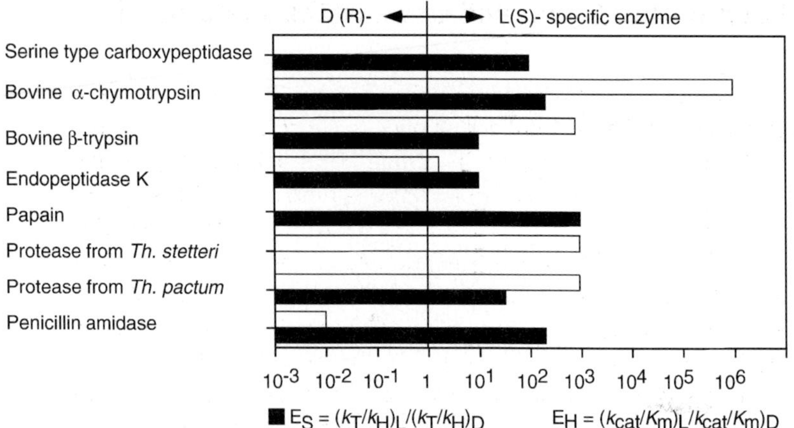

Figure 6 Maximal observed stereoselectivities in the S_1-(E_H) and S_1'-(E_S) binding subsites for the proteases serine type carboxydase (26,42), bovine chymotrypsin (7,28,29), trypsin (40, and unpublished data for E_H), endopeptidase K (29), proteases from the extremely thermophilic organisms *Thermococcus. stetteri* (35) and *Thermococcus pactum* (36) and penicillin amidase (29,47). For E > 1 the enzyme is L-specific, for E < 1 it is D-specific.

quantitative data on their stereoselectivity are missing. Observed stereoselectivities are much larger in the S_1 subsite than the S_1'-subsite. Proteases can therefore be used for the synthesis of L–D peptides. The available data show that the proteases from extremely thermophilic organisms have stereoselectivities of magnitudes similar to most of the proteases from mesophilic organisms. Papain and bovine α-chymotrypsin, where most data are available (7), have, however, much higher stereoselectivities. For penicillin amidase with a D-specific S_1 subsite and an L-specific S_1'-subsite, the stereoselectivities in both subsites allows its use for racemate resolution. This has been applied on an industrial scale (15).

When proteases are used for racemate resolution of the acyl donor P_1 or the nucleophile P_1', a high steric purity of the product requires that the stereoselectivity (E_H, E_S or $E_{S,e}$) is < 0.01 for D-specific or > 100 for L-specific enzymes, respectively.

4 Factors controlling the yield and steric purity in the synthesis of a peptide bond with a given enzyme

Once the enzyme has been selected for the kinetically controlled synthesis, $(k_T/k_H)_{app}$ can be further optimized by changes in the group used to protect the P_1'-carboxyl end, pH, temperature, solvent composition, ionic strength or the use of unconventional systems (reactions in ice or suspensions). For the equilibrium-controlled synthesis, K_{app} can be optimized by the use of P_1-amino-blocked and P_1'-carboxyl-protected amino acids, and changes in pH, ionic strength, solvent composition and temperature.

4.1 Protection of the P_1' and activation of the P_1-carboxyl group

For the proteases where this has been studied $(k_T/k_H)_{app}$ and the product yield in kinetically controlled syntheses has been shown to be much larger for the amino acid amide than for the corresponding ester (25–30, 40). This may be due to the better binding of $-NH_2$ than -O-X ($=P_2'$) to the S_2' binding site. This applies also for serine type carboxypeptidase that can be used for synthesis of peptides with unprotected P_1' amino acids (26,42). The latter applies also for the amidase penicillin amidase that can be used for the production of *N*-phenylacetyl protected amino acids that are used as P_1 groups in peptide synthesis (3). As most proteases that can be used for kinetically controlled synthesis are better esterases than amidases, the P_1-carboxyl group should be esterified for optimal yields. The synthesis then follows scheme (ii) in *Figure 2*, where the product peptide is hardly hydrolysed.

In kinetically controlled peptide synthesis with peptides $P_1'..P_n'$ as nucleophiles, $(k_T/k_H)_{app}$ and the product yield is also influenced by the interactions between P_i' (i = 2,...*n*) and the S_i'-binding subsites on the used protease. With constant P_i' both decreases and increases have been observed for different P_2' amino acids (27,29, 30). The focus in this review is on the P_1 and P_1' specificity, therefore the P_i' specificity ($i \geqslant 2$) will not be analysed further. Data on this have, however, been presented in only a few reports (43,44).

Generally, it has been found that the stereoselectivities E_H and E_S of proteases increase with the specificity constant (k_{cat}/K_m) and the nucleophile specificity $(k_T/k_H)_{app}$, respectively (7,29). From this, it follows that for kinetically controlled peptide synthesis with racemic activated amino acid or P_1'-X, the steric purity of the product increases with E_H and E_S (*Table 1*). Proteases that are more specific for L- than D-amino acids at the S_1' binding site can be used for the kinetically controlled synthesis of L–D peptides (45). The relative importance of a productive binding, when it can deacylate the acyl-enzyme, depends on the relative magnitude of the S_1'-P_1' and S_2'–P_2' interaction. This determines the stereospecificity. When they are of equal magnitude, as for amides, productive binding of D-nucleophiles to L-specific enzymes is less likely. When the S_2'–$P_{2'}'$ interaction is weaker than the S_1'-P_1' interaction, as for esters, the probability of productive binding of D-nucleophiles and the L–D peptide yield should be higher than for the amides. This has been observed in peptide synthesis with α-chymotrypsin (29).

4.2 pH

4.2.1 Equilibrium-controlled synthesis

The pH dependence is given by the concentration ratios in the last factor in Equation 7 (3). The concentration ratios can be derived from the following equilibria for each of the reactants and condensation products that are acids and bases, where the *K*-values are microscopic dissociation constants:

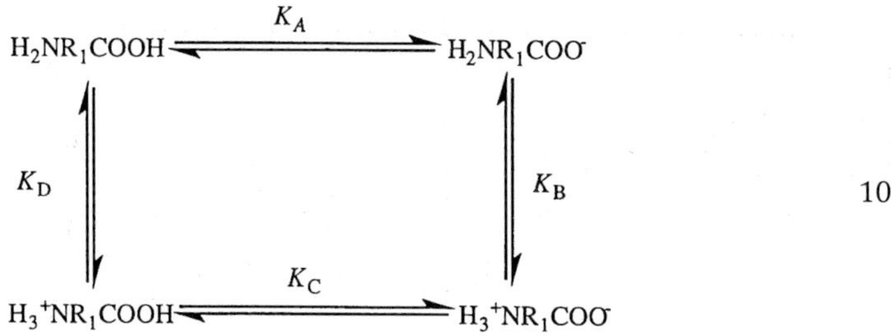

10

With amino-(P_1) and carboxyl-(P_1') protected amino acids the last factor in Equation 5 is

$$\frac{1}{\left(1+\frac{K_A}{[H^+]}\right)\left(1+\frac{[H^+]}{K_D}\right)} \quad 11$$

where the K_A and K_D values for the P_1-carboxyl and P_1'-amino group can be found in ref. (46). The pH-dependence of K_{app} for this and other cases has been calculated and compared with experimental data (*Figure* 7). From these results it is evident that unprotected amino acids (especially P_1) should not be used for peptide synthesis. K_{app} is reduced, owing to charge repulsion, compared with the case where protected amino acids are used. With unprotected amino acids other peptides (P_1-P_1, P_1'-P_1', P_1'-P_1) can be formed in addition to the desired product, P_1-P_1'. This also applies for the kinetically controlled synthesis. An exception here is the synthesis of ampicillin and cephalexin with unprotected D-phenylglycine using penicillin amidase from *E. coli* as a catalyst (3,29,47 and own unpublished data). With D-, but not L-, phenylglycine, side-reactions in which undesired dipeptides were formed have not been observed as this enzyme has a D-specific S_1 binding subsite (*Figure* 6). The calculated pH-dependence in *Figure* 7 agrees well with experimentally determined K_{app} values as a function of pH. For benzylpenicillin, the experimental data at low pH values (< 5) may be perturbed by the reduced activity of the enzyme at low pH and the chemical instability of the β-lactam ring. For peptide synthesis, the experimental data show that the maximum yield can be increased when the difference between the p*K* values for the amino group and carboxyl group involved in the condensation reaction are reduced. This can be achieved by the addition of organic solvents (see Section 4.5). In some cases, enzyme-dependent equilibrium yields in peptide synthesis have been observed. Here, as for penicillin amidase at low pH, this probability results from enzyme instability at high or low pH and/or non-establishment of the equilibrium at pH values far from the pH optimum of the enzyme used, since the latter cannot influence equilibrium yield.

4.2.2 Kinetically controlled synthesis

A pH optimum above the p*K* value for the P_1' amino group (7–8 for amino acid amides or esters; 9–10 for amino acids (46) is generally observed. Below the pH

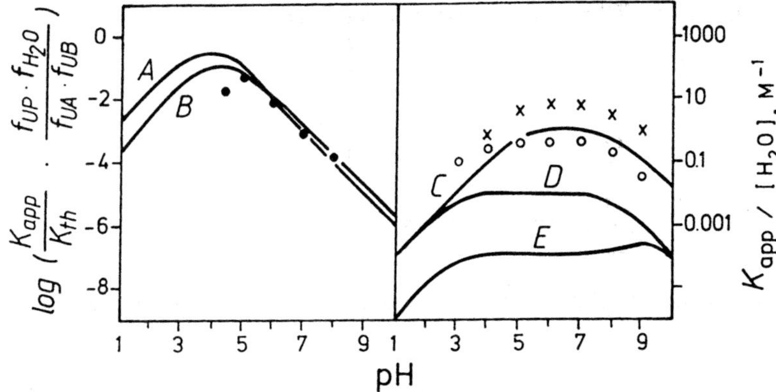

Figure 7 The pH dependence of the association constants for the formation of peptide antibiotics (A, B) and peptides (C, D, E). The last (pH-dependent) factor in Equation 7 was calculated as a function of pH using Equations 10 and 11 and compared with experimental data: pK, values for reactant acid; pK′, values for the reactant base; pK″, values for the product peptide. (A) Ampicillin formation using the values $pK_A = 4.11$; $pK_D = 6.9$, $pK_B = 9.0$; $pK'_A = 2.6$; $pK'_D = 4.6$, $pK'_B = 4.6$; $pK''_A = 2.6$; $pK''_D = 6.9$; $pK''_B = 9.0$. (B) Benzylpenicillin formation using the values $pK_A = 4.2$; $p\ K''_A = 2.6$; $pK'_A = 2.6$; $pK'_D = 4.6$; $pK'_B = 4.6$. Experimental data from ref. (65) (filled circles, right-hand scale). Dipeptide synthesis from amino- and carboxyl-protected amino acids. (C) For $pK_1 = 3.0$; $pK_2 = 8.0$. (D) For $pK_1 = 5.0$; $pK_2 = 8.0$. Experimental data for the synthesis of Cbz-Try-Gly-NH: O, in H_2O ($pK_1 = 3.60$; $pK_2 = 8.20$); x, in 60% (v/v) triethylene glycol ($pK_1 = 4.53$; $pK_2 = 8.17$) from ref. (55) (right-hand scale). (E) Dipeptide synthesis from unprotected amino acids for $pK_1 = pK'_1 = pK''_1 = 3.0$; $pK_2 = pK'_2 = 9.0$; $pK''_2 = 8.0$.

optimum this is due to the decrease in $(k_T/k_H)_{app}$ caused by the decrease in the concentration of the uncharged P'_1 amino group that is required for the nucleophilic deacylation (*Table 1*). Above the pH optimum, the decrease is probably due to the increase in the concentration of OH^- as a nucleophile, although this has not been studied in detail. This will increase k_H and reduce $(k_T/k_H)_{app}$. In cases where Tris has been used as a buffer component, it has been shown that it can deacylate acyl-enzymes (48). This will reduce the yield of the desired peptide. Consequently, here, as in other enzyme kinetic studies, Tris must not be used as a buffering component.

Subtilisin and penicillin amidase (except from *Alcaligenes faecalis*), lack S–S bridges. This reduces their pH stability above pH 9.0 and limits their use for the kinetically-controlled synthesis of peptides in this pH range.

4.3 Temperature

4.3.1 Equilibrium-controlled synthesis

The temperature dependence of K_{th} and the dissociation constants in the last factor of Equation 7 determine the change in K_{app} with temperature. The formation of a peptide bond from uncharged precursors is endothermic (2,49). Thus, K_{th} should increase with temperature. For peptide synthesis from protected amino acids, the temperature dependence of the last factor of Equation 7 is given by Equation 11. This ratio increases with temperature. Thus, in this case, K_{app} and the yield should increase with temperature.

Generally, however, the effect of temperature on K_{app} cannot be predicted but must be studied empirically. Unfortunately, in most studies on protease-catalysed peptide synthesis, the influence of the temperature has not been investigated, although this is of practical and theoretical importance.

4.3.2 Kinetically controlled synthesis

Only a few studies have been published on the influence of the temperature. For proteases from mesophilic and extreme thermophilic enzymes both the yield and $(k_T/k_H)_{app}$ have been found to decrease with temperature (3,34,40,50). This has led to revived interest in studying enzyme-catalysed reactions in ice, where an increase in $(k_T/k_H)_{app}$ and the yields have been found for kinetically controlled peptide synthesis in frozen aqueous systems (51,52).

4.3.3. Steric purity.

As a ratio of rate constants the stereoselectivities in the S_1 (E_H) and S_1'-(E_S) binding subsites of enzymes should depend on temperature. This was recently verified by experimental data (6,7). The steric purity of the products in equilibrium and kinetically controlled peptide synthesis should therefore be temperature dependent. In the few studies on the temperature dependence of the stereoselectivity of proteases it has been observed that E_H generally decreases with temperature, since it is mainly determined by the activation energy of k_{cat}. For E_S an increase with temperature has been observed for chymotrypsin (7). The opposite has been observed for the amidase penicillin amidase in the kinetically controlled synthesis of phenyl acetyl-amino acids (47). This indicates that the binding energy of the nucleophile to the S_1' subsite has a larger influence on the temperature dependence of E_S than the substrate binding energy on E_H. In the temperature interval 5–45 °C the maximum change in these stereoselectivities was found to be about one order of magnitude.

4.4 Ionic strength

4.4.1 Equilibrium-controlled synthesis

Qualitatively, an increased ionic strength should stabilize isolated ions, leading to a decrease in K_{app}. Quantitatively, this can be deduced from the ionic strength dependence of the dissociation constants in the last factor of Equation 7. The other factors do not depend on the ionic strength. The pK values of carboxyl groups decrease and for amino groups are practically constant (or increase for amino acids with unprotected carboxyl groups) with ionic strength. This is in agreement with the qualitative prediction. Thus, for soluble condensation products, increasing the ionic strength should reduce the yield. This implies that the use of ions in buffers, other than those involved in the condensation reaction, should be minimized.

4.4.2 Kinetically controlled synthesis

For both Case i and ii (*Table 1*) systems, the ionic strength influences the binding of the nucleophile to a charged S_1' binding site. When opposite charges are involved, as for the synthesis of β-lactam antibiotics catalysed by *E. coli* penicillin amidase, an increase in the ionic strength causes a decrease in the binding of the nucleophile and the product yield (3). In this case, where both the activated substrate and the nucleophile contribute to the ionic strength, the amount of ions to buffer the reaction medium should be minimized. When equal charges are involved, as in peptide synthesis with negatively charged amino acids (unprotected or as amides) as nucleophiles catalysed by endoproteases with a negative charge in the S_1' or S_2' subsite, an increase in the yield with increasing ionic strength is expected. This has been verified in experimental studies (53).

4.5 Solvent composition

Organic solvents have been used in protease catalysed peptide synthesis to increase the solubility of hydrophobic substrates and products, to perturb K_{app} in equilibrium-controlled synthesis and to isolate the products by solvent partitioning (54). The use of organic solvents here should be limited to environmentally unproblematic solvents that can be easily recycled. Among these is supercritical CO_2 that has recently been extensively used for lipase catalysed reactions. Its possible use for peptide synthesis has not yet been studied in detail.

4.5.1 Equilibrium-controlled synthesis

From Equation 7 it follows that the addition of water-miscible organic solvents influences the activity coefficients of the uncharged solutes and H_2O. This, and their effects on the dissociation constants in the last factor, may change K_{app}. The dissociation constants for carboxyl groups have been found to decrease with the organic solvent content (55). This partly explains the observed increase in K_{app} with organic solvent content shown in *Figure 7*.

Only when the water activity coefficient $\neq 1$ does it influence K_{app} in these systems. This is possible only in non-ideal mixtures of H_2O and organic solvents. The theory of such systems is still not sufficiently developed to allow a more rational analysis of the influence of organic solvents on pK values and activity coefficients. Recently, it has been found that enzymes are more easily inactivated in water-miscible than in water-immiscible organic solvents. With the latter, two-phase systems are formed that are of practical interest, especially when the solubility of the peptides is low in an aqueous system. The reaction rates in these systems are, however, reduced by orders of magnitude compared to the latter system.

4.5.2 Kinetically controlled synthesis

From *Table 1*, it follows that the product yield can be increased when $[H_2O]$ is reduced by the addition of water miscible organic solvents that marginally influence the properties of the enzyme used. Alcohols are not suitable organic solvents here, as transesterification products may be formed (3,48), and as they tend to decrease the stability of the proteases. For peptide synthesis (Case ii), increase in yields with reduced water content have been observed with DMF (50). The data available cannot yet be used to analyse whether these organic solvents influence the P_1'-specificity of these enzymes selectively.

This, however, can be studied with water-immiscible solvents. With these, indirect solvent effects are expected because of the partitioning of the substrates and products in the two liquid phases. Proteases, as other globular proteins, have a surface area of which about 40% consists of hydrophobic groups that may bind organic solvent molecules in the aqueous phase that is saturated with the water-immiscible solvent. (This is the basis for hydrophobic interaction chromatography used to purify proteins.) This gives rise to a direct solvent effect due to bound organic molecules that may change the specificity constants, $(k_T/k_H)_{app}$ and the stereoselectivities E_H and E_S. Large effects due to such direct effects have been observed in the P_1' specificity of chymotrypsin with bound hexane and octane. For hydrophobic amino acids (Leu and Phe) it increased, and for Arg a decrease was observed, whereas for Ser, Val and Ala the changes were within the experimental error. The same applied for the specificity constants for the hydrolysis of *N*-Acetyl-Try-O-Et. No direct solvent effect could be observed for endoproteinase K and penicillin amidase (28,29).

4.6 Peptide synthesis in suspensions with solid product or substrate

Recently, protease-catalysed equilibrium and kinetically controlled peptide synthesis has been carried out in suspensions with solid substrates and/or products as an alternative to the use of organic solvents (56–58). In these systems, high total substrate contents (up to 30% w/v) can be obtained. The reaction rate is limited by the rate of mass transfer from the solid to the aqueous phase, where the protease-catalysed reaction occurs. The latter is increased when the size of the solid substrate particles is decreased (57). In such systems for equilibrium and kinetically controlled peptide synthesis, greater product-formation rates (space–time yields) can be obtained than in systems with organic solvents or in aqueous systems with dissolved substrate (57). Its use is demonstrated in *Protocol 1*.

Protocol 1

Protease-catalysed synthesis of the dipeptide *N*-acetyl-L-tyrosine-L-arginine amide in aqueous two-phase system in suspension with solid substrate (N-acetyl-L-Ty-OEt, ATEE) and soluble nucleophile L-Ar-NH_2 using free or immobilized α-chymotrypsin (CT) (57)

Equipment and reagents

- HPLC-apparatus, pH-meter, thermostatted reaction vessel that can be stirred with a hanging stirrer to avoid mechanical damage of the immobilized enzyme
- Stock solution of CT (10 mg/ml) prepared in 10^{-3}–10^{-2} HCl to avoid autolysis (Section 6.2). It is stable for years when frozen. CT is immobilized in a suitable support as Eupergit C (≈ 1 mg CT/ml settled gel)
- Stock solutions of the nucleophile L-Arg-NH_2 and dissolved substrate ATEE in carbonate buffer, pH 9.0 (I = 0.2 M), the latter are used for calibration in the HPLC analysis

Method

1 *Selection of optimal protease.* From *Figure 4*, it follows that CT is the only enzyme with both the desired P_1 and P_1' specificity. In cases where several enzymes can be used, the one with the highest P_1' specificity (k_T/k_H)app value in *Figure 5* should be selected. A suitable choice is to use CT. For first studies, free enzyme can be used. To avoid disturbance in the HPLC analysis, the free enzyme used must not adsorb to the capillaries or stationary reversed-phase in the columns. Such adsorption has been observed with β-lactamase but, fortunately, not for CT, trypsin, penicillin amidase, endopeptidase K and thermolysin. Commercial bovine CT samples are homogeneous and can easily be immobilized to different supports (61).

2 *Substrate activation and protection of carboxyl group in* P_1'. An ester should be used as activated substrate, as CT is a much better esterase than amidase. From Section 4.1, it follows that a P_1' amide gives higher product yields than a P_1' ester (see also Section 6.3).

3 *Buffer, pH, ionic strength.* An inert buffering system at a pH above the p*K* for L-Arg-NH_2 (8) as carbonate buffer (I = 0.2 M) of pH 9.0 is a suitable choice. The ionic strength (in a homogeneous systems) should not be too high, for immobilized systems it must be selected to minimize possible pH gradients in the support (see Sections 4.4 and 5.3) (60).

4 *HPLC analysis.* Prepare solutions of activated substrate, hydrolysed substrate, nucleophile (and if available of condensation product). These solutions can be used to identify reactants and products and to calibrate the HPLC peaks for quantitative measurements. Reference samples and samples from the reaction mixture (≈ 50 μl)

Protocol 1 continued

are analysed after dilution (≈ 1:10) with the mobile phase using an RP 18 column and as eluent, 70% (v/v) 0.067 M KH_2PO_4, pH 4.7, and 30% (v/v) MeOH). The peak areas at 280 nm (for aromatic amino acids) and 220 nm (for all amino acids) are used to quantify substrates and products.

5 *Synthesis.* Preparing the reaction mixture (1-5 ml): dissolve the nucleophile (400–800 mM) in carbonate buffer, pH 9, (I = 0.2 M), then add the activated substrate (400–750 mM); adjust the pH of the suspension to 9.0 with 2 M NaOH and stir the suspension vigorously to disintegrate particle aggregates of the solid substrate. Add enzyme (10–20 μg free or 100 μg immobilized enzyme per ml reaction suspension) to the thermostatted (25°C) reaction mixture to start the synthesis. The product peptide and hydrolysis product *N*-acetyl-L-Tyr are soluble in this system. Withdraw liquid samples at different times and analyse them as described in step 4. The pH in the suspension should be kept at 9.0 during the reaction with 2 M NaOH or 3 M Na_2CO_3. The latter should be used when the enzyme or products are unstable at pH > 10.

6 From the results in *Figure 8* it follows that this synthesis, as expected, can be described by Case ii (*Table 1* and *Figure 2*). Both the results in solution and in suspension can be used to determine $(k_T/k_H)_{app}$ (Equation 2). No peaks other than these shown were observed and only one condensation product involving the α-amino group (pK ≈ 8) is formed. Condensation products involving other amino groups (ε-amino or guanidino of Lys and Arg, respectively) have been observed in other studies (own unpublished data).

7 *Yields and reaction rates.* The $(k_T/k_H)_{app}$ values calculated from *Figure 8* show that the rates and the product peptide yield are reduced when the enzyme is immobilized. This is in agreement with the predictions that can be made from the reactions in *Table 1*. In the aqueous phase ($[NH]_0 >> [AB]_0$) the yield then increases with $(k_T/k_H)_{app}$. That this ratio and the rates decrease when the enzyme is immobilized is expected from quantitative analysis of such systems (60), and reduction of pH in the immobilized enzyme system cannot be avoided even at high buffer capacities (58, 59) (Section 5.3). At constant $[NH]_0$ the yield can be increased as outlined in Sections 4.1–4.6. The peptide product yields (as % of the activated substrate) in *Figure 8* are 87 % for free CT and 83% for immobilized CT. This, and the high rates and space-time yields demonstrate the potential of enzyme-catalysed reactions in suspensions (56–58).

5 Planning a protease-catalysed synthesis of a peptide bond

5.1 What enzyme?

The proteases that may be used as catalysts can be selected from the data in *Figures 4–6*. Where several proteases may be used, the selection of the most suitable protease depends on whether a dipeptide or an oligopeptide should be synthesized. If an oligopeptide, the protease should have a low specificity for the

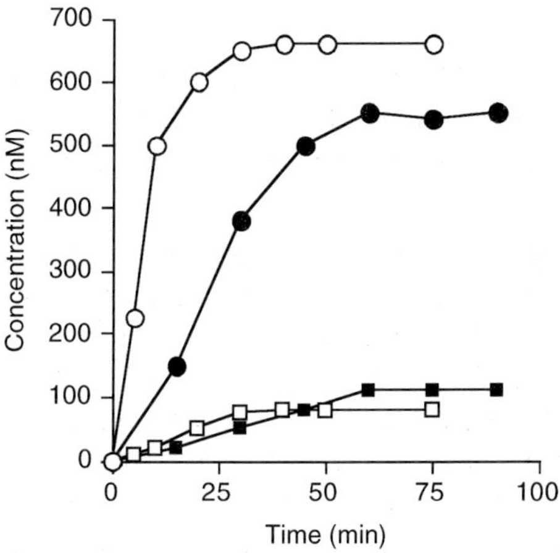

Figure 8 Kinetically controlled synthesis of soluble ATAA (circles) from ATEE as a suspension and soluble L-Arg-NH_2 with free (open symbols) and immobilized (filled symbols) CT according to *Protocol 1*. Conditions: 25 °C, pH 9.0 (carbonate buffer, $I = 0.2$ M), the pH is kept constant during the reaction. Free enzyme: enzyme content 10 μg/ml; starting concentrations: 800 mM L-Arg-NH_2 and 750 mM (total content) ATEE as a suspension. Immobilized enzyme: 200 μl CT immobilized in Eupergit C (0.5 mg CT/ml settled gel) per ml reaction suspension; Starting concentrations were 700 mM L-Arg-NH_2 and 660 mM (total content) ATEE as a suspension; circles, ATAA; squares, the soluble hydrolysis product *N*-acetyl-L-Tyr. The ATEE content in the aqueous system during the reaction with free enzyme was < 1 mM, with immobilized enzyme it decreased from 25 mM at time 0. Without enzyme the dissolved ATEE concentration was 25 mM.

hydrolysis (and synthesis) of the peptide bond(s) in the reactant peptides. This is especially important in equilibrium-controlled synthesis. For kinetically controlled synthesis with amino acid (peptide) esters as activated substrates, these undesired side-reactions are minimized by the use of proteases with large ratios of esterase to amidase activity.

5.2 Equilibrium-controlled or kinetically controlled synthesis?

All proteases can catalyse the equilibrium-controlled synthesis. The kinetically controlled synthesis can only be catalysed by proteases that form covalent acyl-enzyme intermediates. Since much less enzyme is needed, and the reaction times required to reach maximal yields are shorter in the latter case, kinetically controlled synthesis should be used when possible. They should be performed under Case ii conditions (*Figure 2*). This requires the use of esters as activated substrates. In this case, and with an excess of nucleophile, the maximal yield can be estimated using the relations in *Table 1*.

5.3 Free or immobilized enzyme?

For first experiments, low concentrations of free enzymes should be used. With proteases, autolysis cannot be neglected. It is then advantageous to use immobilized enzymes. This also reduces enzyme costs and the costs of separating enzyme from products. The immobilization cannot influence the yields in the equilibrium-controlled synthesis. In kinetically controlled synthesis, the maximal yields and steric purity of the products can be reduced compared with the free enzyme, at low (< 100 mM) nucleophile concentrations because of mass transfer effects (59–61). However, this difference is reduced at the higher concentrations (> 100 mM) that are of practical interest. Changes in enzyme properties due to immobilization may, however, influence the yield and steric purity (*Table 1*).

Peptide synthesis involves the formation of acids and bases. With immobilized proteases this may give rise to considerable pH-gradients (several pH units) in the supports with the enzyme (59,60). This changes the rates and may also lead to pH-induced denaturation of the protease. These negative effects can be minimized when sufficient buffer capacity ($I > 0.2$ M) is used and the system is well stirred to obtain maximal mass transfer rates.

Immobilized proteases can also be used for peptide synthesis when the product precipitates. In such cases a small pore support, where the product cannot precipitate in the support, must be used. Then the immobilized protease can be reused (57).

6 Experimental methods for protease-catalysed peptide synthesis

6.1 Enzyme purity and purification

Commercial protease preparations are generally not pure. They contain different forms of one protease (such as α- and β-trypsin in bovine trypsin) and/or other proteases as impurities. In the former case (enzyme heterogeneity) this only influences the yield in kinetically controlled synthesis. Foreign proteases as impurities may cause undesirable hydrolysis of products and reactants. To obtain reproducible results in kinetically controlled peptide synthesis, pure and homogeneous protease preparations are required. It is thus essential to check the purity of the preparations. A convenient and rapid method is IEF with substrate overlay techniques to visualize the protease (17,20). For fundamental studies, only pure homogeneous protease preparations should be used.

6.2 Enzyme immobilization

Suitable techniques and supports for enzyme immobilization are described in ref. (62). When dissolving proteases, buffers that minimize autolysis must be used. They should ideally be dissolved in buffers at a pH removed from the optimum for activity. During immobilization, the concentration of proteases must be kept below 1 mg/ml to avoid autolysis during the immobilization procedure.

6.3 Substrates and buffers

In equilibrium-controlled synthesis, amino (P_1) and carboxyl (P_1') protected amino acids should be used (*Figure 7*). In kinetically controlled synthesis, the amino group in P_1 must be protected. To obtain the desirable Case ii condition (*Figure 2*) for the synthesis, esterification is better than amidation for the activation of the P_1-carboxyl group. This requires the use of proteases that are better esterases than amidases. The P_1' amino acid carboxyl must be protected (Section 4.1). For the stepwise synthesis of peptides, P_1-...P_n', P_1'-esters are better than amides; then, the resulting P_1-P_1' peptide is also activated.

Buffers that have been shown to have minimal influence on activities should be used at the lowest possible ionic strength (see Sections 4.4 and 5.3). Their p*K*-value should be close to the pH that is optimal for the enzyme catalyzed synthesis. Tris must not be used as a buffer component because it can deacylate acyl-enzymes.

6.4 Monitoring the synthesis; purification of products

High-pressure liquid chromatography is the best analytical technique to monitor reactants and products in protease-catalysed peptide synthesis. Simultaneous detection at several wavelengths (280 nm for aromatic amino acids and about 220 nm for all amino acid) simplifies the identification of peaks. Detection at 220 nm is especially important to identify products from side-reactions such as the hydrolysis of carboxyl-protected P_1' amino acids or formation of P_1'-P_1'-X. For preparative isolation of products, preparative HPLC columns can be used under the same conditions as the analysis.

6.5 Optimizing the yield

Once the optimal system for the synthesis—Case ii—has been selected, the yield can be further increased by changing yield-controlling factors discussed in Section 4. From the determination of $(k_T/k_H)_{app}$ (Equation 2) and the relations in *Table 1*, the P_1' content required for quantitative transfer of P_1 to P_1' in the kinetically controlled synthesis can be calculated.

For equilibrium-controlled synthesis, the yield can be increased by the use of organic solvents. In these cases it is essential to determine the enzyme stability under the conditions used for the synthesis. Only proteases whose activity changes marginally should be selected as catalysts. The application of Sections 4–6 in a specific peptide synthesis is illustrated in *Protocol 1* and *Figure 8*.

7 Proteases in peptide synthesis: limitations and perspectives

For practical purposes, more peptides are synthesized by chemical peptide synthesis than in protease-catalysed processes. For some peptides a combination of chemical and enzymatic condensation steps is used (9,10). The automation of chemical peptide synthesis using peptide synthesizers, and newer developments that reduce the amounts of problematic solvents, still favour this procedure over the enzymatic approach. However, the latter has advantages, including prevention of racemization during the synthesis, reduced use of problematic (toxic) solvents and possible re-use of the catalyst (as an immobilized protease). Every possible peptide can also be synthesized. Finally, proteases can be used for simple amidation of P_1'-carboxyl end groups in peptides (hormones) by transpeptidation with P_1'-NH_2 as a nucleophile.

Since the appearance of the first edition of this handbook some of the main developments in protease catalysed peptide synthesis are:

- the extension of the systems used to aqueous suspensions and frozen systems where higher rates and yields can be obtained than in systems with organic solvents (10, 52, 56–58);
- more systematic studies on the stereoselectivity of enzymes (6, 7);
- new data for proteases including those from extremely thermophilic organisms (22–24, 34–36, 63);
- the introduction of enzyme catalyzed deprotection in peptide synthesis (39);
- the increase in the number of recombinant proteases, especially those of medical interest (63).

These advantages have not yet led to a more extensive use of proteases in peptide synthesis, owing to several factors. These include the lack of systematic studies on the stereospecificity of proteases. Most proteases are not strictly stereospecific (*Figure 6*). Thus D,L mixtures of P_1' cannot generally be used for the synthesis of sterically pure peptides. They can, however, be applied to synthesize L-D (D-L) peptides with L (P_1' or P_1)-specific proteases. There is a paucity of information and sources of (cheap) proteases that can be used for the synthesis of P_1–P_1' bonds where data are missing or the specificity is low (*Figures 4–6*). We also lack data on enzyme properties and stability in organic solvents. Finally, there have been no systematic studies on the use of protected and activated P_1 and P_1' amino acids that can be used both for chemical and enzymatic synthesis.

Even when this information is available, one should not expect that protease-catalysed peptide synthesis would eventually replace chemical peptide synthesis. The protease catalysed synthesis cannot be as readily automated as chemical synthesis, and the latter will remain the first choice for laboratory-scale peptide synthesis. For preparative synthesis of commercially important peptides, protease-catalysed steps will probably increase in importance (64).

The results from studies on the protease-catalysed peptide bond synthesis have been used to analyse the factors that influence product yield. However, they may also be used to analyse more basic enzymological problems. Based on the kinetic data on P_1' specificity reviewed here, systems for direct studies on the P_1'–S_1' interaction by X-ray crystallography can be selected. Such investigations can be used for comparative analysis of the sequence (P_1') specificity of proteinases using natural systems or enzymes designed by site-directed mutagenesis. The results could provide basic information on interactions between substrates and the active site on enzymes that influence their properties as catalysts.

References

1. Van't Hoff, J. H. (1898). *Z. Org. Chem.* **18**, 1.
2. Fruton, J. S. (1982). *Adv. Enzymol. Relat. Areas Mol. Biol.* **53**, 239
3. Kasche, V. (1986). *Enzyme Microb. Technol.* **8,** 5.
4. Freist, W. (1989). *Biochemistry* **28,** 6787
5. Fersht, A. R. (1983). *J. Mol. Biol.* **165**, 655.
6. Phillips, R. S. (1996). *TIBTECH*, **14,** 13.
7. Galunsky, B., Ignatova, Z., and Kasche, V. (1997). *Biochim. Biophys. Acta.* **1343**, 130.
8. Kleinkauf, H., and von Döhren, H (1990). *Eur J. Biochem.*, **192**, 1.
9. Schellenberger, V. and Jakubke, H.-D. (1991). *Angew. Chem. Int. Ed.* **30,** 1437.
10. Gill, I., Lopez-Fandifio, R., Jorba, X., and Vulfson, E. N. (1996). *Enzyme Microb. Technol.* **18**, 162.
11. Bongers, J., and Heimer, E. P. (1994) *Peptides*, **15**, 183.
12. Christensen, T., Dalbøge, H., and Snel, L (1991). In *Drug Biochemistry—Scientific Basis and Practices* (eds. Chiu,Y. Y. H. and Gueriguian J. L.), p. 206, Marcel Dekker, New York.
13. Wallenfels, K. and Malhotra, O. P. (1961). *Adv. Carbohydr. Chem.* **16**, 239.
14. Fink, A. L. and Bender, M. L. (1969). *Biochemistry* **8**, 5109.
15. Zmijewski, Jr, M. J., Briggs, B. S., Thompson, A. R., and Wright, I. G. (1991) *Tetrahedron Lett.* **32**, 1621.
16. Fersht, A. (1985). *Enzyme structure and mechanism*, Ch. 3. W. H. Freeman and Co., New York.
17. Kasche, V., Haufler, U., Markowsky, D., Melnyk, S., and Zeich, A. (1987). *Ann. N. Y. Acad. Sci.*, **501**, 97.
18. Kasche, V., Amneus, H., Gabel, D. and Näslund, L. (1977). *Biochim. Biophys. Acta*, **490**, 1.
19. Kasche, V., Galunsky, B. and Buchholz, K. (1981). *J. Chromatogr.* **216**, 169.
20. Ignatova. Z., Stoeva, S., Galunsky, B., Hörnle, C., Nurk, A., Piotraschke, E. *et al.* (1998). *Biotechnol. Lett.* **20**, 977.
21. Fasman, G. (ed.) (1975). In *Handbook of biochemistry and molecular biology proteins*, Vol. II, p. 211. CRC Press, Cleveland.
22. Schomburg, D. and Salzmann, M. (1991). *Enzyme handbook*, Vol. 5. Springer Verlag, Berlin. Update in Internet: http://www.brenda.uni-koeln.de/php/choose.php3
23. *Enzyme nomenclature* (1992). Academic Press, New York. Update in Internet: http://www.chem.qmw.ac.uk/iubmb/enzyme/EC34/intro.html
24. Lobell, M, and Schneider, M. P. (1998). http://www.ukc.ac.uk/bio/database/proteases/Entry.html
25. Riechmann, L. and Kasche, V. (1984). *Biochem. Biophys. Res. Commun.* **120**, 868.
26. Widmer, F., Breddam, K., and Johansen, J. T. (1981). *Carlsberg Res. Commun.* **46**, 97.
27. Schellenberger, V., Schellenberger, U., Mitin, Y. V., and Jakubke, H.-D. (1991). *Eur. J. Biochem.* **187**, 163.
28. Kasche, V., Michaelis, G., and Galunsky, B. (1991). *Biotechnol. Lett.* **13**, 75.
29. Michaelis, G. (1991). Thesis, Technical University Hamburg-Harburg.
30. Hanisch, U.-K., Könecke, A., Schellenberger, V. and Jakubke, H.-D. (1987). *Biocatalysis* **1**, 129.
31. Schellenberger, V., Schellenberger, U., Mitin, Y. V., and Jakubke, H.-D. (1991). *Eur. J. Biochem.* **179**, 161.
32. Schuster, M., Kasche, V., and Jakubke, H.-D. (1992). *Biochim. Biophys. Acta*, **1121**, 207.
33. Fortier, G. and MacKenzie, S. L. (1986). *Biotechnol. Lett.* **8**, 777.
34. Wilson, S.-A., Daniel, R. M., and Peek, K. (1994). *Biotechnol. Bioeng.*, **44**, 337.
35. Klingeberg, M., Galunsky, B., Sjöholm, C., Kasche, V., and Antranikian, G. (1995). *Appl. Environ. Microbiol.* **61**, 3098.

36. Böckle, B., Galunsky, B., Müller, R. (1995). *Appl. Enmviron. Microbiol.* **61**, 3705.
37. Cho, Y. K., and Northrop, D. B. (1998). *J. Biol. Chem.* **273**, 24305.
38. Stehle, P., Bahsitta, H.-P., Monetr, B., and Fürst, P. (1990). *Enzyme Microb. Technol.* **12**, 56.
39. Waldmann, H. (1994). *Chem. Rev.* **94**, 911.
40. Riechmann, L. and Kasche, V. (1985). *Biochim. Biophys. Acta* **830**, 164.
41. Woodley, J. M., and Tichener-Hooker, N. J. (1996). *Bioprocess Engin.* **14**, 263.
42. Breddam, K. (1985). *Carlsberg Res. Commun.* **50**, 309.
43. Grøn, H. and Breddam, K. (1992). *Biochemistry* **31**, 8967.
44. Schellenberger, V., Turck, C. W., and Rutter, W. J. (1994). *Biochemistry* **33**, 4251.
45. Stoineva, I. B. and Petkov, D. (1985). *FEBS Lett.* **183**, 103.
46. Fasman, G. (ed.). (1975). *Handbook of biochemistry and molecular biology physical and chemical data*, Vol. I, p. 318. CRC Press, Cleveland.
47. Kasche, V., Galunsky, B., Nurk, A., Piotraschke, E., and Rieks, A. (1996). *Biotechnol. Lett.* **18**, 455.
48. Kasche, V. and Zöllner, R. (1982). *Hoppe-Seyler's Z Physiol. Chem.* **363**, 531.
49. Goldberg, R. N., Tewari, Y. B., and Tung, M. (1998). *Thermodynamics of enzyme-catalyzed reactions*, http://imb4.carb.nist.gov:8800/enzyme/intro.html
50. Nilsson, K. and Mosbach, K. (1984). *Biotechnol. Bioeng.* **26**, 1146.
51. Grant, N. H. (1966). *Nature*, **212**, 194.
52. Schuster, M., Aaviksaar, A., and Jakubke, H.-D. (1990). *Tetrahedron* **46,** 8093.
53. Schellenberger, V., Jakubke, H.-D., and Kasche, V. (1991). *Biochim. Biophys. Acta* **1078**, 8.
54. Halling, P. J. (1994). *Enzyme Microb. Technol.* **16**, 178.
55. Homandberg, G. A., Mattis, J. A. and Laskowski, M., (1978). *Biochemistry*, **17**, 5220.
56. Gill, I., and Vulfson, E. N. (1994). *TIBTECH* **12**, 118.
57. Kasche, V., and Galunsky, B. (1995). *Biotechnol. Bioeng.* **45**, 261.
58. Erbeldinger, M., Ni, X., and Halling, P. (1998). *Enzyme Microb. Technol.* **18**, 162.
59. Tischer, W., and Kasche, V. *TIBTECH*, in press.
60. Spieß, A., Schlothauer, R., Hinrichs, J., Scheidat, B., and Kasche, V. (1999). *Biotechnol. Bioeng.*
61. Kasche, V. (1983). *Enzyme. Microb. Technol.* **5**, 2.
62. Woodward, J. (ed.) (1995) *Immobilized Cells and Enzymes: A Practical Approach*. IRL Press, Oxford.
63. Barret, A. J., Rawlings, N. D. and Woessner, J. F. (eds.) (1998) *Handbook of Proteolytic Enzymes*. Academic Press, London.
64 Liese, A., Seelbach, K. and Wanebrey, C. (eds.) (2000) *Industrial Bio-transformations*, Wiley-VCH, Weinheim.

Appendix I

The Schechter and Berger nomenclature for proteinase subsites

R. J. Beynon
Department of Veterinary Preclinical Sciences, University of Liverpool, PO Box 147, Liverpool L69 3BX, UK

J. S. Bond
Department of Biochemistry and Molecular Biology, The Pennsylvania State University, Hershey, PA17033–0850, USA

In 1967, Schechter and Berger (1) introduced a system of nomenclature to describe the interaction of proteases and their substrates which is widely used in the protease literature. In this system, the binding site for a polypeptide substrate on a protease is envisioned as a series of subsites; each subsite interacts with one amino acid residue of the substrate. By convention, the substrate amino acid residues are called P (for peptide); the subsites on the protease that interact with the substrate are called S (for subsite). The subsites are in the catalytic or active site of the protease. The amino acid residues on the amino-terminal side of the scissile bond (bond that is cleaved on the substrate) are numbered P_1, P_2, P_3... counting outwards (*Figure 1*); the residues on the *C*-terminal side of the

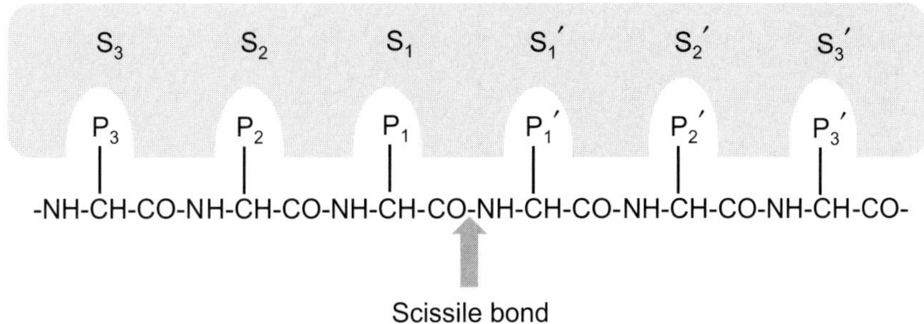

Figure 1 The Schechter and Berger nomenclature for binding of a peptide substrate to a peptidase. The protease is represented as the shaded area. P_1, P_1' are the side-chains of six amino acids, and S_1, S_1' are the corresponding subsites on the enzyme.

scissile bond are numbered P_1', P_2', P_3... The residues can be numbered up to P_6 on each side of the scissile bond. The subsites on the protease are termed S_3, S_2, S_1, S_1', S_2', S_3'... to complement the substrate residues that interact with the enzyme.

For some proteases, the residue interacting with the S_1 binding site is the primary determinant of binding, for example the S_1 subsite of trypsin has a marked preference for the binding of basic amino acid residues. For other proteases (e.g. renin) interactions with other subsites on the enzyme are critical for substrate binding.

Reference

1. Schechter, I. and Berger, A. (1967). *Biochem. Biophys. Res. Commun*, **27**, 157.

Appendix II
Some commercially available proteases

Jill M. Bankus and Judith S. Bond
Department of Biochemistry and Molecular Biology, The Pennsylvania State University, Hershey, PA 17033–0850, USA

Achromopeptidase (1)

Synonyms: lysyl endopeptidase, API, *Achromobacter* proteinase I

EC Number: EC 3.4.21.50

Type: serine endopeptidase (chymotrypsin analogue)

Primary specificity: P_1–P_1' (Lys is preferred for P_1')

Source: *Achromobacter lyticus* M497-1

Substrate: denatured casein, Boc-Val-Leu-Lys-AMC

Specific activity: 4 U/mg

Stability/storage: stable at pH 4.0–10.0 and can be stored at −20 °C in 1–100 mM Tris-HCl buffer at pH 6–10 at enzyme concentrations higher than 0.1 mg/ml

Suggested conditions for use: 50 mM Tris, pH 9.0, at 40 °C

Inhibitors: Zn^{2+}, Ba^{2+}, monovalent cations, DFP, PMSF, Tos-Lys-CH_2Cl (irreversible), 2-Leu-Leu-lysinol (reversible)

Notes: also hydrolyses lysyl amides and esters, stable pH range 4.0–10.0

Acylaminoacyl-peptidase (2,3)

Synonyms: acylamino-acid-releasing enzyme, *N*-acylpeptide hydrolase, *N*-formylmethionine (fMet) aminopeptidase

EC Number: EC 3.4.19.1

Type: ω-peptidase (serine-type)

Primary specificity: Ac-P_1-P_1'- (best if P_1 = Ser, Ala, Met; poor if P_1 = Gly, Tyr, Asp, Asn or Pro) optimal peptide length is 2–3 amino acids

Source: porcine liver

Substrate: Ac-Met-Ala-*p*-nitroanilide (NA), in Tris-HCl, pH 7.5

Specific activity: 13.0 U/mg protein

Stability/storage: store lyophilized at −20 °C

Suggested conditions for use: 0.25 M phosphate buffer or 0.1 M Bis-Tris, pH 7.5 at 37 °C

Inhibitors: DFP (not PMSF), *p*-hydroxymercuribenzoate, diethyl pyrocarbonate, Hg^{2+}, Zn^{2+}, or Cd^{2+}

Notes: liberates *N*-acetyl or *N*-formyl amino acids from proteins and peptides

Aspergillopepsin I (4,5)

Synonyms: *Aspergillus* acid proteinase, Aspergillopeptidase Molsin

EC Number: EC 3.4.23.18

Type: aspartic endopeptidase

Primary specificity: broad specificity, favours hydrophobic residues in P_1 and P_1' but also accepts Lys in P_1

Source: *Aspergillus saitoi*

Substrate: haemoglobin, pH 2.8 at 37 °C

Stability/storage: store at −20 °C

Inhibitors: *Streptomyces* pepsin inhibitor, pepstatin

Calpain (6–8)

Synonyms: CANP

EC Number: EC 3.4.22.17

Type: cysteine endopeptidase

Primary specificity: P_2-P_1-P_1' (P_1 = Tyr, Met or Arg, P_2 = Leu or Val)

Source: rabbit-skeletal muscle

Substrate: *N,N*-dimethylated casein

Stability/storage: store at −20 °C

Suggested conditions for use: pH 7.5, 30 °C

Inhibitors: E-64, IAA, an endogenous inhibitor exists (calpastatin)

Notes: several forms of calpain exist with varying sensitivity to Ca^{2+}. μ-Calpain has high Ca^{2+} sensitivity in the micromolar range; *m*-calpain has low Ca^{2+} sensitivity in the millimolar range; other forms with intermediate sensitivity to Ca^{2+} exist

Carboxypeptidase A (9,10)

Synonyms: carboxypolypeptidase, peptidyl-L-amino acid hydrolase, tissue or pancreatic carboxypeptidase A

EC Number: EC 3.4.17.1

Type: metalloexopeptidase or metallocarboxypeptidase

Primary specificity: -P_1-P_1'- (little or no activity when P_1' is Asp, Glu, Arg, Lys, Pro)

Source: bovine pancreas

Substrate: hippuryl-L-Phe

Specific activity: 840 U/mg protein

Stability/storage: stable at 4°C

Suggested conditions for use: 5 mM HEPES, pH 7.5 at 21°C

Inhibitors: EDTA, other Zn^{2+} chelators, 2-Bz-mercaptopropionic acid, Cu^{2+}, Pb^{2+}, Fe^{3+}, citrate, P_i, oxalate ions

Notes: soluble in 2.4 M LiCl or 1 M NaCl; inactivated by freezing, pH < 6.5, or > 7 M urea; stable in 4 M urea. Forms of the enzyme that have been treated with DFP or PMSF to eliminate trypsin and chymotrypsin are available; insoluble forms of the enzyme attached to agarose are also available from Sigma

Carboxypeptidase B (11–13)

Synonyms: peptidyl-L-lys (L-arg)-hydrolase, tissue or pancreatic carboxypeptidase B, protaminase

EC Number: EC 3.4.17.2

Type: metalloexopeptidase listed as metallocarboxypeptidase

Primary specificity: -P_1-P_1' ($P_{1'}$ = basic amino acids only, e.g. Lys, Arg, ornithine, homoarginine; P_1 = non-specific)

Source: porcine pancreas

Substrate: hippuryl-L-Arg, pH 7.5, 25 °C

Specific activity: 150 U/mg protein

Stability/storage: store at 4 °C

Suggested conditions for use: pH 7–9

Inhibitors: EDTA, other Zn^{2+} chelators, heavy metals

Notes: stable in 1 mg/ml SDS and 1 M urea

Carboxypeptidase C (14,15)

Synonyms: Carboxypeptidase W or Y, Cathepsin A, Peptidyl-L-amino acid hydrolase, Serine carboxypeptidase I, Lysosomal protective protein, Deamidase, Lysosomal carboxypeptidase A, Phaseolin

EC Number: EC 3.4.16.5

Type: serine-type carboxypeptidase

Primary specificity: broad specificity: P_2-P_1-P_1' (P_1 = aromatic or bulky aliphatic residue preferred; $P_{1'}$ not HOPro, Gly disfavoured)

Source: yeast

Substrate: Z-Leu-Phe

Specific activity: 20 U/mg

Stability/storage: store at −20 °C

Suggested conditions for use: 0.1 M sodium citrate, pH 5.3 at 30 °C

Inhibitors: dIFP (diisopropylfluorophosphate), sensitive to thiol blocking reagents

Cathepsin B (16–19)

Synonyms: cathepsin B1

EC Number: EC 3.4.22.1

Type: cysteine endopeptidase

Primary specificity: for synthetic substrates, prefers -Arg-P_1'-; with protein substrates may act as a peptidyldipeptidase, cleaving dipeptides from the *C*-terminus

Source: human liver

Substrate: Bz-Phe-Arg-NMec, pH 5.5, 37 °C

Stability/storage: stable at −80 °C

Suggested conditions for use: pH 3.5–6.0

Inhibitors: sulphydryl reagents, e.g. IAA, Z-Phe-Phe-CHN_2, leupeptin, α_2-macroglobulin, E-64

Notes: a lysosomal and secreted enzyme; a thiol activator, e.g. 2 mM cysteine, should be incubated with the enzyme prior to adding substrate

Cathepsin D (20–22)

EC Number: EC 3.4.23.5

Type: aspartic endopeptidase

Primary specificity: similar to pepsin

Source: bovine spleen

Substrate: denatured haemoglobin, pH 3.0, 37 °C

Specific activity: 10 U/mg protein

Stability/storage: store at −80 °C

Suggested conditions for use: pH 3–5

Inhibitors: pepstatin A

Notes: a lysosomal and granular/secreted enzyme

Cathepsin G (23)

EC Number: EC 3.4.21.20

Type: serine endopeptidase

Primary specificity: homologue of chymotrypsin

Source: human neutrophil

Substrate: MeO-Suc-Ala-Ala-Pro-Phe-4-NA, pH 7.5, 25 °C

Specific activity: 4 U/mg protein

Stability/storage: stable at −80 °C

Suggested conditions for use: pH 7.5

Inhibitors: DFP, Z-Phe-CH_2Br, chymostatin, 3-alkoxy-4-chloro-7-isocoumarin

Notes: from granules of polymorphonuclear leukocytes

Cathepsin H (24,25)

Synonyms: cathepsin B_3

EC Number: EC 3.4.22.16

Type: cysteine endopeptidase

Primary specificity: P_1-P_1' (P_1 = Arg)

Source: human liver

Substrate: Z-Phe-Arg-NMec

Stability/storage: store at −20 °C

Suggested conditions for use: 0.2 M KH_2PO_4, 0.2 M $NaHPO_4$, pH 6.8, 40 mM cysteine (fresh)

Inhibitors: IAA, PCMB, E-64

Chymopapain (26)

Synonyms: papaya proteinase II

EC Number: EC 3.4.22.6

Type: cysteine endopeptidase

Primary specificity: broad-specificity similar to papain

Source: papaya

Substrate: BAEE, pH 6.2 at 25 °C

Specific activity: 0.5–2.0 U/mg

Stability/storage: store at −20 °C

Suggested conditions for use: 0.1 M $NaPO_4$, pH 6/10 mM cysteine/1 mM EDTA at 37 °C

Inhibitors: E-64, IAA

Chymotrypsin (27,28)

Synonyms: chymotrypsin A or B, α-chymotrypsin

EC Number: EC 3.4.21.1

Type: serine endopeptidase

Primary specificity: -P_1-P_1'- (P_1 = aromatic, Trp, Tyr, Phe; P_1' = non-specific)

Source: bovine pancreas

Substrate: BTEE or ATEE substrates; 25 °C, pH 7.8

Specific activity: 40–90 U/mg, 1 U enzyme will hydrolyse 1 μmol BTEE/min

Stability/storage: stable at 4 °C

Suggested conditions for use: pH 7.5–8.5, Ca^{2+}-activated

Inhibitors: DFP, PMSF, TPCK, α_1-antitrypsin, aprotinin, α_2-macroglobulin

Notes: active in 1 mg/ml SDS; Fluka and Sigma also have TLCK-treated forms available (to inactivate traces of trypsin); forms that are attached to agarose beads or carboxymethyl cellulose are also available from Sigma; β-, γ- and δ-chymotrypsin are also available from Sigma

Clostripain (29,30)

Synonyms: clostridiopeptidase B

EC Number: EC 3.4.22.8

Type: Cysteine endopeptidase

Primary specificity: -Arg-P_1'- (P_1', Pro preferred)

Source: *Clostridium histolyticum*

Substrate: BAEE hydrolysis, pH 7.6, 25 °C

Specific activity: 20–100 U/mg protein

Stability/storage: stable at 4 °C

Suggested conditions for use: pH 7.1–7.6

Inhibitors: sulphydryl reagents. iodoacetamide, E-64 (reversible)

Notes: the enzyme must be activated for 2–3 h before use in 2.5 mM DTT containing 1.0 mM calcium choride

Coagulation factor Xa (31–34)

Synonyms: factor $X_{A,}$, thrombokinase, prothrombase, prothrombinase

EC Number: EC 3.4.21.6

Type: serine endopeptidase

Primary specificity: P_2-Arg-P_1'- (cleaves between Arg and P_1'; P_1 = usually Gly; P_1' = non-specific but Ile and Thr preferred)

Source: bovine, human plasma

Substrate: Bz-Ile-Glu-Gly-Arg-4-Na, 25 °C

Specific activity: 1.3 U/mg protein, enzyme stabilized with bovine serum albumin

Stability/storage: stable at 4 °C

Suggested conditions for use: pH 8.3, 25 °C

Inhibitors: antithrombin III (with heparin), DFP, PMSF, soybean trypsin inhibitor

Notes: activates prothrombin, Factor VII, Factor V and Factor IX

Dipeptidyl-peptidase I (35–36)

Synonyms: cathepsin C, dipeptidyl peptidase I, dipeptidyl transferase

EC Number: EC 3.4.14.1

Type: cysteine exopeptidase listed as dipeptidyl peptidase

Primary specificity: P_2-P_1-P_1'-P_2' (P_1 = non-specific, but Phe, Leu, Tyr preferred, P_1' = Gly, Pro preferred; poor if P_1' = Lys, Arg)

Source: beef spleen

Substrate: Gly-Phe-4-NA, 37°C

Specific activity: 3 U/mg

Stability/storage: store at −20°C

Suggested conditions for use: pH 4–6

Inhibitors: IAA, other sulphydryl reagents

Notes: removes *N*-terminal dipeptide at pH 4–6; acts as a transpeptidase at pH 7–8

Dipeptidyl-peptidase IV (37)

Synonyms: lymphocyte antigen CD26, DPP IV, dipeptidyl aminopeptidase IV, Xaa-Pro-dipeptidyl-aminopeptidase, Gly-Pro naphthylamidase, post-proline dipeptidyl aminopeptidase IV, glycoprotein GP110

EC Number: EC 3.4.14.5

Type: dipeptidyl peptidase

Primary specificity: P_2-P_1-P_1' (P_1 = Pro, P_1' not Pro or HOPro, P_2 = *N*-terminus)

Source: porcine kidney, human placenta

Substrate: Gly-Pro-4-nitroaniline, pH 8.0 at37°C

Specific activity: 10 U/mg

Stability/storage: store at −20 °C

Suggested conditions for use: 0.3 M Glycine-NaOH, pH 8.7 at 37 °C

Inhibitors: DFP, PMSF, DCC

Endopeptidase K (38)

Synonyms: proteinase K, *Tritirachium* alkaline proteinase

EC Number: EC 3.4.21.64

Type: serine endopeptidase

Primary specificity: -P_1-P_1'- (P_1 = non-specific but aromatic or hydrophobic amino acids preferred)

Source: *Tritirachium album*

Substrate: ATEE, pH 9.0, 30 °C

Specific activity: 10–300 U/mg protein

Stability/storage: stable at 4 °C. Stored in 50 mM Tris-HCl, pH 8.0, containing 1 mM $CaCl_2$ stable for months at 4 °C

Suggested conditions for use: pH 7.5–12

Inhibitors: PMSF, DFP

Notes: a homologue of subtilisin; enzyme is capable of cleaving native proteins, used to release nucleic acids; enzyme is not inactivated by metal chelators, sulphydryl reagents, TLCK or TPCK

Endoproteinase Arg-C (39)

Synonyms: submandibular proteinase A, esteroprotease

EC Number: no assigned EC number

Type: serine endopeptidase

Primary specificity: -Arg-P_1'-

Source: mouse submaxillaris gland

Substrate: Tos-Arg-OMe, pH 8.0, 37 °C

Specific activity: 100–225 μmol/min.mg protein

Stability/storage: stable dry at 4 °C; in solution at −20 °C

Suggested conditions for use: 0.1 M NH_4CO_3, pH 8.0–8.5, 25–37 °C

Inhibitors: DFP, TLCK, Hg^{2+}, Cu^{2+}, Zn^{2+}, α_2-macroglobulin

Notes: the enzyme loses specificity during long incubations; digestions proceed more rapidly if substrate is denatured in 5 M urea or 1 mg/ml SDS. Endoproteinase Arg-C retains 90% activity after 1 h at 25°C in 1 mg/ml SDS. Unstable in acid

Enteropeptidase (40,41)

Synonyms: enterokinase

EC Number: EC 3.4.21.9

Type: serine endopeptidase

Primary specificity: $(Asp)_4Lys_6$-Ile_7 bond in trypsinogen

Source: recombinant

Substrate: trypsinogen

Stability/storage: store at −20 °C in 50% glycerol

Suggested conditions for use: 20 mM Tris-HCl/pH 7.4, 50 mM NaCl at 23 °C

Inhibitors: PMSF, DFP, not inhibited by protein inhibitors of trypsin

Notes: activates trypsinogen to trypsin

Ficain (42)

Synonyms: ficin

EC Number: EC 3.4.22.3

Type: cysteine endopeptidase

Primary specificity: similar to papain

Source: from fig latex

Substrate: casein, pH 7.0 at 37 °C

Specific activity: 1–2 U/mg

Stability/storage: store at −20 °C. Loses activity after prolonged periods of freezing, lyophilization of ficin in water or volatile salts causes inactivation

Inhibitors: mercuric chloride, NEM, chloroacetamide, IAA

Gelatinase A (43,44)

Synonyms: MMP-2, 72-kDa gelatinase, Type IV collagenase

EC Number: EC 3.4.24.24

Type: metallo-endopeptidase

Primary specificity: Pro-Gln-Gly+Ile-Ala-Gly-Gln

Source: human

Substrate: gelatin type I, azocoll

Specific activity: 12 000 U/mg (1 U cleaves 1 μg gelatin/min at 37°C)

Suggested conditions for use: 50 mM Tris/pH 8.5, 0.15 M NaCl at 37°C

Inhibitors: EDTA, 1,10-phenanthroline, TIMP, glycine, SDS, cysteine

γ-Glu-X carboxypeptidase (45–47)

Synonyms: γ-glutamyl hydrolase, carboxypeptidase G, conjugase, folate conjugase, pteroyl-poly-γ-glutamate hydrolase, lysosomal γ -glutamyl carboxypeptidase

EC Number: EC 3.4.19.9

Type: ω-peptidase

Primary specificity: P_1-P_1' (P_1' = *C*-terminal amino acid, P_1 = γ-glutamic acid)

Source: *Pseudomonas* sp

Substrate: pteroyl triglutamate with *C*-terminal Glu carrying a ^{14}C label

Specific activity: 0.5 Ci/mol

Stability/storage: store at −0 °C (the human form of carboxypeptidase G requires zinc for stability)

Suggested conditions for use: 33 mM sodium acetate, pH 4.5, 37 °C

Inhibitors: Mn^{2+}, Cu^{2+}

Glutamyl endopeptidase (48,49)

Synonyms: endoproteinase Glu-C, Protease V8, staphylococcal serine proteinase

EC Number: EC 3.4.21.19

Type: serine endopeptidase

Primary specificity: P_4-P_3-P_2-P_1-P_1'-P_2' (P_1 = Glu or Asp, Glu preferred; Asp preferred at P_4; Ala or Val preferred at P_3; Phe preferred at P_2; Pro disfavoured at P_3, P_1', P_2'; Asp disfavoured at P_1')

Source: *Staphylococcus aureus* V8

Substrate: oxidized insulin B-chain digestion (HPLC analysis); CBZ-Phe-Leu-Glu-4-NA, pH 7.8, 25 °C (Fluka) *N*-t-BOC-L-Glu-α-phenyl ester, pH 7.8, 37 °C (Sigma)

Specific activity: 15 U/mg protein (Fluka); 500–1000 U/mg solid (Sigma)

Stability/storage: stable at 4 °C

Suggested conditions for use: see notes below

Inhibitors: DFP, α_2-macroglobulin, not inhibited by PMSF

Notes: specificity depends on incubation buffer. In ammonium bicarbonate buffer (50 mM) at pH 7.8 or ammonium acetate buffer (50 mM) at pH 4.0, the protease cleaves only Glu-$P_{1'}$ bonds. In phosphate buffer (50 mM) at pH 7.8, the protease will cleave Asp-$P_{1'}$ bonds as well as Glu-$P_{1'}$ bonds. If $P_{1'}$ = bulky hydrophobic residue, cleavage is slow. M_r of protease = 27 000. The protease is active in 0.1% SDS, 1.0 mol/l urea, 1.0 mol/l guanidine hydrochloride, and 10% acetonitrile. Insoluble enzyme attached to beads also available from Sigma

Leucyl aminopeptidase (50,51)

Synonyms: leucine aminopeptidase, leucyl peptidase, peptidase S, cytosol aminopeptidase, LAP, aminoacyl-peptide hydrolase (cytosol)

EC Number: EC 3.4.11.1

Type: Metallo-exopeptidase

Primary specificity: -P_1-P_1' (P_1, P_1' cannot be Lys, Arg; P_1 cannot be D-amino acid, prefers Leu)

Source: porcine kidney

Substrate: L-leucinamide, pH 8.5, 25–40 °C

Specific activity: 60–100 U/mg protein, 25 °C

Stability/storage: stable at 4 °C

Suggested conditions for use: pH 7.5–9.0

Inhibitors: Cd^{2+}, Hg^{2+}, other heavy metals, EDTA, citrate, bestatin

Notes: membrane alanyl aminopeptidase, EC 3.4.11.2, is available commercially and is listed as Aminopeptidase M in this appendix

Membrane alanyl aminopeptidase (52–54)

Synonyms: aminopeptidase M or N, alanine aminopeptidase, Amino acid arylamidase, microsomal aminopeptidase, particle-bound aminopeptidase, amino-oligopeptidase, membrane aminopeptidase I, pseudo leucine aminopeptidase, peptidase E

EC Number: EC 3.4.11.2

Type: Metallo-exopeptidase, an amino peptidase

Primary specificity: P_1-P_1'- (P_1, P_1' = non-specific; P_1 cannot be Pro or *N*-blocked amino acid)

Source: porcine kidney

Substrate: L-Leu-4-NA, 25 °C

Specific activity: 4 U/mg

Stability/storage: stable at 4 °C

Optimal conditions for use: pH 7.0–7.5, 37 °C, in 60 mM sodium phosphate buffer

Inhibitors: 1,10-phenanthroline (10 μM), acetone, alcohols, guanidinium chloride (0.5 M)

Notes: not inhibited by 6 M urea, DFP, PMSF or PCMB

Membrane Pro-X carboxypeptidase (55,56)

Synonyms: carboxypeptidase P, microsomal carboxypeptidase

EC Number: EC 3.4.17.16

Type: serine exopeptidase

Primary specificity: -P_1-P_1' (P_1' cannot be Ser, Gly)

Source: *Penicillium janthinellum*

Substrate: Z-Glu-Tyr, pH 3.7, 30°C

Specific activity: 50 U/mg protein

Stability/storage: stable at 4 °C

Suggested conditions for use: pH 3.7–5.2; stabilized by non-ionic detergents, Triton X-100, Tween

Inhibitors: PCMB, iodoacetic acid, DFP, SDS, diethylpyrocarbonate, fatty acids

Notes: a renal brush border membrane exopeptidase; stabilized at low protein and acidic pH values by Triton X-100

Microbial collagenase (57–59)

Synonyms: clostridiopeptidase A, collagenase I or A, *Clostridium histolyticum* collagenase, *Achromobacter iophagus* collagenase

EC Number: EC 3.4.24.3

Type: metalloendopeptidase

Primary specificity: -P_3-P_2-P_1-P_1'-P_2'-P_3'- (Gly is preferred for P_3 and P_1'; Pro or Ala is preferred for P_2 and P_2'; Ala, Arg or hydroxy-Pro is preferred for P_3')

Source: *Clostridium histolyticum*

Substrate: native collagen

Specific activity: 150–300 U/mg

Stability/storage: store at −20 °C

Suggested conditions for use: pH 7.4, 37 °C, in the presence of calcium ions

Inhibitors: EDTA

Notes: variants of this enzyme have been purified from several bacteria

Multicatalytic endopeptidase complex (60–62)

Synonyms: proteasome, MCP, Ingensin, macropain, prosome, a component of lens neutral proteinase

EC Number: EC 3.4.99.46

Type: threonine endopeptidase

Primary specificity: P_1-P_1' (P_1 = hydrophobic, basic, or acidic side-chain)

Source: *M. thermophila* recombinant enzyme

Substrate: Suc-Ala-Ala-Phe-AMC (chymotrypsin-like activity), CBZ-Leu-Leu-Glu-β-NA (peptidylglutamyl-peptide hydrolase activity)

Specific activity: 1 mg protein yields 1.2 nmol AMC/min and 8.9 nmol β-naphthylamine/min

Stability/storage: store at 0 °C

Suggested conditions for use: 10 mM Tris-HCl, pH 7.5/1 mM EDTA at 25 °C

Inhibitors: lactacystin, mercurial reagents, some inhibitors of serine proteases

Notes: a multisubunit protein abundant in the cytoplasm of most eukaryotic cells containing at least three distinct catalytic sites: trypsin-like, chymotrypsin-like, and peptidylglutamyl peptide-hydrolysing activities; this particular product from ICN is prokaryotic in origin and contains only two enzymatic activities: chymotrypsin-like, and peptidylglutamyl peptide-hydrolysing activities

Pancreatic elastase (63,64)

Synonyms: elastase, pancreatopeptidase e, pancreatic elastase I

EC Number: EC 3.4.21.36

Type: serine endopeptidase

Primary specificity: $-P_1-P_1'-$(P_1 = uncharged, non-aromatic, e.g. Ala, Val, Leu, Ile, Gly, Ser; P_1' = non-specific)

Source: porcine pancreas, human neutrophil

Substrate: *N*-acetyl-trialanyl-methyl ester, or Suc-(Ala)3–4-NA, pH 7.8, 25 °C

Specific activity: 12–70 U/mg

Stability/storage: stable at 4 °C

Suggested conditions for use: pH 7.8–8.5

Inhibitors: DFP, α_2-macroglobulin, elastinal, α_1-antitrypsin

Notes: the enzyme sticks to glass; use plastic vessels wherever possible and add 0.05% Triton X-100 to incubation solutions

Papain (65,66)

Synonyms: papaya peptidase I

EC Number: EC 3.4.22.2

Type: cysteine endopeptidase

Primary specificity: $P_2-P_1-P_1'-$ (P_1 = non-specific but Arg, Lys preferred; P_1' = not Val; P_2 prefers bulky hydrophobic)

Source: *Carica papaya*

Substrate: B_3-L-Arg-ethyl ester, pH 6.2, 25 °C

Specific activity: 3–30 U/mg protein

Stability/storage: stable at 4 °C; activity decreases about 20% in 6 months

Suggested conditions for use: pH 6.0–7.0

Inhibitors: sulphydryl-blocking reagents, heavy metal ions, carbonyl derivatives and ascorbic acid, leupeptin, PMSF, TPCK, TLCK, E-64, α_2-macroglobulin

Notes: also available as insoluble enzyme attached to agarose or carboxymethyl cellulose from Sigma

Pepsin A (67,68)

Synonyms: pepsin

EC Number: EC 3.4.23.1

Type: aspartic endopeptidase

Primary specificity: -P_1-P_1'- (P_1 = non-specific but aromatic and other hydrophobic residues preferred, especially Phe, Leu; P_1, P_1' cannot be Val, Ala, Gly)

Source: porcine gastric mucosa

Substrate: denatured haemoglobin (69)

Specific activity: 2500–3800 U/mg; 1 U liberates sufficient tyrosine to produce an increase in absorbance of TCA-soluble products 0.001/min at 280 nm

Stability/storage: stable at 4 °C

Suggested conditions for use: pH 2–4

Inhibitors: pepstatin A, diazoketones, phenylacyl bromides, aliphatic alcohols, pH > 6

Peptidyl-Asp metalloendopeptidase (70,71)

Synonyms: endoproteinase Asp-N

EC Number: EC 3.4.24.33

Type: metallo-endopeptidase

Primary specificity: P_1-P_1' (P_1' = Asp or Cys)

Source: *Pseudomonas fragi*

Substrate: digestion of glucagon, azocoll, casein or haemoglobin

Stability/storage: stable in lyophilized form at 4 °C; solution stable 1 month at −20 °C; 1 week at 4°C; freeze/thawing is not recommended

Suggested conditions for use: for glucagon: 0.05 M sodium phosphate, pH 8.0, 37 °C; for proteins: 0.2 M Tris/20 mM $CaCl_2$, pH 7.8, 37 °C

Inhibitors: EDTA, 1,10-phenanthroline

Notes: the enzyme is active in 0.01% (w/v) SDS, 1.0 M urea, 0.1 M guanidine hydrochloride, 10% (v/v) acetonitrile; inactivate the enzyme by boiling at 100 °C for 5 min

Peptidyl-dipeptidase A (72)

Synonyms: ACE, kininase II, dipeptidyl carboxypeptidase I; peptidase P, carboxycathepsin, dipeptide hydrolase

EC Number: EC 3.4.15.1

Type: peptidyl dipeptidase

Primary specificity: P_1-P_1'-P_2' (P_1' not Pro, P_2' not Asp, Glu)

Source: porcine and human kidney

Substrate: 5 mM Hippuryl-His-Leu

Stability/storage: store at −20 °C

Suggested conditions for use: 100 mM potassium phosphate, pH 8.3, 0.3 M NaCl, 37 °C

Inhibitors: captopril, lisinopril, *o*-phenanthroline, EDTA

Notes: removes *C*-terminal dipeptides from substrates such as angiotensin I and bradykinin; Cl- and Zn-dependent

Plasma kallikrein (73,74)

Synonyms: kininogenin, serum kallikrein

EC Number: EC 3.4.21.34

Type: serine endopeptidase

Primary specificity: P_1-P_1' (P_1 = Arg, Lys where Arg is preferred over Lys, in small molecular substrates)

Source: human plasma, porcine pancreas

Substrate: BAEE, pH 8.7, 25 °C

Specific activity: 5 U/mg

Stability/storage: store at −20 °C

Suggested conditions for use: pH 8.7, 25 °C

Inhibitors: selective synthetic inhibitor D-Ile-Phe-Arg-CH_2Cl

Notes: formed from plasma prokallkrein by Factor XIIA; activates Factors XII, VII and plasminogen; cleaves kininogen to bradykinin

Plasmin (75–77)

Synonyms: fibrinolysin, fibrinase

EC Number: EC 3.4.21.7

Type: serine endopeptidase

Primary specificity: -P_1-P_1' (P_1 = Lys or Arg)

Source: bovine and human plasma

Substrate: α-casein, pH 7.5 at 37 °C

Stability/storage: lyophilized form stable at least 3 months at −20 °C

Suggested conditions for use: pH 8.9

Notes: converts fibrin into soluble products, also active in other systems

t-Plasminogen activator (78–81)

Synonyms: tissue plasminogen activator, tPA

EC Number: EC 3.4.21.68

Type: serine endopeptidase

Primary specificity: P_1-P'_1 (P_1 = Arg, P'_1 = Val)

Source: human melanoma cells

Substrate: plasminogen, D-Val-Leu-Lys-pNA

Stability/storage: 1000 U/ml stable for months at −70°C, use fresh working dilutions daily

Suggested conditions for use: pH 7.5

Inhibitors: DFP, DCC, PMSF

Notes: activity is enhanced by fibrin

u-Plasminogen activator (82–84)

Synonyms: urinary plasminogen activator, uPA, urokinase, cellular plasminogen activator

EC Number: EC 3.4.21.73

Type: serine endopeptidase

Primary specificity: P_1-P'_1 (P_1 = Arg, P'_1 = Val)

Source: human kidney cells

Substrate: plasminogen

Stability/storage: store at 4 °C

Suggested conditions for use: pH 7.5, 37 °C

Inhibitors: DFP, DCC < PMSF

Notes: cleaves plasminogen to form plasmin; does not bind fibrin as does tPA

Pyroglutamyl-peptidase I (85,86)

Synonyms: pyroglutamyl aminopeptidase, 5-oxoprolyl-peptidase, pyrrolidone-carboxylate peptidase

EC Number: EC 3.4.19.3

Type: cysteine exopeptidase

Primary specificity: P_1-P'_1 (P_1 = 5-oxoproline or pyroglutamic; P'_1 = non-specific but Ala preferred; no hydrolysis if P'_1 = Pro)

Source: calf and porcine liver

Substrate: pyroGlu-NHNap

Specific activity: 4 mU/mg, 1 U hydrolyses 1 μmol of substrate/min

Stability/storage: stable at 4 °C dry; stability enhanced by addition of sucrose and EDTA; reconstitute in solutions containing 5 mM DTT and 10 mM EDTA and store frozen

Suggested conditions for use: pH 7–9, 37 °C in phosphate buffer containing 2 mM DTT and 2 mM EDTA

Inhibitors: IAA, PCMB, *p*-chloromercuribenzoate, DTNB (5,5′-dithiobis(2-nitrobenzoic acid)

Notes: requires at least one free sulphydryl group for activity; formerly this enzyme was classified as EC 3.4.11.8

Renin (87)

Synonyms: angiotensin-forming enzyme, angiotensinogenase

EC Number: EC 3.4.23.15

Type: aspartic endopeptidase

Primary specificity: P_1-P_1' (P_1 = Leu in angiotensin)

Source: porcine kidney

Substrate: angiotensinogen, *N*-acetyltetradecapeptide

Specific activity: 1000 U/mg where 1 unit = amount renin required to raise dog blood pressure by 30 mM Hg

Stability/storage: stable at −70°C for over 1 year

Suggested conditions for use: pH 6–7

Inhibitors: pepstatin

Notes: cleaves angiotensinogen to generate angiotensin I; formed from prorenin in plasma and kidney

Retropepsin (88,89)

Synonyms: HIV-1 protease, AIDS protease

EC Number: EC 3.4.23.16

Type: aspartic endopeptidase

Primary specificity: P_1-P_1' (P_1 = hydrophobic, P_1' = variable, but often Pro)

Source: recombinant

Substrate: HIV substrate III

Stability/storage: −70°C

Suggested conditions for use: 50 mM PIPES, pH 6.5/150 mM NaCl/0.2% Triton X-100/1 mM EDTA at 37°C

Inhibitors: pepstatin A

Stem bromelain (90,91)

Synonyms: bromelain

EC Number: EC 3.4.22.32

Type: cysteine endopeptidase

Primary specificity: P_1 and P_2 = Arg for small substrates

Source: *Ananas comosas* (pineapple stem)

Substrate: Z-Arg-Arg-NHMec

Specific activity: $K_m = 15.4\ \mu M$

Stability/storage: stable lyophilized and stored at −20 °C

Suggested conditions for use: 0.2 M Na_2HPO_4/0.2 M NaH_2PO_4/4 mM EDTA pH 6.8 (8 mM DTT or 16 mM cysteine added fresh) at 40 °C

Inhibitors: Hg^{2+}, NEM, IAA, other sulphydryl reagents; not inhibited by cystatin, slow inactivation by E-64

Subtilisin (92)

Synonyms: subtilopeptidase A or B, subtilisin Carlsberg, subtilisin BPN, Nagarse

EC Number: EC 3.4.21.62

Type: serine endopeptidase

Primary specificity: -P_1-P_1' (P_1 = non-specific but neutral and acidic amino acid preferred; P_1' = non-specific)

Source: *Bacillus subtilis*

Substrate: denatured haemoglobin degradation

Specific activity: 5–20 U/mg, 37 °C, casein or haemoglobin as substrate, pH 8.0

Stability/storage: stable at 4 °C, dry

Suggested conditions for use: pH 7.0–8.0

Inhibitors: PMSF, DFP, indole, aprotinin, α_2-macroglobulin

Notes: Ca^{2+} is an activator; stable at 100–200 mg/ml solution in 0.1 M borate, pH 8, containing 0.1 M $CaCl_2$ for 1–2 days at 4 °C; avoid phosphate buffers, as Ca^{2+} will precipitate as calcium phosphate; use Tris-HCl for higher buffering capacity. Subtilisin B (Nagarse proteinase, or subtilisin BPN) is another non-specific proteinase from *B. subtilis* and is available from Sigma

Thermolysin (93–95)

Synonyms: *Bacillus thermoproteolyticus* neutral proteinase

EC Number: EC 3.4.24.27

Type: metalloendopeptidase

Primary specificity: P_1-P_1'-P_2' (P_1' = Leu, Phe, Ile, Val, Met, Ala; P_1 = non-specific; P_2' cannot be Pro)

Source: *Bacillus thermoproteolyticus*

Substrate: casein, pH 7.2, 37°C

Specific activity: 40–100 U/mg protein

Stability/storage: stable at 4°C if dry

Suggested conditions for use: pH 7–9

Inhibitors: EDTA, 1,10-phenanthroline, Hg^{2+}, $AgNO_3$, α_2-macroglobulin, oxalate, citrate, P_i

Notes: stabilized by 2 mM Ca^{2+}, retains activity at 80°C

Thrombin (96,97)

Synonyms: coagulation factor IIa, fibrinogenase

EC Number: EC 3.4.21.5

Type: serine endopeptidase

Primary specificity: -P_1-P_1' (P_1 = Arg; P_1' = Gly preferred)

Source: human plasma

Substrate: fibrinogen

Stability/storage: stable at −70 °C for 1 year, susceptible to deterioration caused by repeated freeze–thaw cycles

Suggested conditions for use: 0.1 M Tris pH 7.5–8.5/10% DMSO at 25 °C

Inhibitors: antithrombin III, hirudin, PMSF, DFP

Notes: biotinylated thrombin also available from Novagen (for easy removal by binding to immobilized streptavidin); activates fibrinogen to fibrin

Tissue kallikrein (98,99)

Synonyms: urinary kallikrein, glandular kallikrein, pancreatic kallikrein, submandibular kallikrein, submaxillary kallikrein, kidney kallikrein

EC Number: EC 3.4.21.35

Type: serine endopeptidase

Primary specificity: P_1-P_1' (P_1 = Lys, P_1' = Arg OR P_1 = Met, Leu, P_1' = Lys

Source: porcine pancreas

Substrate: BAEE, pH 8.7, 25 °C

Specific activity: 40 U/mg

Stability/storage: stable for 2 years. at −20 °C, 6 months at 4 °C with 0.02% (w/v) sodium azide

Suggested conditions for use: pH 8.7, 25 °C

Notes: formed from tissue prokallikrein by activation with trypsin

Trypsin (100–102)

Synonyms: α- and β-trypsin

EC Number: EC 3.4.21.4

Type: serine endopeptidase

Primary specificity: P_1-P_1'- (P_1 = Lys, Arg; P_1' = non-specific)

Source: bovine pancreas

Substrate: oxidized insulin B-chain digestion (HPLC); BAEE, pH 7.6, 25 °C

Specific activity: 1000–13 000 U/mg protein with BAEE as substrate

Stability/storage: stable at 4 °C in lyophilized form

Suggested conditions for use: pH 8.5–8.8, stabilized at 20 mM Ca^{2+}

Inhibitors: TLCK, DFP, PMSF, leupeptin, soybean trypsin inhibitor, aprotinin, α_2-macroglobulin, APMSF, α_1-antitrypsin

Notes: active in protein denaturants such as SDS (0.1%), urea (1 M), guanidine hydrochloride (1 mol/l), acetonitrile (10%); TPCK-treated forms are available to inhibit trace amounts of chymotrypsin. Insoluble forms attached to polyacrylamide, agarose or glass mesh are also available from Sigma

References

1. Sakiyama, F. and Masaki, T. (1994). *Methods Enzymol.* **244**, 126.
2. Kobayushi, K. and Smith, J. A. (1987). *J. Biol. Chem.* **262**, 11435.
3. Jones, W. M., Scaloni, A., and Manning, J. M. (1994). *Methods Enzymol.* **244**, 227.
4. Tanaka, N., Takeuchi, M. and Ichishima, E. (1977). *Biochim. Biophys. Acta* **485**, 406.
5. Shintani, T., Kobayashi, M., and Ichishima, E. (1996). *J. Biochem.* **120**, 974.
6. Suzuki, K., Imajoh, S., Emori, Y., Kawasaki, H., Minami, Y., and Ohno, S. (1987). *FEBS Lett.* **220**, 271.
7. Suzuki, K., Sorimachi, H., Yoshizawa, T., Kinbara, K., and Ishiura, S. (1995). *Biol. Chem. Hoppe-Seyler* **376**, 523.
8. Saido, T. C., Sorimachi, H., and Suzuki, K. (1994). *FASEB J.* **8**, 814.
9. Petra, P. H. (1970). In *Methods in enzymology* (ed. Colowick, S. P. and Kaplan, N. O.), Vol. 19, p. 460. Academic Press, New York.
10. Bodwell, J. E. and Meyer, W. L. (1981). *Biochemistry* **20**, 2767.
11. Folk, J. E. (1970). In *Methods in enzymology* (ed. Colowick, S. P. and Kaplan, N. O.), Vol. 19, p. 504. Academic Press, New York.
12. Wallace, E. F., Evans, C. J., Jurik, S. M., Mefford, I. N., and Barchas J. D. (1982). *Life Sci.* **31**, 1793.
13. Bradley, G., Naude, R. J., Muramoto, K., Yamauchi, F., and Oelofsen, W. (1996). *Int. J. Biochem. Cell Biol.* **28**, 521.
14. Miller, J. J., Changaris, D. G. and Levy, R. S. (1992). *J. Chromatogr.* **627**, 153.
15. Satake, A., Itoh, K., Shimmoto, M., Saido, T. C., Sakuraba, H., and Suzuki, Y. (1994). *Biochem. Biophys. Res. Comm.* **205**, 38.
16. Barrett, A. J. and Kirschke, H. (1981). In *Methods in enzymology* (ed. Colowick, S. P. and Kaplan, N. O.), Vol. 80, p. 535. Academic Press, New York.
17. Barrett, A. J., Buttle, D. J., and Mason, R. W. (1988). *ISI Atlas Sci. Biochem.* **1**, 256.
18. Kirschke, H., Wikstrom, P., and Shaw, E. (1988). *FEBS Lett.* **228**, 128.
19. Berquin, I. M. and Sloane, B. F. (1996). *Adv. Exp. Med. Biol.* **389**, 281.
20. Takayuki, T. and Tang, J. (1980). In *Methods in enzymology* (ed. Colowick, S. P. and Kaplan, N. O.), Vol. 80, p. 565. Academic Press, New York.
21. Conner, G. E. (1989). *Biochem. J.* **263**, 601.
22. Yamamoto, K. (1995). *Adv. Exp. Med. Biol.* **362**, 223.
23. Powers, J. C. and Kam, C.-M. (1994). *Methods Enzymol.* **244**, 442.
24. Bromme, D. Bescherer, K., Kirsche, H., and Fittkau, S. (1987). *Biochem. J.* **245**, 381.
25. Sivaparvathi, N., Sawaya, R., Gokaslan, Z. L., Chintala, S. K., Rao, J. S., and Chintala, S. K. (1996). *Cancer Lett.* **104**, 121.

26. Buttle, D. J., Dando, P. M., Coe, P. F., Sharp, S. L., Shepherd, S. T., and Barrett, A. J. (1990). *Biochem. Hoppe-Seyler* **371**, 1083.
27. Okamoto, Y. and Sekine, T. (1985). *J. Biochem.*, **98**, 1143.
28. Lesk, A. M. and Fordham, W. D. (1996). *J. Mol. Biol.* **258**, 501.
29. Gilles, A.-M., Imhoff, J.-M., and Keil, B. (1979). *J. Biol. Chem.* **254**, 1462.
30. Rawlings, N. D. and Barrett, A. J. (1994). *Methods Enzymol.* **244**, 479.
31. Owen, W. G., Esmon, C. T., and Jackson, C. M. (1974). *J. Biol. Chem.* **249**, 594.
32. Cho, K., Tanaka, T., Cook, R. R., Kisiel, W., Fujikawa, K., and Powers, J. C. (1984). *Biochemistry* **23**, 644.
33. Walker, R. K. and Krishnaswamy, S. (1994). *J. Biol. Chem.* **269**, 27441.
34. Krishnaswamy, S., Nesheim, M. E., Pryzdial, E. L., and Mann, K. G. (1993). *Methods Enzymol.* **222**, 260.
35. Lynn, K. R. and Labow, R. S. (1985). *Can. J. Biochem. Cell Biol.* **62**, 1301.
36. Dolenc, I., Turk, B., Pungercic, G., Ritonja, A., and Turk, V. (1995). *J. Biol. Chem.* **270**, 21626.
37. Ikehara, Y., Ogata, S., and Misumi, Y. (1994). *Methods Enzymol.* **244**, 215.
38 Lebherz, H. G., Burke, T., Shackelford, J. E., Strickler, J. E., and Wilson, K. J. (1986). *Biochem. J.* **233**, 51.
39. Ahmed, S. A., Fairwell, T., Dunn, S., Kirschner, K., and Miles, E. W. (1986). *Biochemistry* **25**, 3118.
40. LaVallie, E. R., Rehemtulla, A., Racie, L. A., DiBlasio E. A., Ferenz, C., Grant, K. L. *et al.* (1993). *J. Biol. Chem.*, **268**, 23311.
41. Light, A. and Janska, H. (1989). *Trends Biochem. Sci.* **14**, 110.
42. Liener, I. E. and Friedenson, B. (1970). *Methods Enzymol.* **19**, 261.
43. Okada, Y., Morodomi, T., Enghild, J. J., Suzuki, K., Yasui, A., Nakanishi, I. *et al.* (1990). *Eur. J. Biochem.* **194**, 721.
44. Corcoran, M. L., Hewitt, R. E., Kleiner Jr, D. E., and Stetler-Stevenson, W. G. (1996). *Enzyme Protein* **49**, 7.
45. Silink, M., Reddel, R., Bethel, M., and Rowe, P. B. (1975). *J. Biol. Chem.* **250**, 5982.
46. Elsenhans, B., Ahmad, O., and Rosenberg, J. H. (1984). *J. Biol. Chem.* **259**, 6364.
47. Yao, R., Schneider, E., Ryan, T. J., and Galivan, J. (1996). *Proc. Natl. Acad. Sci. USA* **93**, 10134.
48. Houmard, J. and Drapeau, G. R. (1972). *Proc. Natl. Acad. Sci USA* **69**, 306.
49. Birktoft, J. J. and Breddam, K. (1994). *Methods Enzymol.* **244**, 114.
50. Bergmeyer, H. U. (1983). *Methods in enzymatic analysis*, Volume II, 3rd edn, p.234. Academic Press, New York.
51. Taylor, A (1993). *FASEB J.* 7, 290.
52. Pfleiderer, G. (1970). In *Methods in enzymology* (ed. Colowick, S. P. and Kaplan, N. O.), Vol. 19, p.514. Academic Press, New York.
53. Feracci, H. and Maroux, S. (1980). *Biochim. Biophys. Acta* **599**, 448.
54. Lendeckel, U., Wex, T., Reinhold, D., Kahne, T., Frank, K., Faust, J. *et al.* (1996). *Biochem. J.* **319**, 817.
55. Yokoyama, S., Oobayashi, A., Tanabe, O., and Ichishima, E. (1975). *Agr. Biol. Chem.* **39**, 1211.
56. Yokoyama, S., Oobayashi, A., Tanabe, O., and Ichishima, E. (1981). *Agr. Biol. Chem.* **45**, 311.
57. Bond, M. D. and van Wart, H. D. (1984). *Biochemistry* **23**, 3085.
58. VanWart, H. D. and Steinbrink, D. R. (1985). *Biochemistry* **24**, 6520.
59. Makinen, K. K. and Makinen, P.-L. (1987). *J. Biol. Chem.* **262**, 12488.
60. Fenteany, G., Standaert, R. F., Lane, W. S., Choi, S., Corey, E. J., and Schreiber, S. L. (1995). *Science* **268**, 726.

61. Savory, P. J., Djaballah, H., Angliker, H., Shaw, E., and Rivett, A. J. (1993). *Biochem. J.* **296**, 601.
62. Maupin-Furlow, J. A., and Ferry, J. G. (1995). *J. Biol. Chem.* **270**, 28617.
63. Grunnet, I. and Knudsen, J. (1983). *Biochem. J.* **209**, 215.
64. Bode, W., Meyer Jr, E. and Powers, J. C. (1989). *Biochemistry* **28**, 1951.
65. Cleveland, D. W., Fischer, S. G., Kirschner, M. W., and Laemmli, U. K. (1977). *J. Biol. Chem.*, **252**, 1102.
66. Rawlings, N. D. and Barrett, A. J. (1994). *Methods Enzymol.* **244**, 461.
67. Bergmeyer, H. U. (1984). *Methods enzymatic analysis*, Volume V, 3rd edn, p. 232. Academic Press, New York.
68. Hirasawa, A., Athauda, S. B., and Takahashi, K. (1996). *J. Biochem.* **120**, 407.
69. Anson, M. L. (1938). *J. Gen. Physiol.* **22**, 79.
70. Drapeau, G. R. (1980). *J. Biol. Chem.* **255**, 839.
71. Hagmann, M.-L., Geuss, U., Fischer, S, and Kresse, G.-B. (1995). *Methods Enzymol.* **248**, 782.
72. Corvol, P., Williams, T. A., and Soubrier, F. (1995). *Methods Enzymol.* **248**, 283.
73. Heimark, R. L. and Davie, E. W. (1981). *Methods Enzymol.* **80**, 157.
74. Silverberg, M. and Kaplan, A. P. (1988). *Methods Enzymol.* **163**, 85.
75. Robbins, K. C., Summaria, L., and Wohl, R. C. (1981). *Methods Enzymol.* **80**, 379.
76. Castellino, F. J. and Powell, J. R. (1981). *Methods Enzymol.*, **80**, 365.
77. Kolev, K., Lerant, I., Tenekejiev, K., and Machovich, R. (1994). *J. Biol. Chem.* **269**, 17030.
78. Loskutoff, D. J. and Schleef, R. R. (1988). *Methods Enzymol.* **163**, 293.
79. Verheijen, J. H. (1988). *Methods Enzymol.* **163**, 302.
80. Pryzdial, E. L., Bajzar, L., and Nesheim, M. E. (1995). *J. Biol. Chem.* **270**, 17871.
81. Mori, K., Dwek, R. A., Downing, A. K., Opdenakker, G., and Rudd, P. M. (1995). *J. Biol. Chem.* **270**, 3261.
82. Loskutoff, D. J. and Schleef, R. R. (1988). *Methods Enzymol.* **163**, 293.
83. Ragno, P., Montuori, N., Vassalli, J. D., and Rossi, G. (1993). *FEBS Lett.* **323**, 279.
84. Liu, J. N. and Gurewich, V. (1995). *J. Biol. Chem.* **270**, 8408.
85. Armentrout, R. W. and Doolittle, R. F. (1969). *Arch. Biochem. Biophys.* **132**, 80.
86. Cummins, P. M. and O'Connor, B. (1996). *Int. J. Biochem. Cell Biol.* **28**, 883.
87. Slater, E. E. (1981). *Methods Enzymol.* **80**, 427.
88. Seelmeier, S., Schmidt, H., Turk, V., and von der Helm, K. (1988). *Proc. Natl. Acad. Sci. USA* **85**, 6612.
89. Lin, X. L., Lin, Y. Z., and Tang, J. (1994). *Methods Enzymol.* **241**, 3.
90. Rowan, A. D. and Buttle, D. J. (1994). *Methods Enzymol.* **244**, 555.
91. Rowan, A. D., Buttle, D. J., and Barrett, A. J. (1990). *Biochem. J.* **266**, 869.
92. Serrano, L., Avila, J., and Maccioni, R. B. (1984). *Biochemistry* **23**, 4675.
93. Price, N. C., Duncan, D., and McAlister, J. W. (1985). *Biochem. J.* **229**, 167.
94. Matthews, B.W (1988). *Acc. Chem. Res.* **21**, 333.
95. Yang, J. J. and van Wart, H. E. (1994). *Biochemistry* **33**, 6508.
96. Lawson, J. H., Kalafatis, M., Stram, S. and Mann, K. G. (1994). *J. Biol. Chem.* **269**, 23357.
97. Okamoto, S. and Hijikata-Okunomiya, A. (1993). *Methods Enzymol.* **222**, 328.
98. Geiger, R. and Miska, W. (1988). *Methods Enzymol.* **163**, 102.
99. Fiedler, F. (1987). *Eur. J. Biochem.* **163**, 303.
100. Gonias, S. L. and Pizzo, S. V. (1983). *J. Biol. Chem.* **258**, 14682.
101. Fiedler, F. (1987). *Eur. J. Biochem.* **163**, 303.
102. Hedstrom, L. (1996). *Biol. Chem. Hoppe-Seyler* **377**, 465.

Appendix III
Commercially available proteinase inhibitors

Robert J. Beynon
Department of Veterinary Preclinical Sciences, University of Liverpool, PO Box 147, Liverpool L69 3BX

Guy Salvesen
The Burnham Institute, 10901 North Torrey Pines Road, La Jolla, CA 92037, USA

This appendix lists some of the properties of proteinase inhibitors that will be useful, either as analytical reagents or as additives to prevent or diminish proteolysis. Low molecular weight inhibitors are listed separately from the proteinaceous inhibitors as they are often used in quite different applications. The list of inhibitors is not meant to be exhaustive, but is nonetheless a reasonable reflection of those that are cited most commonly in the literature. Suppliers are not indicated, except where unusual inhibitors are from one company. Other suppliers may also offer similar products. The effective concentration ranges cannot be guaranteed to be optimal for every proteinase or system (see also Chapters 5 and 6).

Table 1 summarizes the inhibitors that are included in this compilation. Many specialized inhibitors that exhibit reasonable selectivity for some serine or cysteine proteinases are not listed. Suppliers of inhibitors include Bachem, Calbiochem and Enzyme Systems Products (see Core List).

Table 1 Overview

Name	Class	Mode	Notes	Page
Low molecular weight inhibitors				
AEBSF	Serine	Irr	Less toxic alternative to PMSF	319
Ac-YVAD-CHO	Cysteine	Rev	Caspase inhibitor	319
Amastatin	Metallo-	Rev	Aminopeptidase inhibitor	319
Antipain	Serine/ cysteine	Rev	Trypsin-like and many cysteine proteinases	319
APMSF	Serine	Irr	Unstable, prepare fresh	320
Bestatin	Metallo-	Rev	Aminopeptidase	320
Chymostatin	Serine/ Cysteine	Rev	Chymotrypsin-like and cysteine proteinases	320
CS-peptide	Cysteine	Rev	Calpain inhibitor, through calcium site	320
Dansyl-pepstatin	Aspartic	Rev	Active-site titrant	320
3,4-DCI	Serine	Irr	Fast inhibitor of serine proteinases	321
DFP	Serine	Irr	Very toxic, requires special care	321

Table 1 Continued

Name	Class	Mode	Notes	Page
Low molecular weight inhibitors				
Diprotin A	Metallo-	Rev	Aminopeptidase inhibitor	321
Diprotin B	Metallo-	Rev	Aminopeptidase inhibitor	321
320E-64	Cysteine	Irr	Very specific, active site titrant	322
E-64d	Cysteine	Irr	Cell permeant	322
EDTA	Metallo-	Rev	Chelator	322
Elastatinal	Serine	Rev	Elastase-like serine proteinases	322
Elastase inhibitor 1	Serine	Irr		323
Iodoacetate/ iodoacetamide	Cysteine	Irr	Not specific for proteinases	323
Leupeptin	Serine/ Cysteine	Rev	Trypsin-like and many cysteine proteinases	323
NCO-700	Cysteine	Irr	Epoxide inhibitor	324
PD 150606	Cysteine	Irr	Calpain inhibitor	324
Pepstain	Aspartic	Rev	Light binding	324
1,10-Phenathroline	Metallo-	Rev	Chelator	324
Phosphoramidon	Metallo-	Rev	Not all metallo-endopeptidases	324
PMSF	Serine	Irr	Slower than DCI	325
Subtilisn inhibitor 1	Serine	Irr		325
TLCK	Serine	Irr	Trypsin-like serine proteinases	325
TPCK	Serine	Irr	Chymotrypsin-like serine proteinases	325
Z-Phe-Ala-CHN_2	Cysteine	Irr	Cathepsin B, L, papain	326
Z-Leu-Leu-$B(OH)_2$	Threonine	Rev	Proteasome inhibitor	326
Z-Leu-Leu-Leu-CHO	Threonine/ Other	Rev	Proteasome inhibitor, also cathepsins, calpain	326
Z-Leu-Leu -Leu-CHO	Threonine/ Other	Rev	proteasome inhibitor, also cathepsin	326
ZINCOV	Metallo	Rev	Thermolysin, Neprilysin	327
High molecular weight (proteinaceous) inhibitors				
α_2-Antiplasmin	Serine	Rev		
Antithrombin III	Serine	Rev		327
Aprotinin	Serine	Rev	Very stable	327
Cystatin	Cysteine	Rev	Very stable	327
Calpastatin	Cysteine	Rev	Calpain inhibitor	327
Chymotrypsin inhibitor	Serine	Rev		328
Corn trypsin inhibitor	Serine	Rev	Inhibits Factor XIIa	328
Lima bean trypsin inhibitor	Serine	Rev	Dissociated at low pH	328
α_2-Macroglobulin	All classes	Irr	Forms 'trap' around most proteinases	329
α_2-Proteinase inhibitor	Serine	Rev		329
Soybean trypsin inhibitor	Serine	Rev	Dissociated at low pH	329

Low molecular weight inhibitors

AEBSF (1)

Target proteinases: serine proteinases

Synonyms: Pefabloc SC; 4-(2-aminoethyl)benzenesulphonyl fluoride·HCl; *p*-aminoethylbenzenesulphonyl fluoride·HCl

Molecular weight: 239.5

Effective concentration: 0.1–1.0 mM

Stock solution: 10 mM in water. Stable for 1-2 months at 4 °C

Notes: Some hydrolysis at alkaline pH values. Non-toxic alternative to PMSF. Low activity towards proteinase K

Ac-YVAD-CHO

Target proteinases: capsases 1 and 3

Synonyms: Ac-Tyr-Val-Ala-Asp-aldehyde

Molecular weight: 475.6

Effective concentration: 10–100 nM

Notes: Calbiochem produce a large range of caspase inhibitors, including aldehydes, chloromethyl ketones, fluoromethylketones and fluoroacyloxymethyl ketones. This a representative example of a reversible aldehyde inhibitor

Amastatin (2,3)

Target proteinases: aminopeptidases, notably alanyl-aminopeptidase

Synonyms:

Molecular weight: 474.6

Effective concentration: 1–10 μM. Stable for 1 day

Stock solution: 1 mM in methanol. Stable for at least 1 month at −20 °C

Antipain (4)

Target proteinases: trypsin-like serine and some cysteine proteinases

Synonyms: [(5)-l-carboxy-2-phenylethyl]-CarbamoylArgValArgal

Molecular weight: antipain, 604.7 antipain dihydrochloride: 677.6

Effective concentration: 1–100 μM. Stable for several hours

Stock solution: 10 mM in water or buffer. Stable for 1 week at 4 °C, 1 month at −20 °C

Notes: antipain is an amino acid aldehyde, the aldehyde being contributed by an arginine residue. It has a similar specificity to leupeptin

APMSF (5,6)

Target proteinases: trypsin-like serine proteinases

Synonyms: *p*-APMSF,4-(amidinopheny1) methanesulphonyl fluoride

Molecular weight: APMSF, 216.2; APMSF·HCl·H_2O, 270.7

Effective concentration: 10–100 μM. Half-life = 6 min in pH 7.0 buffer systems

Stock solution: 50 mM in water. Stable when aliquoted at −20°C

Notes: not as effective as DipF, but more effective than PMSF. Specific for trypsin-like serine proteinases. No effect on acetylcholinesterase

Bestatin (3,7)

Target proteinases: aminopeptidases, notably alanyl aminopeptidase

Synonyms:

Molecular weight: 308.4

Effective concentration: 1–10 μM. Stable for 1 day

Stock solution: 1 mM in methanol. Stable for at least 1 month at −20°C

Chymostatin (4)

Target proteinases: chymotrypsin-like serine proteinases and some cysteine proteinases

Synonyms: Phe-(Cap)-Leu-Phe-al, N-[(S)-1-carboxy-isopentyl-carbamoyl-α-(2-iminohexahydro-4(S)-pyrimidyl]-L-glycyl-L-phenylalaninal

Molecular weight: 604.7

Effective concentration: 10–100 μM. Stable for several hours

Stock solution: 10 mM in DMSO. Stable for months at −20°C

Notes: chymostatin is an amino acid aldehyde, the aldehyde being contributed by a phenylalanine residue. It inhibits chymotrypsin-like serine proteinases and most cysteine proteinases

CS-peptide (8)

Target proteinases: calpains I and II

Synonyms:

Molecular weight: 3177.7

Effective concentration: 10–100 nM. Stable for several hours

Stock solution: soluble in water. Stable for 1 month at −20°C

Notes: the 27-residue peptide encoded by exon 1B of human calpastatin. It is cell permeant and able to inhibit calpain I and calpain II (Calbiochem, Sigma)

Dansyl-pepstatin (9)

Target proteinases: aspartic proteinases

Synonyms: *N*-pepstatinyl-*N′*-dansyldiaminopropane

Molecular weight: 975.3

Effective concentration: K_i = 37 pM towards pepsin

Stock solution: DMSO or warm ethanol. Stable

Notes: An active site titrant for pepstain inhibitable aspartic proteinases. The fluorescence is enhanced on binding

3,4-DCI (10)

Target proteinases: serine proteinases

Synonyms: 3,4-dichloroisocoumarin

Molecular weight: 215.0

Effective concentration: 5–100 μM. Half-life = 20 mm at pH 7.5

Stock solution: 10 mM in DMF or DMSO. Stable for months at −20 °C

Notes: active towards wide range of serine proteinases, not active towards β-lactamases

DFP (11)

Target proteinases: serine proteinases

Synonyms: di-isopropylfluorophosphate, Dip-F

Molecular weight: 184.2

Effective concentration: 0.1 mM. Half-life = 1 h at pH 7.5

Stock solution: 200–500 mM in dry propan-2-ol. Stable for several months at −70 °C

Notes: highly toxic. Stock solvent can be dried by storage over 'molecular sieve'. [^{14}C]DFP and [^{3}H]DFP can be used as affinity labels and for autoradiography

Diprotin A (12)

Target proteinases: inhibitor of dipeptidyl aminopeptidase IV

Synonyms: H-Ile-Pro-Ile-OH

Molecular weight: 359.3 (monohydrate)

Effective concentration: 10–50 μM. Stable for 1 day

Stock solution: 1 mM in water, methanol or ethanol. Stability unknown

Diprotin B (12)

Target proteinases: inhibitor of dipeptidyl aminopeptidase IV

Synonyms: H-Val-Pro-Leu-OH

Molecular weight: 327.4

Effective concentration: 50–100 μM. Stable for 1 day

Stock solution: 1–10 mM in water, methanol or ethanol. Stability unknown

E-64 (13,14)

Target proteinases: cysteine proteinases

Synonyms: L-*trans*-epoxysuccinyl-leucylamide-(4-guanidino)-butane, *N-[N'*-(L-3-transcarboxyirane-2-carbonyl)-L-leucyl]-agmatine

Molecular weight: 357.4

Effective concentration: 1–10 μM. Stable for days at neutral pH

Stock solution: 1 mM in aqueous solution. Stable for months at −20 °C

Notes: E-64 is an effective irreversible inhibitor of cysteine proteinases that does not affect cysteine residues in other enzymes or reacts with low molecular weight thiols such as 2-mercaptoethanol. It is an excellent active-site titrant

E-64d (15)

Target proteinases: cysteine proteases

Synonyms: loxistatin, EST, (2S,3S)-*trans*-epoxysuccinyl-L-leucylamido-3-methylbutane ethyl ester

Molecular weight: 342.4

Effective concentration: 10–100 μM. Stable for several hours

Stock solution: 1 mM in chloroform or ethanol. Stable for 1 week at 4 °C, 1 month at −20 °C

Notes: similar to E64, but devoid of charged functionalities. The ester is hydrolysed inside the cell, and the free acid product, E64c, is considered to be active

EDTA (16)

Target proteinases: metalloproteinases, metal-activated proteinases

Synonyms: ethylenediaminetetraacetic acid

Molecular weight: disodium salt, dihydrate: 372.24

Effective concentration: 1–10 mM. Stable in aqueous solution

Stock solution: 0.5 M in water, pH 8.5. Stable for months at 4 °C

Notes: EDTA acts as a chelator of the active site zinc ion in metalloproteinases but can also inhibit other metal ion-dependent proteinases such as the calcium-dependent cysteine proteinases. EDTA may interfere with other metal-dependent biological processes

Elastatinal (4)

Target proteinases: elastase-like serine proteinases

Synonyms: Leu-(Cap)-Gln-Ala-al, *N*-[(S)-1-carboxy-isopentyl)-carbamoyl-α-(2-iminohexahydro-4(S)-pyrimidyl]-L-glycyl-L-glutaminyl-L-alaninal

Molecular weight: 512.6

Effective concentration: 10–100 μM in water. Stable for several hours

Stock solution: 10 mM in water. Stable for 1 week at 4 °C, months at −20 °C

Notes: elastatinal is an amino acid aldehyde, the aldehyde being contributed by an alanine residue

Elastase inhibitor I (17)

Target proteinases: elastase-like serine proteinases

Synonyms: Boc-Ala-Ala-Ala-NHO-Bz, PPE inhibitor

Molecular weight: 450.5

Effective concentration: Irreversible inhibitor; pancreatic elastase 128/M.s

Stock solution: soluble in DMSO, ethanol. Unstable in aqueous solution, with a half-life of 4 h at 30 °C and 50 h at 0 °C

Notes: Calbiochem

Iodoacetic acid/iodoacetamide

Target proteinases: cysteine proteinases

Synonyms: IAA, IAN; iodoacetate (usually as sodium salt)

Molecular weight: iodoacetic acid, 185.9; sodium iodoacetate, 207.9; iodoacetamide, 184.9

Effective concentration: 10–100 μM; use freshly prepared

Stock solution: 10–100 mM in water; prepare fresh as required

Notes: IAA is not specific for the active site cysteine residue of cysteine proteinases and can inhibit many other enzymes. It also reacts with low molecular weight thiols such as 2-mercaptoethanol. Radiolabelled IAA can be used as an active-site titrant provided that the concentration is kept low, so that the nucleophilic active site cysteine residue will react preferentially; this interaction will be E64-inhibitable

Leupeptin (4)

Target proteinases: trypsin-like serine and some cysteine proteinases

Synonyms: *N*-acetyl-Leu-Leu-Arg-al

Molecular weight: leupeptin, 426.6; leupeptin hemisulphate monohydrate: 542.7

Effective concentration: 10–100 μM. Stable for several hours

Stock solution: 10 mM in water. Stable for 1 week at 4 °C, 1 month at −20 °C

Notes: leupeptin is an amino acid aldehyde, the aldehyde being contributed by an arginine residue. It inhibits trypsin-like serine proteinases and most cysteine proteinases. Because the aldehyde is hydrated the active concentration may be as low as 2% of total. The inhibitor is susceptible to the action of peptidases

NCO-700 (18)

Target proteinases: cysteine proteinases

Synonyms:

Molecular weight: 1141.2 (sulphate)

Effective concentration: 10–100 μM

Stock solution: soluble in water. Stable for month at 4 °C. Protect from light

Notes: an epoxysuccinic acid derivative, active towards calpains, cathepsin B and L and papain

PD 150606 (19)

Target proteinases: calpains

Synonyms: 3-(4-iodophenyl)-2-mercapto-(Z)-2-propenoic acid

Molecular weight: 306.1

Effective concentration: 100 nM–1 μM

Stock solution: soluble in DMSO or methanol. Stable at −20 °C

Notes: a cell-permeant inhibitor of calpains, directed towards the calcium binding sites, rather than the active site nucleophilic cysteinyl group

Pepstatin (4)

Target proteinases: some aspartic proteinases

Synonyms: pepstatin A, isovaleryl-Val-Val-AHMHA-Ala-AHMHA [AHMHA = (3S,4S)4-amino-3-hydroxy-6-methyl-heptanoic acid]

Molecular weight: 685.9

Effective concentration: 1 μM. Stable for at least 1 day

Stock solution: 1 mM in MeOH or DMSO. Stable for months at −20 °C

Notes: pepstatin is a transition-state analogue that is a potent inhibitor of cathepsin D, pepsin, renin and many microbial aspartic proteinases. Immobilized pepstatin is available from Pierce

1,10-Phenanthroline (16)

Target proteinases: metalloproteinases, metal-activated proteinases

Synonyms: orthophenanthroline

Molecular weight: 198.2

Effective concentration: 1–10 mM. Stable for days

Stock solution: 200 mM in methanol. Stable for months at −20 °C

Notes: 1,10-phenanthroline has a strong UV absorbance and may interfere with spectrophotometric assays

Phosphoramidon (4,16)

Target proteinases: some metallo-endopeptidases

Molecular weight: 543.6

Effective concentration: 1–10 μM. Stable for 1 day

Stock solution: 1 mM in water. Stable for 1 month at −20 °C

Notes: phosphoramidon is an inhibitor of many bacterial metallo-endopeptidases but few of mammalian origin. Neprilysin is very susceptible to this inhibitor

PMSF (20,21)

Target proteinases: all serine proteinases

Synonyms: phenylmethanesulphonyl fluoride, phenylmethylsulphonyl fluoride, α-toluenesulphonyl fluoride

Molecular weight: 174.2

Effective concentration: 0.1–1 mM. Half-life = 1 h at pH 7.5

Stock solution: 200 mM in dry solvents (propan-2-ol, MeOH, EtOH). Stable for at least 9 months at 4 °C

Notes: not as effective nor as toxic as Dip-F, also inhibits cysteine proteinases (reversible by reduced thiols)

Subtilisin inhibitor 1 (22)

Target proteinases: subtilin, thermitase and related enzymes

Synonyms: Boc-Ala-Ala-NHO-Bz

Molecular weight: 379.4

Effective concentration: Irreversible inhibitor; subtilisin 226/M.s

Stock solution: soluble in DMSO, ethanol or acetonitrile

Notes: unstable in aqueous solution. Half-life is 4 h at 30 °C or 60 h at 0 °C. Calbiochem

TLCK (23)

Target proteinases: trypsin-like serine proteinases

Synonyms: L-1-chloro-3-[4-tosylamido]-7-amino-2-heptanone HCl, tosyl lysyl chloromethyl ketone, Tos-Lys-CH_2Cl

Molecular weight: TLCK, 332.5; TLCK.HCl, 369.4

Effective concentration: 10–100 μM. Very unstable above pH 6.0

Stock solution: 10 mM in aqueous solution (1 mM HCl, pH 3.0). Prepare fresh as needed

Notes: active towards some trypsin-like serine proteinases and has some activity against cysteine proteinases. Has an offensive odour. Used in the preparation of chymotrypsin free of contaminating trypsin activity

TPCK (24)

Target proteinases: chymotrypsin-like serine proteinases

Synonyms: L-1-chloro-3-[4-tosylamido]-4-phenyl-2-butanone, tosyl phenylalanyl chloromethyl ketone, Tos-Phe-CH_2Cl

Molecular weight: 351.5

Effective concentration: 10–100 μM. Stable for several hours

Stock solution: 10 mM in methanol or ethanol. Stable for several months at 4 °C

Notes: active towards some chymotrypsin-like serine proteinases. Used in the preparation of trypsin free of contaminating chymotrypsin activity

Z-Phe-Ala-CHN_2 (25)

Target proteinases: cysteine proteinases

Synonyms: *N*-Cbz-Phe-Ala-diazomethane

Molecular weight: 395

Effective concentration: 10 μM, optimally active at pH 6.0. Use freshly diluted

Stock solution: 10 mM in DMSO or acetonitrile. Stable for 1 week at −20 °C

Notes: a fast, irreversible inhibitor of cysteine proteinases

Z-Leu-Leu-$B(OH)_2$

Target proteinases: proteasome

Synonyms:

Molecular weight: 491.4

Effective concentration: K_i = 0.023 nM versus SDS-activated 20S proteasome

Stock solution: soluble and stable in DMSO

Notes: structurally similar to MG-132, cell-permeant

Z-Leu-Leu-Leu-aldehyde (26)

Target proteinases: proteasome

Synonyms: MG-132

Molecular weight: 475.6

Effective concentration: K_i = 4.2 nM vs. SDS-activated 20S proteasome; IC50 in cell culture 5 μM

Stock solution: Soluble and stable in DMSO

Notes: Also inhibits calpains and cathepsins

Z-Leu-Leu-Leu-Nva-aldehyde (27)

Target proteinases: proteasome

Synonyms: Z-Leu-Leu-Leu-norvaline aldehyde, MG-115

Molecular weight: 461.6

Effective concentration: K_i = 20 nM vs SDS-activated 20S proteasome; IC_{50} in cell culture 5 μM

Stock solution: soluble and stable in DMSO

Notes: also inhibits calpains and cathepsins.

ZINCOV (28)

Target proteinases: thermolysin-like metalloproteinases, neprilysin

Synonyms: 2-(*N*-hydroxycarboxyamido)-4-methylpentanoyl-L-Ala-Gly-amide

Molecular weight: 302.3

Effective concentration: K_i = 480 nM vs thermolysin, 57 nM vs. *Pseudomonas aeruginosa* metalloendopeptidase and 3.1 nM vs. mouse neprilysin

Stock solution: soluble in water, and stable for months at 4 °C

Notes:

High molecular weight (proteinaceous) inhibitors

α_2-Antiplasmin (29)

Target proteinases: plasmin, trypsin, chymotrypsin

Synonyms: α_2-2-AP, α_2-2-plasmin inhibitor

Molecular weight: 67 000

Effective concentration: equimolar with proteinase

Stock solution: stable at −20 °C at pH 6–7.5. Unstable below pH 5.7

Antithrombin III (30)

Target proteinases: thrombin and Factor Xa, trypsin, other trypsin-like serine proteinases

Synonyms: ATIII, antithrombin–heparin cofactor

Molecular weight: 67 000

Effective concentration: equimolar with proteinase

Stock solution: stock solutions are stable at −20 °C

Notes: the reaction between ATIII and thrombin is accelerated by heparin

Aprotinin (31)

Target proteinases: serine proteinases – not thrombin and Factor Xa

Synonyms: Trasylol, bovine pancreatic trypsin inhibitor (Kunitz), BPTI

Molecular weight: 6500

Effective concentration: equimolar with proteinase

Stock solution: very soluble in water; very stable

Notes: available immobilized from Sigma

Calpastatin (32)

Target proteinases: calpains

Synonyms:

Molecular weight: 14 000

Effective concentration: equimolar with proteinase

Stock solution: stable to freezing

Notes:

Chymotrypsin inhibitor II, potato (33)

Target proteinases: serine proteinases

Synonyms:

Molecular weight: 20 000

Effective concentration: equimolar with proteinase

Stock solution: stable to freezing

Notes:

Corn trypsin inhibitor (34)

Target proteinases: serine proteases, especially factor XIIa

Synonyms:

Molecular weight: 14 000

Effective concentration: equimolar with proteinase

Stock solution: avoid freeze-thaw cycles

Notes:

Cystatin (35,36)

Target proteinases: cysteine proteinases, including dipeptidyl peptidase III

Synonyms: stefins, low and high molecular weight kininogens, -ã-trace

Molecular weight: 12 000

Effective concentration: equimolar with proteinase

Stock solution: stable to heat. Unstable to freezing unless in the presence of 20% glycerol or buffered to pH 7.5

Notes: data here refers to hen egg white cystatin; other members of the cystatin superfamily have markedly different properties. See (36) for a discussion of the cystatin superfamily

Lima bean trypsin inhibitor (37)

Target proteinases: serine proteinases

Synonyms: LBTI

Molecular weight: 9000

Effective concentration: equimolar with proteinase

Stock solution: stable at −20 °C at pH 6–7.5. Stable at 4 °C at pH 3.0

Notes: suitable for immobilization and as an affinity ligand. LBTI has separate binding sites for trypsin and chymotrypsin, and can form a ternary complex with these two proteinases

α_1-Macroglobulin (38)

Target proteinases: most endopeptidases

Synonyms: α_1-M

Molecular weight: 725 000

Effective concentration: equimolar with proteinase

Stock solution: stable at −20 °C at pH 6–7.5

Notes: inactivated by small primary amines (ammonia, hydroxylamine, methylamine). Titrate to determine activity if purchased commercially (see Chapter 5). Available in immobilized form from Boehringer

α_1-Proteinase inhibitor (39)

Target proteinases: trypsin, plasma proteinases, elastase, most mammalian serine proteinases

Synonyms: α_1-P1, -antitrypsin

Molecular weight: 52 000

Effective concentration: equimolar with proteinase

Stock solution: stable for several months in solutions containing 0.01% NaN_3. Can be stored at −80 °C but should not be refrozen. Unstable below pH 5.5

Notes: inactivated by some non-serine proteinases. Inactivated by oxidation of active site methionine residue

Soybean trypsin inhibitor (40)

Target proteinases; serine proteinases

Synonyms; STI, SBTI

Molecular weight; 20 100

Effective concentration: equimolar with proteinase

Stock solution: stock solutions are stable at −20 °C

Notes: the reaction between STI and serine proteinases is reversed by low pH (3.0) which does not inactivate the inhibitor. Thus, STI is an effective affinity matrix if the proteinase is tolerant to low pH. Immobilized STI is available from Pierce and Sigma

References

1. Baker, B. R. and Cory, M. (1971). *J. Med. Chem.* **14**, 119.
2. Tobe, H., Morishima, H., Naganawa, T., Takita, T., and Umezawa, H. (1979). *Agric. Biol. Chem.* **43**, 591.
3. Umezawa, H. (1982). *Annu. Rev. Microbiol.* **36**, 75.
4. Umezawa, H. (1976). In *Methods in enzymology* (ed. Lorand, L.), Vol. 45, p. 678. Academic Press, New York.
5. Laura, R. (1980). *Biochemistry* **19**, 4859.

6. Powers, J. C. and Harper, J. W. (1986). In *Proteinase inhibitors.* (ed. Barrett, A. J. and Salvesen, G.), p. 55. Elsevier, Amsterdam.
7. Umezawa, H., Aoyagi, T. Suda, H., Hamada, M., and Takeuchi. T. (1976). *J. Antibiot.* **29**, 97.
8. Eto, A. *et al.* (1995). *J. Biol. Chem.* **270**, 25115.
9. Yonezawa, H. *et al.* (1997). *J. Biochem.* **122**, 294.
10. Harper, J. W., Hemmi, K., and Powers, J. C. (1985). *Biochemistry* **24**, 1831.
11. Beynon, R. J. (1989). In *Protein purification methods: a practical approach.* IRL Press, Oxford.
12. Umezawa, H., Aoyagi, T., Ogawa, K., Naganawa, N. Hamada, M., and Takeuchi, T. (1984). *J. Antibiot.* **37**, 422.
13. Hanada, K., Tamai, M., Yamagishi, M., Ohmura, S., Sawada, J., and Tanaka, I. (1978). *Agric. Biol. Chem.* **42**, 523.
14. Barrett, A. J., Kembhavi, A. A., Brown, M. A., Kirschke, H., Knight, C. G., Tamai, M., and Hanada, K. (1982). *Biochem. J.* **201**, 189.
15. Tamai *et al.* (1987). *J. Pharmacobiodyn.* **10**, 678.
16. Powers, J. C. and Harper, J. W. (1986). In *Proteinase Inhibitors* (ed. Barrett, A. J. and Salvesen, G.), p. 219. Elsevier, Amsterdam.
17. Schmidt, C. *et al.* (1991). *Peptides* **100**, 761.
18. Hara, K. and Takahashi, K. (1983). *Biomed. Res.* **4**, 121.
19. Wang, K. K *et al.* (1996). *Proc. Natl. Acad. Sci. USA* **93**, 6687.
20. Fahrney, D. E. and Gold, A. M. (1963). *J. Am. Chem. Soc.* **85**, 997.
21. James, G. T. (1978). *Anal. Biochem.* **86**, 574.
22. Bromme, D. and Demuth, H-U (1994). *Methods Enzymol.* **244**, 671.
23. Shaw, E. (1965). *Biochemistry* **4**, 2219.
24. Schoellman, G. and Shaw, E. (1963). *Biochemistry* **2**, 252.
25. Shaw, E. and Green, G. (1981). In *Methods in enzymology* (ed. Colowick, S. P. and Lorand, L.), Vol. 80, p. 820. Academic Press, New York.
26. Adams, J. and Stein, R. (1996). *Annu. Rep. Med. Chem.* **31**, 279.
27. Palombella, V. J. (1994). *Cell* **78**, 773.
28. Hudgin, R. L. (1981). *Life. Sci.* **29**, 2593.
29. Lijnen, H. R. and Collen, D. (1986). In *Proteinase inhibitors.* (ed. Barrett, A. J. and Salvesen, G.), p. 457. Elsevier, Amsterdam.
30. Damus, P. S. and Rosenberg, R. D. (1976). In *Methods in enzymology* (ed. Lorand, L.), Vol. 45, p. 653. Academic Press, New York.
31. Gebhard, W., Tschesche, H., and Fritz, H. (1986). In *Proteinase inhibitors* (ed. Barrett, A. J. and Salvesen, G.), p. 375. Elsevier, Amsterdam.
32. Salamino, F. *et al.* (1994). *Biochem. Biophys. Res. Commun.* **199**, 1326.
33. Bryant, J. *et al.* (1976). *Biochemistry* **15**, 3418.
34. Swartz, M. J. *et al.* (1977). *J. Biol. Chem.* **252**, 8105.
35. Anastasi, A., Brown, M. A., Kembhavi, A. A., Nicklin, M. J. H., Sayers, C. A., Sunter, D. C., and Barrett, A. J. (1982). *Biochem. J.* **211**, 129.
36. Barrett, A. J. (1986). In *Proteinase Inhibitors* (ed. Barrett, A. J. and Salvesen, G.), p. 515. Elsevier, Amsterdam.
37. Birk, Y. (1976). In *Methods in enzymology* (ed. Lorand, L.), Vol. 45, p. 707. Academic Press, New York.
38. Barrett, A. J. (1980). In *Methods in enzymology.* (ed. Colowick, S. P. and Lorand. L.), Vol. 80, p. 737. Academic Press, New York.
39. Travis, J. and Johnson, D. (1981). In *Methods in enzymology.* (ed. Packer, L.), Volume 81, p. 754. Academic Press, New York.
40. Birk, Y. (1976). *Methods in enzymology* (ed. Lorand, L.), Vol. 45, p. 700. Academic Press, New York.

List of suppliers

Agilent Technologies
395 Page Mill Road,
PO Box 10395,
Palo Alto, CA 94304, USA
Web site: www.agilent.com

Amersham Pharmacia BioTech
Pharmacia Biotech (Biochrom) Ltd., Unit 22, Cambridge Science Park, Milton Road, Cambridge CB4 0FJ, UK.
Tel: 01223 423723
Fax: 01223 420164
Web site: www.biochrom.co.uk
Pharmacia and Upjohn Ltd., Davy Avenue, Knowlhill, Milton Keynes, Buckinghamshire MK5 8PH, UK.
Tel: 01908 661101
Fax: 01908 690091
Web site: www.eu.pnu.com

Anderman and Co. Ltd., 145 London Road, Kingston-upon-Thames, Surrey KT2 6NH, UK.
Tel: 0181 541 0035
Fax: 0181 541 0623

Beckman Coulter Inc.
Beckman Coulter Inc., 4300 N Harbor Boulevard, PO Box 3100, Fullerton, CA 92834-3100, USA.
Tel: 001 714 871 4848
Fax: 001 714 773 8283
Web site: www.beckman.com
Beckman Coulter (UK) Ltd., Oakley Court, Kingsmead Business Park, London Road, High Wycombe, Buckinghamshire HP11 1JU, UK.
Tel: 01494 441181 Fax: 01494 447558
Web site: www.beckman.com

Becton Dickinson and Co.
Becton Dickinson and Co., 21 Between Towns Road, Cowley, Oxford OX4 3LY, UK.
Tel: 01865 748844 Fax: 01865 781627
Web site: www.bd.com
Becton Dickinson and Co., 1 Becton Drive, Franklin Lakes, NJ 07417-1883, USA.
Tel: 001 201 847 6800
Web site: www.bd.com

Bio 101 Inc.
Bio 101 Inc., c/o Anachem Ltd., Anachem House, 20 Charles Street, Luton, Bedfordshire LU2 0EB, UK.
Tel: 01582 456666 Fax: 01582 391768
Web site: www.anachem.co.uk
Bio 101 Inc., PO Box 2284, La Jolla, CA 92038-2284, USA. Tel: 001 760 598 7299
Fax: 001 760 598 0116
Web site: www.bio101.com

Bio-Rad Laboratories Ltd.
Bio-Rad Laboratories Ltd., Bio-Rad House, Maylands Avenue, Hemel Hempstead, Hertfordshire HP2 7TD, UK.
Tel: 0181 328 2000 Fax: 0181 328 2550
Web site: www.bio-rad.com

Bio-Rad Laboratories Ltd., Division Headquarters, 1000 Alfred Noble Drive, Hercules, CA 94547, USA.
Tel: 001 510 724 7000
Fax: 001 510 741 5817
Web site: www.bio-rad.com
Bio-Rad Micromeasurements Ltd., Haxby Road York, North Yorkshire YO3 7SD, UK.
Web site: www.bio-rad.com/index1.html

Biosepra Inc.
111 Loch Drive, Marlborough MA 01752, USA
Tel: 001 508 357 7525
Fax: 001 508 357 7595
Web site: www.biosepra.com

Biosoft, 37 Cambridge Place, Cambridge CB2 1NS
Web site: www.biosoft.com

Boehringer Mannheim Corporation
please see Roche Diagnostics

Calbiochem–Novabiochem Corporation, P.O. Box 12087, La Jolla, CA 92039-2087, USA.
Web site: www.calbiochem.com/

CP Instrument Co. Ltd., PO Box 22, Bishop Stortford, Hertfordshire CM23 3DX, UK.
Tel: 01279 757711 Fax: 01279 755785
Web site: www.cpinstrument.co.uk

Dupont
Dupont (UK) Ltd., Industrial Products Division, Wedgwood Way, Stevenage, Hertfordshire SG1 4QN, UK.
Tel: 01438 734000
Fax: 01438 734382
Web site: www.dupont.com
Dupont Co. (Biotechnology Systems Division), PO Box 80024, Wilmington, DE 19880-002, USA.
Tel: 001 302 774 1000
Fax: 001 302 774 7321
Web site: www.dupont.com

Eastman Chemical Co., 100 North Eastman Road, PO Box 511, Kingsport, TN 37662-5075, USA.
Tel: 001 423 229 2000
Web site: www.eastman.com

Fisher Scientific
Fisher Scientific UK Ltd., Bishop Meadow Road, Loughborough, Leicestershire LE11 5RG, UK.
Tel: 01509 231166
Fax: 01509 231893
Web site: www.fisher.co.uk
Fisher Scientific, Fisher Research, 2761 Walnut Avenue, Tustin, CA 92780, USA.
Tel: 001 714 669 4600
Fax: 001 714 669 1613
Web site: www.fishersci.com

Fluka
Fluka, PO Box 2060, Milwaukee, WI 53201, USA.
Tel: 001 414 273 5013
Fax: 001 414 2734979
Web site: www.sigma-aldrich.com
Fluka Chemical Co. Ltd., PO Box 260, CH-9471, Buchs, Switzerland.
Tel: 0041 81 745 2828
Fax: 0041 81 756 5449
Web site: www.sigma-aldrich.com

Hewlett-Packard Company,
please see Agilent Technologies

Hybaid
Hybaid Ltd., Action Court, Ashford Road, Ashford, Middlesex TW15 1XB, UK.
Tel: 01784 425000
Fax: 01784 248085
Web site: www.hybaid.com
Hybaid US, 8 East Forge Parkway, Franklin, MA 02038, USA.
Tel: 001 508 541 6918
Fax: 001 508 541 3041
Web site: www.hybaid.com

HyClone Laboratories, 1725 South HyClone Road, Logan, UT 84321, USA.
Tel: 001 435 753 4584
Fax: 001 435 753 4589
Web site: www.hyclone.com

Invitrogen
Invitrogen BV, PO Box 2312, 9704 CH Groningen, The Netherlands.
Tel: 00800 5345 5345 Fax: 00800 7890 7890
Web site: www.invitrogen.com
Invitrogen Corp., 1600 Faraday Avenue, Carlsbad, CA 92008, USA.
Tel: 001 760 603 7200
Fax: 001 760 603 7201
Web site: www.invitrogen.com

Life Technologies
Life Technologies Ltd., PO Box 35, Free Fountain Drive, Incsinnan Business Park, Paisley PA4 9RF, UK.
Tel: 0800 269210 Fax: 0800 838380
Web site: www.lifetech.com
Life Technologies Inc., 9800 Medical Center Drive, Rockville, MD 20850, USA.
Tel: 001 301 610 8000
Web site: www.lifetech.com

Merck Sharp & Dohme
Merck Sharp & Dohme Research Laboratories, Neuroscience Research Centre, Terlings Park, Harlow, Essex CM20 2QR, UK.
Web site: www.msd-nrc.co.uk
MSD Sharp and Dohme GmbH, Lindenplatz 1, D–85540, Haar, Germany.
Web site: www.msd-deutschland.com

Michrom BioResources, Inc., 5673 W. Las Positas Boulevard, Suite 291, Pleasanton, CA 94566, USA.

Millipore
Millipore (UK) Ltd., The Boulevard, Blackmoor Lane, Watford, Hertfordshire WD1 8YW, UK.
Tel: 01923 816375 Fax: 01923 818297
Web site: www.millipore.com/local/UK.htm
Millipore Corp., 80 Ashby Road, Bedford, MA 01730, USA.
Tel: 001 800 645 5476
Fax: 001 800 645 5439
Web site: www.millipore.com

Molecular Probes
Molecular Probes, Inc., 4849 Pitchford Avenue, Eugene, OR, 97402-9165, USA
Web site: www.probes.com/

New England Biolabs, 32 Tozer Road, Beverley, MA 01915-5510, USA.
Tel: 001 978 927 5054

Nikon
Nikon Corp., Fuji Building, 2-3, 3-chome, Marunouchi, Chiyoda-ku, Tokyo 100, Japan.
Tel: 00813 3214 5311
Fax: 00813 3201 5856
Web site: www.nikon.co.jp/main/index_e.htm
Nikon Inc., 1300 Walt Whitman Road, Melville, NY 11747-3064, USA.
Tel: 001 516 547 4200
Fax: 001 516 547 0299
Web site: www.nikonusa.com

Novex, 11040 Roselle Street, San Diego, California, 92121 USA.
Web site: www.novex.com/

Nycomed
Nycomed Amersham plc, Amersham Place, Little Chalfont, Buckinghamshire HP7 9NA, UK.
Tel: 01494 544000
Fax: 01494 542266
Web site: www.amersham.co.uk
Nycomed Amersham, 101 Carnegie Center, Princeton, NJ 08540, USA.
Tel: 001 609 514 6000
Web site: www.amersham.co.uk

Perkin Elmer Ltd., Post Office Lane, Beaconsfield, Buckinghamshire HP9 1QA, UK.
Tel: 01494 676161
Web site: www.perkin-elmer.com

Perseptive Biosystems, Inc., 500 Old Connecticut Path, Framingham, MA 01701, USA.
Web site: www.pbio.com/

Pharmacia (please see Amersham Pharmacia Bio Tech)

Phenomenex, 2320 W. 205th Street, Torrance, CA 90501-1456, USA.

Pierce, PO Box 117, Rockford, IL 61105, USA.
Web site: www.piercenet.com/

PolyLC Inc., 9151 Rumsey Road, Suite 180, Columbia, MD 21045, USA.
E-mail; polylc@aol.com

Polymicro Technologies, Inc., 18109 N 25th Avenue, Phoenix, AZ 85023-1200, USA
Web site: www.polymicro.com

Promega
Promega UK Ltd., Delta House, Chilworth Research Centre, Southampton SO16 7NS, UK.
Tel: 0800 378994 Fax: 0800 181037
Web site: www.promega.com
Promega Corp., 2800 Woods Hollow Road, Madison, WI 53711-5399, USA.
Tel: 001 608 274 4330
Fax: 001 608 277 2516
Web site: www.promega.com

Protana A/S, Staermosegaardsvej 16, DK-5230 Odense M, Denmark

Qiagen
Qiagen UK Ltd., Boundary Court, Gatwick Road, Crawley, West Sussex RH10 2AX, UK.
Tel: 01293 422911 Fax: 01293 422922
Web site: www.qiagen.com
Qiagen Inc., 28159 Avenue Stanford, Valencia, CA 91355, USA.
Tel: 001 800 426 8157
Fax: 001 800 718 2056
Web site: www.qiagen.com

Roche Diagnostics
Roche Diagnostics Ltd., Bell Lane, Lewes, East Sussex BN7 1LG, UK.
Tel: 01273 484644 Fax: 01273 480266
Web site: www.roche.com
Roche Diagnostics Corp., 9115 Hague Road, PO Box 50457, Indianapolis, IN 46256, USA.
Tel: 001 317 845 2358
Fax: 001 317 576 2126
Web site: www.roche.com
Roche Diagnostics GmbH, Sandhoferstrasse 116, 68305 Mannheim, Germany.
Tel: 0049 621 759 4747
Fax: 0049 621 759 4002
Web site: www.roche.com

Schleicher and Schuell Inc., Keene, NH 03431A, USA.
Tel: 001 603 357 2398

Shandon Scientific Ltd., 93-96 Chadwick Road, Astmoor, Runcorn, Cheshire WA7 1PR, UK.
Tel: 01928 566611
Web site: www.shandon.com

Sigma-Aldrich
Sigma-Aldrich Chemie GmbH, Reidstrasse 2, D-89555 Steinheim, Germany
Sigma-Aldrich Co. Ltd., The Old Brickyard, New Road, Gillingham, Dorset XP8 4XT, UK.
Tel: 01747 822211
Fax: 01747 823779
Web site: www.sigma-aldrich.com
Sigma-Aldrich Co. Ltd., Fancy Road, Poole, Dorset BH12 4QH, UK.
Tel: 01202 722114
Fax: 01202 715460
Web site: www.sigma-aldrich.com

Sigma Chemical Co., PO Box 14508, St Louis, MO 63178, USA.
Tel: 001 314 771 5765
Fax: 001 314 771 5757
Web site: www.sigma-aldrich.com

Stratagene
Stratagene Europe, Gebouw California, Hogehilweg 15, 1101 CB Amsterdam Zuidoost, The Netherlands.
Tel: 00800 9100 9100
Web site: www.stratagene.com
Stratagene Inc., 11011 North Torrey Pines Road, La Jolla, CA 92037, USA.
Tel: 001 858 535 5400
Web site: www.stratagene.com

United States Biochemical, PO Box 22400, Cleveland, OH 44122, USA.
Tel: 001 216 464 9277

Vydac/The Separations Group, Inc., 17434 Mojave Street, Hesperia, CA 92345 USA.
Web site: www.vydac.com/

Wako Chemicals USA Inc., BioProducts Div., 1600 Bellwood Road.,Richmond , VA 23237, USA.
Web site: www.wako-chem.co.jp

Worthington Biochemical Corp.,
18109 N 25th Avenue,
Phoenix, AZ 85023-1200, USA
Web site: www.worthington-biochem.com

Index